Frontiers in Pure and Applied Probability

FRONTIERS IN PURE AND APPLIED PROBABILITY

Volume 1

Proceedings of the Third Finnish-Soviet Symposium on Probability Theory and Mathematical Statistics

Turku, Finland, August 13 - 16, 1991

Editors

Niemi H., Högnas G.,
Shiryaev A.N., and Melnikov A.V.

///VSP///
Utrecht, The Netherlands
Tokyo, Japan

TVP
Science Publishers
Moscow, Russia

VSP BV
P.O. Box 346
3700 AH Zeist
The Netherlands

TVP Science Publishers
Vavilov st. 42
117966 Moscow GSP-1
Russia

First published in 1993
ISBN 90-6764-156-1
NUGI 811

Printed in Russia by Novosti, Moscow

CONTENTS

Preface vii

Renovation, Regeneration, and Coupling in Multiple-Server Queues in Continuous Time
S. Asmussen and S. G. Foss 1

On Functionals of Branching Brownian Motion
A. N. Borodin and P. H. Salminen 7

Convergence Rates in Transient Phenomena for Branching Processes
K. A. Borovkov 22

The Wick Product
H. Gjessing, H. Holden, T. Lindstrøm, B. Øksendal, J. Ubøe, and T.-S. Zhang 29

Modelling VARMAX-Processes by Extended Sample Autocorrelation and Linear Regression Techniques
R. Höglund and R. Östermark 68

A Brief Account of the Theory of Homogeneous Gaussian Diffusions in Finite Dimensions
M. Jacobsen 86

A Note on First Passages in Branching Brownian Motions
I. Kaj and P. Salminen 95

Random Graphs with Marked Cycles
V. F. Kolchin 103

Local Limit Theorem on Convergence of Markov Chains to Diffusion Processes
V. D. Konakov 112

A Note on the Stationary Modification of the Loglikelihood Function in Asymptotic Continuous Time System Identification
T. Koski 135

Convergence of Filtered Experiments to the Experiment Generated by a Semimartingale
D. O. Kramkov 145

Martingale Approach to the Procedures of Stochastic Approximation
A. V. Melnikov and A. E. Rodkina 165

On a General Class of Stochastic Approximation Algorithms
A. V. Melnikov, A. E. Rodkina, and E. Valkeila 183

Stochastic Approximations in the Maximum Likelihood Inference for Markov Random Fields
A. Penttinen 197

The Edgeworth Expansion in $\mathbf{R}^k$
Yu. V. Prokhorov 206

An Improved Nonuniform Convergence Rate Estimate in the Central Limit Theorem in $\mathbf{R}^k$
V. V. Sazonov 209

Asymptotic Properties of the Maximum Likelihood Estimators under Random Normalization for a First Order Autoregressive Model
A. N. Shiryaev and V. G. Spokoiny 223

Infinite Systems of Dissimilar Particles with Random Interactions
A. V. Skorokhod 228

On Sequential Detection of a Small Disorder
V. G. Spokoiny 238

Branching Processes and Random Trees
V. A. Vatutin 256

On the Marcinkiewicz Weak Laws of Large Numbers in Banach Spaces
A. I. Volodin 269

Estimating Functions and Efficiency in a Filtered Model
W. Wefelmeyer 287

List of Contributors 296

PREFACE

During last decade Finland and Soviet Union were constantly extending their scientific collaboration, largely in the form of contacts between Academies of Sciences of these countries. Intensification of the existing scientific contacts became possible in the framework of the Perspective Program of Collaboration in Sciences and Humanities, which included in particular concentration of joint efforts in two traditional to our countries branches of mathematics, namely probability theory and mathematical statistics. The problem of establishing more close relations between Soviet and Finnish specialists in probability theory was repeatedly discussed, and the both parties agreeded in necessity of periodically organizing (with 2-3 year intervals) joint symposia. In above mentioned Program this idea was formally realised as one of the forms of our collaboration.

The first symposium took place in Helsinki and Lahti, August 31–September 5, 1987, and 26 scientists participated in it. As all following symposia, it was conducted under the aegis of the Soviet-Finnish Commission on Scientific and Technical Collaboration, and concrete organization work was done by Helsinki University. Scientific meetings were held in Mukkula-Conference-Hall in Lahti and were devoted to such topics as martingales, Markov processes, limit theorems in statistics and stochastic analysis.

The second symposium was organized in Leningrad, October 16–20, 1989, in Leningrad department of Steklov Mathematical Institute of the USSR Academy of Sciences; more then 30 scientists participated in it. The topics discussed included such branches of modern probability theory and mathematical statistics as martingales and their applications in statistics of stochastic processes, limit theorems in statistics, stochastic analysis.

The third symposium took place in Åbo Academy in Turku, August 13–16, 1991, with participation of about 50 scientists, and among them there were specialists from several Scandinavian and Baltic countries. The symposium had gathered a number of famous specialists in probability theory and mathematical statistics. Accordingly, a wide range of actual topics in probability theory, diffusion and branching processes, and stochastic analysis was discussed in the reports of its participants.

As expected, these symposia that have been prepared up till now by A.N. Shiryaev and A.V. Melnikov (Mathematical Institute of Russian Academy of Sciences) and H. Niemi and E. Valkeila (Helsinki University) are becoming more and more representative in the level of reports, the number of participants and countries involved. So it was agreed that the future Russian-Finnish symposia will be held with participation of specialists from Scandinavian and Baltic countries on permanent basis. The proceedings of the first two symposia were published in *Acad. Sci. Fennicae. Mathematica* in 1989 and 1992.

The present volume is a collection of papers presented by the participants of the third symposium. It encompasses the whole range of the topics discussed and is a tangible result of our collaboration. The editors hope that it will find readers in other countries as well.

It is a good opportunity for the editors to thank University of Helsinki, Åbo Academy, Academy of Finland, and Russia Academy of Sciences for their significant support of the symposium. TVP Science Publishers agreed to publish the Proceedings of the Third Symposium in a special volume and realized it in rather short time with high quality. And we are grateful to TVP Science Publishers for this work.

H. Niemi — *A.N. Shiryaev*

G. Högnas — *A.V. Melnikov*

Frontiers in Pure and Appl. Probab. 1, pp. 1 – 6
H. Niemi et al. (Eds.)
1993 TVP/VSP

RENOVATION, REGENERATION, AND COUPLING IN MULTIPLE–SERVER QUEUES IN CONTINUOUS TIME

S. ASMUSSEN and S. G. FOSS

Institute of Electronic Systems, Fr. Bajersv. 7, DK 9220 Aalborg, Denmark
Novosibirsk State University, Pirogov Street 2, 630090 Novosibirsk, Russia

ABSTRACT

This paper demonstrates some different, though related, approaches for obtaining ergodicity results for queues on a $GI/GI/p$ example, where the weak convergence of the workload- and queue length vectors in continuous time to their stationary distributions is derived under minimal assumptions. The spectrum of methods incorporate concepts like renovating events, regeneration (in a broad sense) and ε-coupling.

1. INTRODUCTION

We consider the $GI/GI/p$ queue with service times $U_0, U_1, \ldots$ and interarrival times $T_0, T_1, \ldots$ with common distribution $A(dt)$. We denote by $\vec{W}_n = (W_n^{(1)}, \ldots, W_n^{(p)})$ the Kiefer–Wolfowitz vector of workloads at the p stations just before the arrival of the nth customer ($n = 0, 1, \ldots$), with the components ordered in ascending order, and by $\vec{V}_t = \left(V_t^{(1)}, \ldots, V_t^{(p)}\right)$ the (ordered) vector of workloads at time t. We allow general initial conditions so that $\vec{V}_0$ is not necessarily the zero vector and that customer 0 arrives at time $\tau(0) = D$, where D needs not be zero (the nth customer arrives at time $\tau(n) = D + T_0 + \cdots + T_{n-1}$). The connection between these two processes in the customer time scale and the physical time scale, respectively, is

$$\vec{W}_n = \vec{V}_{D+T_0+\cdots+T_{n-1}-}. \tag{1.1}$$

We assume throughout that $\rho = \mathbf{E}U/(p\mathbf{E}T) < 1$. Then the following result is well-known (see, e.g., (Borovkov, 1984; Asmussen, 1987)).

THEOREM 1.1. *Without conditions beyond $\rho < 1$, there exists a random vector $\vec{W}$ such that $\vec{W}_n \to \vec{W}$ in total variation as $n \to \infty$.*

(Actually, it is even known that $\{W_n\}$ is a Harris chain, see (Charlot et al., 1978).) From this and analogues with single-server queues, one expects the following result to be true:

THEOREM 1.2. *If $\rho < 1$ and A is nonlattice, then there exists a random vector $\vec{V}$ such that $\vec{V}_t \xrightarrow{D} \vec{V}$ as $t \to \infty$.*

In fact, it is sometimes claimed in the literature that this can be proved exactly as for single-server queues. In our opinion, the problem is, however, more delicate than it may appear at first glance, and it seems that Theorem 1.2 has only been proved with some additional assumptions like regeneration in the idle state (see the discussion in §2) or A being spread-out as in (Asmussen, 1987, Chapter XI). Therefore, our purpose here is therefore of two kinds: to provide rigorous proofs without assumptions beyond those of Theorem 1.2 and to give some general discussion of the problem which we believe has a wider methodological interest. In this respect, the aim is the same as in (Asmussen, 1992; Foss and Kalashnikov, 1991). In fact, §5 of the present note and (Foss and Kalashnikov, 1991), which exploit the method of renovating events originating from the Russian literature (e.g., (Borovkov, 1984; Foss, 1986)) and discuss its relation to the concept of regeneration, are parallel outcomes of discussions at a January 1990 conference in Karpacz, Poland; similarly, §4 provides one of the main applications of the method of ε-coupling as discussed in (Asmussen, 1992) (and originating from (Lindvall, 1977)).

2. EXISTENCE OF LIMITS IN DISCRETE TIME

The problem in question is easy if $\mathbf{P}(U < T) > 0$: The system then empties infinitely often (i.o.), and the instants of customers entering an the empty system serve as regeneration points (both in discrete and continuous time) in the sense that the process is split up into i.i.d. cycles. Unfortunately, $\mathbf{P}(U < T) > 0$ is not automatic from $\rho < 1$ as in the single-server case, and what can be inferred is only that $\mathbf{P}(U <!pT) > 0$. These points are discussed, e.g., in (Whitt, 1972). In discrete time it turns out, however, that the relation $\mathbf{P}(U < pT) > 0$ can be exploited in a more subtle way to prove Theorem 1. The idea goes back at least to (Gnedenko and Kovalenko, 1968) (see also (Borovkov, 1978)) and is used, e.g., in (Borovkov, 1984) and (Asmussen, 1987; Chapter XI) in slightly different ways. An outline of the argument is as follows. Choose u_+, t_- with $u_+ < pt_-$, $\mathbf{P}(U < u_+) > 0$, $\mathbf{P}(T > t_-) > 0$, let K, L be large numbers to be defined later on (with L integer), and define $\varepsilon = pt_- - u_+$,

$$A_n = \{W_n^{(p)} \le K\} \bigcap_{j=n}^{n+L} \{U_j < u_+\} \cap \{T_j > t_-\}. \tag{2.1}$$

If A_n occurs, then at least the amount ε of work is cleared at each service station between each arrival $j = n, \ldots, n + L$ and the following one. Thus, if $L > pK/\varepsilon$, then the $\vec{W}_{n+L}$, $\vec{W}_{n+L+1}$, ... are completely independent of the value of $\vec{W}_n$. Equivalently,

$$\vec{W}_{n+k} = \varphi_k(T_n, \ldots, T_{n+k-1}, U_n, \ldots, U_{n+k-1}), \qquad k \ge L, \tag{2.2}$$

which in the terminology of (Borovkov, 1984; Foss, 1986) means that the event A_n is *renovating* on $\{n, \ldots, n + L\}$. Note that this line of thought is similar to the splitting argument for Harris chains (Nummelin, 1978; Athreya and Ney, 1978), the set $S_K = \{\vec{w} : w_n^{(p)} \leq K\}$ serves as a splitting set from which there is a uniform probability of having the "minorant measure" after L steps. Moreover, it was shown in (Borovkov and Foss, 1992) that these two concepts, positive recurrence in the Harris sense and the existence of a stationary sequence of renovating events, are equivalent in the Markov case.

Total variation convergence of $\vec{W}_n$ can now be deduced, for example, along the following lines. First show that if K is large enough, then $\mathbf{P}(A_n \text{ i.o.}) = 1$ and then let $\alpha(0) = \inf\{n \geq 0: A_n\}$, $\alpha(k+1) = \inf\{n \geq \alpha(k) + L: A_n\}$. Then $\{\alpha(k)\}$ is a renewal process, and $\vec{W}_{\alpha(k)+L}, \vec{W}_{\alpha(k)+L+1}, \ldots$ is independent of $\{\alpha(0), \ldots, \alpha(k)\}$ with distribution not depending on k. This means that $\left\{\vec{W}_n\right\}$ is regenerative (in a wide sense, cf. (Asmussen, 1987, Chapter V)) with respect to the renewal process $\{\alpha(k) + L\}$, and the checking that the cycle length $\alpha(k+1) - \alpha(k)$ is aperiodic with finite mean establishes Theorem 1.

3. RENOVATION AND REGENERATION IN CONTINUOUS TIME

In continuous time, it is tempting to infer from the imbedding (1.1) of $\left\{\vec{V}_t\right\}$ in $\left\{\vec{W}_n\right\}$ that $\left\{\vec{V}_t\right\}$ is regenerative with respect to $\{\tau(\alpha(k)+L)\}$. However, a closer inspection reveals some difficulties: for example, the post-$\tau(\alpha(k)+L)$ process is independent of $D, T_0, \ldots, T_{\alpha(k)-1}$ but not necessarily of $T_{\alpha(k)}, \ldots, T_{\alpha(k)+L-1}$ and, therefore, may depend on $\tau(\alpha(k)+L)$.

To overcome this difficulty, we need a modification of our definition of renovating events. Let u_+, t_- be as before, but choose now $t_+ > t_-$ with $\mathbf{P}(t_- < T < t_+) > 0$ and define

$$B_n \;=\; \{W_n^{(p)} \leq K\} \bigcap_{j=n}^{n+L} \{U_j < u_+\} \cap \{t_- < T_j < t_+\}. \tag{3.1}$$

Then B_n is renovating on $[\tau(n), \tau(n) + Lt_+)$ in the sense that the post-$\tau(n) + Lt_+$ process does not depend on what happened before $\tau(n)$. Thus, if we put $\vec{U}_t = \vec{V}_{t+Lt_+}$,

$$\beta(0) = \inf\{n \geq 0 : B_n\}, \quad \beta(k+1) = \inf\{n \geq \beta(k) + L : B_n\},$$

then $\left\{\vec{U}_t\right\}$ is regenerative (in the wide sense) w.r.t. $\{\tau(\beta(k))\}$ and $\{\tau(\beta(k))\}$ is readily checked to be a renewal process. Thus to obtain weak convergence of $\left\{\vec{U}_t\right\}$ and thereby $\left\{\vec{V}_t\right\}$, we have only to check that $\{\tau(\beta(k))\}$ is nonlattice and has finite mean interarrival time. The first of these statements is a general property of thinning of a nonlattice renewal process (Asmussen, 1987, p.187). For the second, according to Wald's identity it is sufficient to check that $\{\beta(k)\}$ has finite mean interarrival time. However, from the positive recurrence in discrete time we know that the limiting proportion of time $\left\{\vec{W}_n\right\}$ visits S_K is strictly positive. Hence by the geometric trials argument, the limiting proportion of instants of the form $\beta(k)$ is also positive.

4. ε-COUPLING

Our second proof of Theorem 1.2 proceeds by ε-coupling, in the form of the following result from (Asmussen, 1992).

LEMMA 4.1. *Let $\{Z_t\}_{0\le t<\infty}$, $\{Z'_t\}_{-\infty<t<\infty}$ be right-continuous stochastic processes with the same metric state space E, and assume that*

(a) *$\{Z'_t\}$ is strictly stationary with invariant distribution π,*

(b) *for each $\varepsilon > 0$ one can find versions of $\{Z_t\}$, $\{Z'_t\}$ defined on a common probability space and a.s. finite random times $D = D_\varepsilon$, $T = T_\varepsilon$ such that*

$$|D| \le \varepsilon, \quad Z_t = Z'_{t+D}, \qquad t \ge T. \tag{4.1}$$

Then $Z_t \xrightarrow{D} \pi$, $t \to \infty$.

Thus, to prove Theorem 1.2, it is sufficient to find random times δ, T and a stationary version $\left\{\vec{V}'_t\right\}$ of $\left\{\vec{V}_t\right\}$ such that

$$\vec{V}_t = \vec{V}'_{t+\delta}, \qquad t \ge T, \tag{4.2}$$

where ε is a given positive number.

The existence of a stationary version can be inferred in various ways, say (a) from the existence of a stationary version of $\left\{\vec{W}_n\right\}$ and the theory of Palm measures (see, e.g., (Rolski, 1981)) or (b) by adding the elapsed time A_t since the last arrival time to obtain the Markov process $\left\{(\vec{V}_t, A_t)\right\}$ with the imbedded Harris chain $\left\{(\vec{W}_n, 0)\right\}$. Then a unique invariant measure π exists ((Asmussen, 1992, Corollary 3.1)), and $|\pi| < \infty$ follows, say by tightness, which in turn is a consequence of comparisons with single server queues (see (Asmussen, 1987, Chapter XI) for details).

Now since renewal processes can be ε-coupled (see (Lindvall, 1977; Asmussen, 1992)), we can construct $\left\{\vec{V}'_t\right\}$, $\left\{\vec{V'}_t\right\}$ on a common probability space such that for suitable l, l' we have $|\delta| \le \varepsilon$, where $\delta = \tau(l) - \tau(l')$ and then take

$$T_{l+k} = T'_{l'+k}, \quad U_{l+k} = U'_{l'+k}, \qquad k = 0, 1, \ldots. \tag{4.3}$$

By Proposition 3.1 of (Asmussen, 1992), there is a K such that $\left\{(\vec{W}_{l+n}, \vec{W}'_{l'+n})\right\}$ visits $S_K \times S_K$ i.o. By the splitting (renovation) property surveyed in §2, there is the uniform lower bound $\mathbf{P}(U < u_+)^L \mathbf{P}(T > t_-)^L$ of the probability that the coupling of $\left\{\vec{W}_{l+n}\right\}$, $\left\{\vec{W}'_{l'+n}\right\}$ is obtained within L steps from a visit to $S_K \times S_K$. Hence for some m we have $\vec{W}_{l+k} = \vec{W}'_{l'+k}$ for all $k \ge m$, and combining with (4.3), we see that (4.2) holds with $T = \tau(m)$.

Remark. From the fact that Harris chains can be coupled (see, e.g., (Asmussen and Thorisson, 1987)) it is tempting to infer without further arguing that coupling of $\left\{\vec{W}_{l+n}\right\}, \left\{\vec{W}'_{l'+n}\right\}$ occurs. However, the two processes are dependent because of the ε-coupling of the arrival processes, and it is certainly not true that any two dependent Harris chains eventually meet. Thus, some explicit knowledge of the coupling structure must enter the argument at some stage. This problem is sometimes overlooked (e.g., (Asmussen, 1987, p. 253; Sigman, 1989, proof of Theorem 5.1)), but is easily repaired along the above lines.

5. EXTENSIONS

The weak convergence of a number of related functionals follows easily by entirely similar arguments. In particular, let $\vec{Z}_t$ be the $3p+2$ dimensional vector

$$\vec{Z}_t = \left(A_t, B_t, \vec{V}_t, R_t^{(1)}, \ldots, R_t^{(p)}, Q_t^{(1)}, \ldots, Q_t^{(p)}\right),$$

where A_t is the time since the last arrival, B_t the time until the next arrival, $R_t^{(i)}$ the residual service time at the ith station and $Q_t^{(i)}$ the number in line at the ith station.

COROLLARY 5.1. *If $\rho < 1$ and A is nonlattice, then there exists a random vector $\vec{Z}$ such that $\vec{Z}_t \xrightarrow{D} \vec{Z}$ as $t \to \infty$.*

Applying the continuous mapping theorem, say to the total number in system, we get the following result of (Borovkov, 1990).

COROLLARY 5.2. *If $\rho < 1$ and A is nonlattice, then there exists a random variable Q such that $Q_t = Q_t^{(1)} + \cdots + Q_t^{(p)} \xrightarrow{D} \vec{Q}$ as $t \to \infty$.*

A regenerative structure similar to that of the present paper, the regeneration occuring as a result of short service times and long interarrival times following visits to compact sets, is a quite common feature of queueing models, and obviously then weak convergence to stationarity can be obtained by one or another of the present methods.

Acknowledgements
We would like to thank V.V. Kalashnikov and H. Thorisson for stimulating discussions.

REFERENCES

Asmussen, S. (1987). *Applied Probability and Queues.* John Wiley, Chichester, New York.

Asmussen, S. (1992). On coupling and weak convergence to stationarity. *Ann. Appl. Probab.* (to appear).

Asmussen, S. and Thorisson, H. (1987). A Markov chain approach to periodic queues. *J. Appl. Probab.* **24**, 215–225.

Athreya, K.B. and Ney, P. (1978). A new approach to the limit theory of recurrent Markov chains. *Trans. Amer. Math. Soc.* **245**, 493–501.

Borovkov, A.A. (1978). Ergodicity and stability results for a class of stochastic systems. *Theory Probab. Appl.* **23**, 227–247.

Borovkov, A.A. (1984). *Asymptotic Methods in Queueing Theory.* John Wiley, Chichester, New York.

Borovkov, A.A. (1990). *Stochastic Processes in Queueing Theory.* Springer-Verlag, New York.

Borovkov, A.A. and Foss, S.G. (1992). Stochastically recursive sequences and their generalization. *Siberian Math. J.* **2**, 16–81.

Charlot, F., Ghidouche, M., and Hamami M. (1978). Irreducibilite et recurrence au sens de Harris des "Temps d'attente" des files $GI/G/q$. *Z. Wahrscheinlichkeitstheor. verw. Geb.* **43**, 187–203.

Foss, S.G. (1986). The method of renovating events and its applications in queueing theory. In: *Proceedings of the International Symposium on Semi-Markov Processes and Their Applications, June* 4–7, (1984), Brussels, Belgium. Ed. by J. Janssen, Plenum Press, New York, pp. 337–350.

Foss, S.G. (1991). Ergodicity of queueing networks. *Siberian Math. J.* **32**, 184–203.

Foss, S.G. and Kalashnikov, V.V. (1991). Regeneration and renovation in queues. *Queueing Systems* **8**, 211–223.

Gnedenko, B. and Kovalenko, I.N. (1968). *An Introduction to Queueing Theory.* Israel Program for Scientific Translations, Jerusalem.

Lindvall, T. (1977). A probabilistic proof of Blackwell's renewal theorem. *Ann. Probab.* **5**, 482–485.

Miyazawa, M. (1977). Time and customer processes in queues with stationary inputs. *J. Appl. Probab.* **14**, 349–357.

Nummelin, E. (1978). A splitting technique for Harris recurrent Markov chains. *Z. Wahrscheinlichkeitstheor. verw. Geb.* **43**, 309–318.

Rolski, T. (1981). *Stationary Random Processes Associated with Point Processes.* Springer-Verlag, New York.

Sigman, K. (1989). One-dependent regenerative processes and queues in continuous time. *Math. Operat. Res.* **15**, 175–189.

Sigman, K. (1990). The stability of open queueing networks. *Stoch. Proc. Appl.* **35**, 11–25.

de Smit, J.H.A. (1973). Some general results for many-server queues. *Adv. Appl. Probab.* **5**, 153–169.

Whitt, W. (1972). Embedded renewal processes in the $GI/G/s$ queue. *J. Appl. Probab.* **9**, 650–658.

Frontiers in Pure and Applied Probab. 1, pp. 7 – 21
H. Niemi et al. (Eds.)
1993 TVP/VSP

ON FUNCTIONALS OF BRANCHING BROWNIAN MOTION

A. N. BORODIN* and P. H. SALMINEN
Steklov Mathematical Institute, St. Petersburg Branch, Naberezhnaya Fontanki 27, 191011 St. Petersburg, Russia
Åbo Akademi, Mathematical Institute, SF-20500 Åbo 50, Finland

ABSTRACT

In this paper we consider additive functionals of integral type of a branching Brownian motion. Differential equations are derived for Laplace transforms of such functionals. Distributions of local times, spatial maximum and minimum of a branching Brownian tree are also treated. Our results generalise some results obtained by Sevast'yanov in 1958. Furthermore, explicit examples are presented.

1. INTRODUCTION

Let $X = \{X_t\colon t \geq 0\}$ be a branching Brownian motion (BBM) in $\mathbf{R}$ with constant creation rate α and offspring distribution $p = \{p_k\colon k = 0, 1, 2, \ldots\}$. By

$$G(z) := \sum_{n=0}^{\infty} p_n z^n, \qquad |z| \leq 1,$$

we denote the generating function of p. Assume that $G'(1) < \infty$. The process starts with one particle located at some given point x at time 0. The probability measure associated with X is denoted by $\mathbf{P}_x$. Let N_t be the number of particles in the process X at time t and, in the case $N_t \geq 1$, let $(\omega_1(t), \omega_2(t), \ldots, \omega_{N_t}(t))$ be their locations, i.e., the state of X at time t. Here, the order of the particles plays no role. We refer to (Ikeda, Nagasawa, and Watanabe, 1968a,b; 1969) or (Asmussen and Hering, 1983) for a rigorous definition of the state space of a branching Markov process. Given the state of X at the "last-before-t" branching time the locations $\omega_i(t)$ vary until the next branching time independently of each other according to the law of a standard Brownian motion (BM) denoted by $\mathbf{P}^B$. For $t > 0$ and a given non-negative measurable function f we are interested in the distribution of the additive functional

$$I_t(f) := \int_0^t \sum_{k=1}^{N_s} f(\omega_k(s))ds.$$

*Research supported in part by internal funds of Åbo Akademi and by the Research Institute of the Foundation of Åbo Akademi.

More generally, introducing the local time of X at level y via

$$L_t(y) := \lim_{\varepsilon \downarrow 0} \frac{1}{\varepsilon} \int_0^t \sum_{k=1}^{N_s} 1_{[y,y+\varepsilon)}(\omega_k(s))ds, \tag{1.1}$$

where $1_{\{\cdot\}}$ is the indicator function of a set $\{\cdot\}$, we consider for $n \in \mathbf{N}$, $\lambda_i \geq 0$, and $y_i \in \mathbf{R}$, $i = 1, 2, \cdots, n$, the functional

$$A_t := I_t(f) + \sum_{i=1}^{n} \lambda_i L_t(y_i). \tag{1.2}$$

The fact that the limit in (1.1) exists $\mathbf{P}$-a.s. follows easily from the corresponding result for BM.

In the next section it is shown (Theorem 2.1) that

$$u(t,x) := \mathbf{E}_x(\exp(-I_t(f)))$$

solves an integral equation which, in the case when f is locally Hölder continuous, is equivalent to a nonlinear partial differential equation. This result could be called Feynman–Kac's formula for BBM. It may also be viewed as a particular example of a multiplicative functional of branching type as introduced in (Ikeda, Nagasawa, and Watanabe, 1968a,b; 1969, §5, especially Example 5.2). However, we feel that it is worthwhile in this important specific case to give a short proof.

In Theorems 2.2 and 2.3 ordinary differential equations are derived for the Laplace transform of

$$A_\infty := \lim_{t \to \infty} A_t$$

for (i) a killed BBM and (ii) in the presence of a condition on maximum and minimum of BBM, respectively. These are generalizations of some classical results of Sevast'yanov (Sevast'yanov, 1958); see also (Watanabe, 1965).

In §3 we give some remarks concerning the moments of A_t and A_∞ and their computation. Finally, in §4 some explicit computations are presented. We are looking there at the critical case $p_0 = p_2 = 1/2$ and compute, e.g., the Laplace transform of the total time of positivity of this BBM, i.e., $f = 1_{[0,\infty)}$ in $I_\infty(f)$.

2. MAIN RESULTS

Let $(X, \mathbf{P}_x)$ be a BBM as introduced above.

THEOREM 2.1. *Let f and g be bounded measurable functions. Moreover, assume that $|g| \leq 1$. Then the function*

$$u(t,x) := \mathbf{E}_x\left(\prod_{k=1}^{N_t} g(\omega_k(t)) \exp\left(-\int_0^t \sum_{k=1}^{N_s} f(\omega_k(s))ds\right)\right) \tag{2.1}$$

satisfies the integral equation

$$u(t,x) = e^{-\alpha t}u_0(t,x) + \alpha \int_0^t ds \left(e^{-\alpha s} \int_{\mathbf{R}} K(s;x,dy)G(u(t-s,y)) \right), \quad (2.2)$$

where for $A \in \mathcal{B}(\mathbf{R})$

$$K(t;x,A) := \mathbf{E}_x^B \left(\exp\left(-\int_0^t f(\omega(s))ds \right); \quad \omega(t) \in A \right),$$

$$u_0(t,x) := K(t;x,g) := \mathbf{E}_x^B \left(g(\omega(t)) \exp\left(-\int_0^t f(\omega(s))ds \right) \right).$$

In the case where f is locally Hölder continuous, (2.2) *is equivalent to the* PDE

$$\frac{\partial u}{\partial t} = \frac{1}{2}\frac{\partial^2 u}{\partial x^2} - (\alpha + f)u + \alpha G(u) \quad (2.3)$$

with the initial condition

$$u(0,x) = g(x). \quad (2.4)$$

Proof. Let β be the first branching time. Then $\beta \sim \exp(\alpha)$ and we have

$$u(t,x) = \mathbf{E}_x \left(g(\omega(t)) \exp\left(-\int_0^t f(\omega(s))ds \right); \ t < \beta \right)$$

$$+\mathbf{E}_x \left(\prod_{k=1}^{N_t} g(\omega_k(t)) \exp\left(-\int_0^\beta f(\omega(s))ds - \int_\beta^t \sum_{k=1}^{N_s} f(\omega_k(s))ds \right); \ t > \beta \right).$$

From this using the branching property, we obtain equation (2.2). Assume now that f is locally Hölder continuous. Then it is well known (the Kac's formula) that $(t,x) \mapsto u_0(t,x)$ solves

$$\begin{cases} \dfrac{\partial u_0}{\partial t} = \dfrac{1}{2}\dfrac{\partial^2 u_0}{\partial x^2} - f u_0, \\ u_0(0+,x) = g(x). \end{cases} \quad (2.5)$$

Let

$$V(s, t-s, x) := K(s;x,G(u(t-s,\cdot))).$$

Since $(t,x) \mapsto G(u(t,x))$ is bounded and measurable (in fact, jointly continuous), it follows that $(s,x) \mapsto V(t,s,x)$ is smooth and solves (the Kac's formula)

$$\begin{cases} \dfrac{\partial V}{\partial s} = \dfrac{1}{2}\dfrac{\partial^2 V}{\partial x^2} - fV, \\ V(t,0+,x) = G(u(t,x)). \end{cases} \tag{2.6}$$

Differentiating in (2.2) with respect to t and taking into account (2.5) and (2.6), we have

$$\frac{\partial}{\partial t}u(t,x) = \frac{\partial}{\partial t}(e^{-\alpha t}u_0(t,x)) + \alpha\int_0^t \frac{\partial}{\partial t}(e^{-\alpha(t-s)}V(s,t-s,x))ds + \alpha V(t,0+,x)$$

$$= \frac{1}{2}\frac{\partial^2}{\partial x^2}u(t,x) - (\alpha + f(x))u(t,x) + \alpha G(u(t,x)).$$

Initial condition (2.4) is satisfied by the definition (2.1) of the function u.

Let A_t be a functional of type (1.2) with $f \geq 0$. Since an ordinary BM may be seen as a BBM, we use the notation A_t also for the Brownian functional

$$A_t := \int_0^t f(\omega(s))ds + \sum_{k=1}^n \lambda_k l_t(y_k),$$

where $l_t(y)$ is the Brownian local time defined by

$$l_t(y) := \lim_{\varepsilon\downarrow 0}\frac{1}{\varepsilon}\int_0^t 1_{[y,y+\varepsilon)}(\omega(s))ds.$$

This should not cause any ambiguity because the measure under which A_t is considered is always explicitly given. Further, let

$$A_\infty := \lim_{t\to\infty} A_t,$$

and assume that $\mathbf{P}_x\{A_\infty > 0\} > 0$. It is obvious that in the subcritical case, i.e., $G'(1) < 1$, we have $A_\infty < \infty$ a.s. In the (super)critical case, i.e., $G'(1) \geq 1$, it is easily seen that

$$\{A_\infty < \infty\} = \{\zeta < \infty\} \quad \mathbf{P}.\text{-a.s.},$$

where $\zeta := \inf\{t\colon N_t = 0\}$ is the extinction time of X. Consequently,

$$\mathbf{E}.(\exp(-A_\infty)) = \mathbf{E}.(\exp(-A_\infty);\ \zeta < \infty). \tag{2.7}$$

Let now $(\widehat{X}, \widehat{\mathbf{P}}.)$ be a BBM living inside a given interval (a,b), $-\infty < a < b < \infty$. By this we mean a branching process having an ordinary BM living in (a,b) and killed when it hits a or b as the nonbranching part. Informally, this means that a particle in the process X hitting a or b is immediately at that time removed from

the process. The extinction of $\widehat{X}$ (the extinction time is defined as above) can happen in two different ways:
1) The last particle dies without offspring.
2) The last particle is killed at the boundary point a or b.

THEOREM 2.2. *Let $\widehat{X}$ and A_∞ be as above. Then the function*

$$\hat{u}(x) := \widehat{\mathbf{E}}_x(\exp(-A_\infty))$$

is the smallest non-negative solution of the integral equation

$$\hat{u}(x) = \hat{u}_0(x) + \int_{\mathbf{R}} \widehat{K}_\alpha(x, dy) G(\hat{u}(y)), \tag{2.8}$$

where for $E \in \mathcal{B}((a,b))$, $\tau := \inf\{t\colon \omega(t) = a \text{ or } b\}$, and $\eta \sim \exp(\alpha)$

$$\widehat{K}_\alpha(x, E) := \widehat{\mathbf{E}}_x^B(\exp(-A_\eta);\ \omega(\eta) \in E),$$
$$\hat{u}_0(x) := \widehat{\mathbf{E}}_x^B(\exp(-\alpha\tau - A_\infty)).$$

In the case where f is piecewise continuous in $[a,b]$ the function $\hat{u}$ is the unique solution of the problem

$$\begin{cases} \frac{1}{2}\hat{u}'' - (\alpha + f)\hat{u} + \alpha G(\hat{u}) = 0 & \text{on } (a,b) \setminus \{y_1, y_2, \dots, y_n\}, \\ \hat{u}'(y_i+) - \hat{u}'(y_i-) = 2\lambda_i \hat{u}(y_i), & \\ \hat{u}(a+) = \hat{u}(b-) = 1, & \\ 0 \le \hat{u} \le 1, & \text{on } (a,b). \end{cases} \tag{2.9}$$

Proof. To get (2.8) consider

$$\begin{aligned} \hat{u}(x) &= \widehat{\mathbf{E}}_x(\exp(-A_\infty);\ \tau \le \beta) + \widehat{\mathbf{E}}_x(\exp(-A_\infty);\ \tau > \beta) \\ &= \hat{u}_0(x) + \widehat{\mathbf{E}}_x(\exp(-A_\beta - (A_\infty - A_\beta));\ \tau > \beta) \\ &= \hat{u}_0(x) + \widehat{\mathbf{E}}_x\left(\exp(-A_\beta) \sum_{n=0}^{\infty} p_n (\widehat{\mathbf{E}}_{\omega_\beta}(\exp(-A_\infty)))^n;\ \tau > \beta\right) \\ &= \hat{u}_0(x) + \int_0^\infty dt \alpha e^{-\alpha t} \int_{\mathbf{R}} \widehat{\mathbf{E}}_x^B(\exp(-A_t);\ \omega_t \in dy) G(\hat{u}(y)) \\ &= \hat{u}_0(x) + \int_{\mathbf{R}} \widehat{K}_\alpha(x, dy) G(\hat{u}(y)). \end{aligned}$$

To prove that $\hat{u}$ is the smallest solution we use the idea of Sevast'yanov (Sevast'yanov, 1958), see also (Watanabe, 1965). For this define a sequence of random times via

$$T_0 := 0, \qquad T_k := \inf\{t > T_{k-1};\ N_t \ne N_{t-}\}$$

and introduce a sequence of functions via

$$\hat{u}^{(0)}(x) = 0,$$
$$\hat{u}^{(k)}(x) = \hat{\mathbf{E}}_x(\exp(-A_\infty);\ N_{T_k} = 0).$$

Using the convention "if $N_t = 0$, then $N_{t+s} = 0$ for all $s \geq 0$", and (2.7) we have $\hat{u}^{(k)}(x) \uparrow \hat{u}(x)$ for every x as $k \to \infty$. Furthermore, we have for $k \geq 2$

$$\hat{u}^{(k)}(x) = \widehat{\mathbf{E}}_x(\exp(-A_\infty);\ N_{T_1} = 0,\ N_{T_k} = 0) + \widehat{\mathbf{E}}_x(\exp(-A_\infty);\ N_{T_1} \neq 0,\ N_{T_k} = 0)$$
$$= \hat{u}_0(x) + p_0 \widehat{K}_\alpha(x, \mathbf{R}) + \int_{\mathbf{R}} \widehat{K}_\alpha(x, dy) \sum_{n=1}^{\infty} p_n \widehat{\mathbf{E}}_{\bar{y}_n}(\exp(-A_\infty);\ N_{T_{k-1}} = 0),$$

where $\bar{y}_n = (y, \ldots, y) \in \mathbf{R}^n$. Denote by $\omega^{(i)}$ the subtree starting at the ith coordinate in $\bar{y}_n$. Then, taking into account that

$$\{X_0 = \bar{y}_n;\ N_{T_{k-1}} = 0\} \subset \{X_0 = \bar{y}_n,\ N_{T_{k-1}}(\omega^{(1)}) = 0, \ldots, N_{T_{k-1}}(\omega^{(n)}) = 0\},$$

we have

$$\mathbf{E}_{\bar{y}_n}(\exp(-A_\infty);\ N_{T_{k-1}} = 0) = \widehat{\mathbf{E}}_{\bar{y}_n}\left(\exp\left(-\sum_{i=1}^{n} A_\infty(\omega^{(i)})\right);\ N_{T_{k-1}} = 0\right)$$
$$\leq \widehat{\mathbf{E}}_{\bar{y}_n}\left(\exp\left(-\sum_{i=1}^{n} A_\infty(\omega^{(i)})\right);\ N_{T_{k-1}}(\omega^{(1)}) = 0, \ldots, N_{T_{k-1}}(\omega^{(n)}) = 0\right)$$
$$= \left(\widehat{\mathbf{E}}_y(\exp(-A_\infty);\ N_{T_{k-1}} = 0)\right)^n.$$

Consequentely,

$$\hat{u}^{(k)}(x) \leq \hat{u}_0(x) + \int \widehat{K}_\alpha(x, dy) G(\hat{u}^{(k-1)}(y)). \tag{2.10}$$

Letting now $\hat{v}$ be an arbitrary non-negative solution of (2.8) by induction it follows easily from (2.10) that $\hat{u}(x) \leq \hat{v}(x)$. Next assume that f is piecewise continuous on $[a, b]$, and introduce

$$\hat{u}_1(x) := \frac{1}{\alpha} \int \widehat{K}_\alpha(x, dy) G(\hat{u}(y)).$$

Since $x \mapsto \hat{u}(x)$ is continuous the mapping $x \mapsto G(\hat{u}(x))$ is also continuous. It follows from Theorem 2.1 in (Borodin, 1989) that $\hat{u}_1$ is the solution of

$$\begin{cases} \frac{1}{2}\hat{u}_1'' - (\alpha + f)\hat{u}_1 = -G(\hat{u}) & \text{on} \quad (a, b) \backslash \{y_1, y_2, \ldots, y_n\}, \\ \hat{u}_1'(y_i+) - \hat{u}_1'(y_i-) = 2\lambda_i \hat{u}_1(y_i), & \\ \hat{u}_1(a+) = \hat{u}_1(b-) = 0. & \end{cases}$$

Further, since $\hat{u}_0$ satisfies

$$\begin{cases} \frac{1}{2}\hat{u}_0'' - (\alpha + f)\hat{u}_0 = 0, & \text{on} \quad (a,b)\backslash\{y_1, y_2, \dots, y_n\}, \\ \hat{u}_0'(y_i+) - \hat{u}_0'(y_i-) = 2\lambda_i \hat{u}_0(y_i), & \\ \hat{u}_1(a+) = \hat{u}_1(b-) = 1, & \end{cases}$$

it follows that $\hat{u} = \hat{u}_0 + \alpha\hat{u}_1$ solves (2.9). To prove the uniqueness assume that $\hat{v}$ is a solution such that $0 \le \hat{u} \le \hat{v} \le 1$. Set $\check{u} := 1 - \hat{u}$ and $\check{v} := 1 - \hat{v}$, then $0 \le \check{v} \le \check{u} \le 1$ and they solve the problem

$$\begin{cases} \mathcal{A}\varphi + \alpha(G(1-\varphi) - (1-\varphi)) = f & \text{on} \quad (a,b) \setminus \{y_1, y_2, \cdots, y_n\}, \\ \varphi'(y_i+) - \varphi'(y_i-) = -2\lambda_i(1 - \varphi(y_i)), & \\ \varphi(a+) = \varphi(b-) = 0, & \\ 0 \le \varphi \le 1, & \text{on} \quad (a,b), \end{cases}$$

where

$$\mathcal{A}\varphi := -\frac{1}{2}\varphi'' + f\varphi. \tag{2.11}$$

From this we obtain easily that

$$\int(\check{v}\mathcal{A}\check{u} - \check{u}\mathcal{A}\check{v}) + \int f(\check{u} - \check{v}) + \alpha\int\big(\check{v}(G(1-\check{u}) - (1-\check{u}))$$

$$-\check{u}(G(1-\check{v}) - (1-\check{v}))\big) = 0, \tag{2.12}$$

where the integration is carried over $[a,b]$. Using the conditions at $a, y_1, \dots, y_n$ and b given in (2.9) it is easily seen that

$$\int(\check{v}\mathcal{A}\check{u} - \check{u}\mathcal{A}\check{v}) = \sum_{i=1}^{n}\lambda_i(\check{u}(y_i) - \check{v}(y_i)).$$

Therefore, since $f \ge 0$ and $\check{u} \ge \check{v}$ the first and the second terms in (2.12) are non-negative. For the third term consider

$$\int\Big(\check{v}(G(1-\check{u}) - (1-\check{u})) - \check{u}(G(1-\check{v}) - (1-\check{v})big)\Big)$$

$$= \int\check{u}\check{v}\left(\frac{G(1-\check{u}) - (1-\check{u})}{\check{u}} - \frac{G(1-\check{v}) - (1-\check{v})}{\check{v}}\right)$$

$$= \int\check{u}\check{v}\left(\frac{1 - G(1-\check{v})}{\check{v}} - \frac{1 - G(1-\check{u})}{\check{u}}\right).$$

The claim is that

$$x \mapsto R(x) := \frac{1 - G(1-x)}{x}, \qquad 0 < x \le 1,$$

is decreasing. From this it follows that even the third integral is non-negative, and this implies that $\check{u} = \check{v}$. To prove the claim consider the function

$$x \mapsto H(x) := R(1-x), \qquad 0 \le x < 1.$$

An easy computation shows that

$$H(x) = \sum_{n=0}^{\infty} x^n \left(1 - \sum_{m=0}^{n} p_m\right).$$

Since $\{p_m;\ m = 0, 1, 2\ldots\}$ is a probability distribution it is obvious that H is increasing, and hence R is decreasing. The proof is complete.

In the next result the event $\{\inf X_s > a\}$ means that no particle in the process X enters the set $(-\infty, a]$. The event $\{\sup X_s < a\}$ is defined analogously. Denote by q the extinction probability of X, i.e., q is the smallest non-negative solution of the equation $x = G(x)$.

THEOREM 2.3. *Let X be a BBM and $A_\infty := \lim A_t$, where A_t is as in (1.2). Then the function*

$$\tilde{u}(x) := \mathbf{E}_x(\exp(-A_\infty);\ a < \inf X_s,\ \sup X_s < b)$$

is the smallest non-negative solution of the integral equation

$$\tilde{u}(x) = \int \widehat{K}_\alpha(x, dy) G(\tilde{u}(y)), \tag{2.13}$$

where the operator $\widehat{K}_\alpha$ is as in Theorem 2.2. In the case when f is piecewise continuous on $[a,b]$ the function $\tilde{u}$ is the unique solution of the problem

$$\begin{cases} \frac{1}{2}\tilde{u}'' - (\alpha + f)\tilde{u} + \alpha G(\tilde{u}) = 0 & \text{on} \quad (a,b) \setminus \{y_1, y_2, \ldots, y_n\}, \\ \tilde{u}'(y_i+) - \tilde{u}'(y_i-) = 2\lambda_i \tilde{u}(y_i), & \\ \tilde{u}(a+) = \tilde{u}(b-) = 0, & \\ 0 \le \tilde{u} \le q, & \text{on} \quad (a,b). \end{cases} \tag{2.14}$$

Proof. Let

$$S_t := \{a < \inf_{0 \le s < t} X_s,\ \sup_{0 \le s < t} X_s < b\}.$$

To get the equation (2.13) consider

$$\tilde{u}(x) = \mathbf{E}_x(\exp(-A_\infty);\ S_\infty)$$

$$= \mathbf{E}_x\left(\exp(-A_\beta) \sum_{n=0}^{\infty} p_n (\mathbf{E}_{\omega_\beta}(\exp(-A_\infty);\ S_\infty))^n;\ S_\beta\right)$$

$$= \int \mathbf{E}_x^B(\exp(-A_\beta);\ S_\beta,\ \omega_\beta \in dy) G(\tilde{u}(y))$$

$$= \int \widehat{K}_\alpha(x, dy) G(\tilde{u}(y)).$$

To prove that $\tilde{u}$ is the smallest non-negative solution we may proceed as in the proof of Theorem 2.2 after having introduced

$$\tilde{u}^{(0)}(x) = 0,$$
$$\tilde{u}^{(k)}(x) = \mathbf{E}_x(\exp(-A_\infty);\ S_\infty,\ N_{T_k} = 0), \qquad k = 1, 2, \ldots$$

The fact that $\tilde{u}$ solves the problem (2.14) is also obtained as in the proof of Theorem 2.2. However, note that

$$\begin{aligned}\tilde{u}(x) &= \mathbf{E}_x(\exp(-A_\infty);\ S_\infty)\\ &= \mathbf{E}_x(\exp(-A_\infty);\ S_\infty,\ \zeta < \infty)\\ &\leq \mathbf{P}_x\{\zeta < \infty\} =: q.\end{aligned}$$

To see the uniqueness assume that $\tilde{v}$ is a solution such that $\tilde{u} \leq \tilde{v}$, and let $\mathcal{B} = -\mathcal{A}$, where $\mathcal{A}$ is as in (2.11). Proceeding as in the proof of Theorem 2.2, we see that

$$\int (\tilde{v}\mathcal{B}\tilde{u} - \tilde{u}\mathcal{B}\tilde{v}) + \alpha \int \tilde{v}\tilde{u} \left(\frac{G(\tilde{u})}{\tilde{u}} - \frac{G(\tilde{v})}{\tilde{v}} \right) = 0. \tag{2.15}$$

It is easily seen that the first term in (2.15) equals zero. To study the second term we may assume that $p_0 > 0$; otherwise, $\tilde{u} \equiv 0$ which is the unique solution of (2.14) in the case $q = 0$. Now it is quite easily to show that the function

$$x \mapsto \frac{G(x)}{x}$$

is decreasing on $(0, q)$. Hence, the second term in (2.15) is non-negative implying that $\tilde{u} = \tilde{v}$.

Remarks.

(i) Equations (2.8) and (2.13) are also valid in the case $a = -\infty$ and/or $b = \infty$. However, to get enough smoothness to be able to state problems (2.9) and (2.14) we have to assume, e.g., that f is bounded and locally Hölder continuous.

(ii) In the case where f is discontinuous the ODE's in (2.9) and (2.14) should be treated in the following sense: they hold at all points of continuity of f and at points of discontinuity the solutions from both sides of such a point must be matched so that they form a continuously differentiable function.

3. MOMENTS OF FUNCTIONALS

Using the well-known fact that

$$\mathbf{E}_x \left(\sum_{n=1}^{N_t} f(\omega^{(k)}(t)) \right) = \exp(-\gamma t) \mathbf{E}_x^B (f(\omega(t))),$$

where $\gamma := \alpha(1 - G'(1))$, we see that

$$v(t,x) := \mathbf{E}_x(I_t(f)) = \int_0^t e^{-\gamma s}\mathbf{E}_x^B(f(\omega(s)))ds, \tag{3.1}$$

and, moreover, for $y \in \mathbf{R}$

$$\mathbf{E}_x(L_t(y)) = \int_0^t e^{-\gamma s}p(s;x,y)ds, \tag{3.2}$$

where p is the transition density (with respect to the Lebesgue's measure) of the standard BM. Assuming that f is bounded and locally Hölder continuous we see from (3.1) that v solves

$$\frac{\partial v}{\partial t} = \frac{1}{2}\frac{\partial^2 v}{\partial x^2} - \gamma v - f. \tag{3.3}$$

Equation (3.3) is also obtained formally from (2.3) by considering the corresponding equation for the functional $\gamma I_t(f)$. Differentiating in this equation with respect to γ and setting $\gamma = 0$, we have (3.3). PDE (3.3) is solved in the standard way by taking the Laplace transforms. Indeed, letting

$$v_\eta(x) := \int_0^\infty e^{-\eta t}v(t,x)dt$$

we see that v_η solves

$$\frac{1}{2}v_\eta'' - \gamma v_\eta - \eta v_\eta = -\frac{1}{\eta}f \tag{3.4}$$

and for $\eta > -\gamma$ it has only one bounded solution. Inverting the Laplace transform gives (uniquely) the function v.

Assume that X is subcritical and consider the expectation of the functional A_∞ introduced in (1.2). From (3.1) and (3.2), we obtain immediately

$$\mathbf{E}_x(A_\infty) = \int g_\gamma(x,y)f(y)dy + \sum_{i=1}^n \lambda_i g_\gamma(x,y_i),$$

where g_γ is the Green's function of BM and $\gamma := \alpha(1-G'(1)) > 0$. Higher moments can be obtained similarly by differentiating in (2.9). For example, the Laplace transform s_η of $s(t,x) := \mathbf{E}_x(I_t^2(f))$ is the unique bounded solution of

$$\frac{1}{2}s_\eta'' - (\gamma+\eta)s_\eta = -2fv_\eta - \alpha G''(1)\int_0^\infty e^{-\eta t}v^2(t,x)dt.$$

Next consider a killed BBM and an additive functional A_t defined in (1.2) with a bounded and piecewise continuous f. We are interested of computing $\hat{v}(x) := \widehat{\mathbf{E}}_x(A_\infty)$. Let λ_0 be the smallest eigenvalue of the problem

$$\begin{cases} \frac{1}{2}u'' + \lambda u = 0, \\ u(a+) = u(b-) = 0. \end{cases}$$

Then $\lambda_0 > 0$, and it is well known (see (Sevast'yanov, 1958) or (Watanabe, 1965, Theorem 2.2)) that $\mathbf{P}_x\{\zeta < \infty\} = 1$ for all x if and only if $\lambda_0 + \alpha(1 - G'(1)) \geq 0$. In fact, this is proved in (Watanabe, 1965) under the assumption that $p_0 = 0$. However, it is easily seen that the same proof is valid also in the case $p_0 > 0$. Therefore, $\hat{v} \equiv \infty$ if $\lambda_0 + \alpha(1 - G'(1)) < 0$. This is also true if $\lambda_0 + \alpha(1 - G'(1)) = 0$, as can be seen using in (3.1) and (3.2) the spectral expansion of the transition density of this killed BM. If $\lambda_0 + \alpha(1 - G'(1)) > 0$ the function $\hat{v}$ is the unique continuous solution of

$$\begin{cases} \frac{1}{2}\hat{v}'' - \alpha(1 - G'(1))\hat{v} = -f & \text{on} \quad (a,b) \setminus \{y_1, y_2, \dots, y_n\}, \\ \hat{v}'(y_i+) - \hat{v}'(y_i-) = 2\lambda_i \hat{v}(y_i), \\ \hat{v}(a+) = \hat{v}(b-) = 0. \end{cases} \tag{3.5}$$

One can obtain this by considering for $\gamma > 0$ the function $\widehat{\mathbf{E}}_x(\exp(-\gamma A_\infty))$ which solves (2.9) with γf instead of f. Differentiating (2.9) with respect to the γ and setting $\gamma = 0$ we arrives at (3.5).

Finally, we consider the expectation

$$\tilde{v}(x) := \mathbf{E}_x(A_\infty;\ a < \inf X_s,\ \sup X_s < b).$$

Proceeding as above we see that $\tilde{v}$ solves, if finite,

$$\begin{cases} \frac{1}{2}\tilde{v}'' - \alpha(1 - G'(r))\tilde{v} = -fr & \text{on} \quad (a,b) \setminus \{y_1, y_2, \dots, y_n\}, \\ \tilde{v}'(y_i+) - \tilde{v}'(y_i-) = 2\lambda_i \tilde{v}(y_i), \\ \tilde{v}(a+) = \tilde{v}(b-) = 0, \end{cases} \tag{3.6}$$

where the function r is given by

$$r(x) := \mathbf{P}_x\{a < \inf X_s,\ \sup X_s < b\}.$$

We remark that (3.6) has a solution, which is unique and continuous, if and only if the first eigenvalue of the problem

$$\begin{cases} \frac{1}{2}u'' - \alpha(1 - G'(r))u + \lambda u = 0, \\ u(a+) = u(b-) = 0. \end{cases}$$

is positive.

4. EXAMPLES

4.1. *Total length of a branching Brownian tree*

Let X be a BBM, and, for $f \equiv 1$, consider the functional

$$I_\infty := \int_0^\infty \sum_{k=1}^{N_s} f(\omega_k(s))ds. \tag{4.1}$$

We call I_∞ the total length of a branching Brownian tree or (due to the fact that X is spatially homogeneous) a (continuous time) Galton–Watson tree with creation parameter α and the generating function $G(u)$, $|u| \le 1$. It follows from Theorem 2.2 that

$$u(x) := \mathbf{E}_x(\exp(-\gamma I_\infty)), \qquad \gamma > 0,$$

is the unique solution of the problem

$$\begin{cases} \alpha(G(u) - u) - \gamma u = 0, \\ 0 \le u \le 1. \end{cases} \tag{4.2}$$

In the case $\gamma = 0$, we have the classical result concerning the extinction probability of a Galton–Watson process.
In particular, if $p_0 = p_2 = 1/2$,

$$u(x) = \frac{\alpha+\gamma}{\alpha} - \sqrt{\left(\frac{\alpha+\gamma}{\alpha}\right)^2 - 1} =: B. \tag{4.3}$$

This Laplace transform can be inverted (see (Erdélyi, 1954; 5.3.49, p. 237)) giving the density, g_α, of the random variable I_∞ :

$$g_\alpha(t) = \frac{e^{-\alpha t}}{t}\mathcal{I}_1(\alpha t), \qquad t > 0,$$

where $\mathcal{I}_1$ is the modified Bessel function of order 1.

4.2. *Total length of the positive part of a critical branching Brownian tree*

Let X be a BBM with $p_0 = p_2 = 1/2$ and consider the functional

$$I_\infty^+ := \int_0^\infty \sum_{k=1}^{N_s} \mathbf{1}_{[0,\infty)}(\omega_k(s))ds.$$

Then, with I_∞ as in (4.1), we have

$$\lim_{x\uparrow\infty} \mathbf{E}_x(\exp(-\gamma I_\infty^+))$$

$$= \lim_{x\uparrow\infty}(\mathbf{E}_x(\exp(-\gamma I_\infty);\ \inf X_s > 0) + \mathbf{E}_x(\exp(-\gamma I_\infty^+);\ \inf X_s < 0))$$

$$= \lim_{x\uparrow\infty}(\mathbf{E}_x(\exp(-\gamma I_\infty)) - \mathbf{E}_x(\exp(-\gamma I_\infty);\ \inf X_s < 0)$$

$$+\mathbf{E}_x(\exp(-\gamma I_\infty^+);\ \inf X_s < 0)) = B,$$

where B is as in (4.3) and we have used the fact (see Example 4.3)

$$\lim_{x\uparrow\infty} \mathbf{P}_x(\inf X_s < 0) = 0.$$

Analogously,

$$\lim_{x\downarrow-\infty} \mathbf{E}_x(\exp(-\gamma I_\infty^+)) = 1.$$

According to Theorem 2.2 or 2.3 (see Remark (i) at the end of §2), the function

$$\hat{u}(x) := \mathbf{E}_x(\exp(-\gamma I_\infty^+))$$

is the unique continuously differentiable solution (≤ 1) of the problem

$$\begin{cases} \frac{1}{2}\hat{u}'' - (\alpha+\gamma)\hat{u} + \alpha(\frac{1}{2}\hat{u}^2 + \frac{1}{2}) = 0, & \text{for} \quad x > 0, \\ \frac{1}{2}\hat{u}'' - \alpha\hat{u} + \alpha(\frac{1}{2}\hat{u}^2 + \frac{1}{2}) = 0, & \text{for} \quad x < 0. \end{cases}$$

To solve this consider first the case $x < 0$. Setting $\hat{v} := 1 - \hat{u}$ we see that $\hat{v}$, satisfies

$$\begin{cases} \frac{1}{2}\hat{v}'' = \alpha\hat{v}^2, \\ \lim_{x\downarrow-\infty}\hat{v}(x) = \lim_{x\downarrow-\infty}\hat{v}'(x) = 0. \end{cases} \tag{4.4}$$

Multiplying the ODE in (4.4) by $\hat{v}'$ and integrating from $-\infty$ to $x < 0$ gives the result

$$\hat{v}' = \left(\frac{2\alpha}{3}\hat{v}^3\right)^{1/2}.$$

From this we deduce for $x < 0$

$$\hat{u}(x) = 1 - C_1\left(1 - \sqrt{\frac{\alpha C_1}{6}}x\right)^{-2},$$

where the unknown constant C_1 is such that $0 < C_1 < 1$. In the case $x > 0$ introduce $\hat{t} := \hat{u} - B$, where B is as in (4.3). Using the fact that B solves the equation in (4.2), we see that $\hat{t}$ satisfies

$$\begin{cases} \hat{t}'' + \alpha\hat{t}^2 + 2(\alpha B - (\alpha+\gamma))\hat{t} = 0, \\ \lim_{x\downarrow\infty}\hat{t}(x) = \lim_{x\downarrow\infty}\hat{t}'(x) = 0. \end{cases} \tag{4.5}$$

The problem (4.5) is solved analogously to (4.4). The result is

$$\hat{t}(x) = A\left(1 - \left(\frac{\sqrt{A-C_2} + \sqrt{A} + (\sqrt{A-C_2} - \sqrt{A})e^{-Dx}}{\sqrt{A-C_2} + \sqrt{A} - (\sqrt{A-C_2} - \sqrt{A})e^{-Dx}}\right)^2\right), \tag{4.6}$$

where $A := 3\sqrt{(\alpha+\gamma/\alpha)^2 - 1}$, $D = \sqrt{2\alpha A/3}$, and C_2 is an unknown constant such that $\hat{t}(0+) = C_2$. The constants C_1 and C_2 are determined from the matching conditions

$$\begin{cases} \hat{u}(0-) = \hat{u}(0+), \\ \hat{u}'(0-) = \hat{u}'(0+). \end{cases}$$

This leads to the system

$$\begin{cases} C_1 + C_2 = 1 - B \\ C_1^3 + C_2^3 = AC_2^2, \end{cases}$$

from which $C_1$and C_2 can be explicitly obtained. However, expressions being quite messy, we do not give any further details.

4.3. *Local time at zero of a critical branching Brownian tree*

Let X be as in Example 4.2 and $t \mapsto L_t$ be its local time at 0 at time t. Then

$$\hat{u}(x) := \mathbf{E}_x(\exp(-\gamma L_\infty))$$

is the unique solution of the problem

$$\begin{cases} \frac{1}{2}\hat{u}'' - \alpha\hat{u} + \alpha(\frac{1}{2}\hat{u}^2 + \frac{1}{2}) = 0, & x \neq 0, \\ \hat{u}'(0+) - \hat{u}'(0-) = 2\gamma\hat{u}(0), \\ \lim_{x\downarrow-\infty} \hat{u}(x) = \lim_{x\uparrow\infty} \hat{u}(x) = 1, \\ 0 \leq \hat{u} \leq 1. \end{cases} \tag{4.7}$$

Introducing $\hat{v} := 1 - \hat{u}$ and proceeding as above, we see that

$$\hat{u}(x) = 1 - C\left(1 + \sqrt{\frac{\alpha C}{6}}|x|\right)^{-2} \tag{4.8}$$

The constant C is determined from the condition at 0 given in (4.7). Having the explicit form (4.8), we see that this condition is equivalent to

$$\sqrt{\frac{2\alpha}{3}}C^{3/2} = \gamma(1 - C).$$

Unfortunately, in this case the solution (there is only one real solution) is also complicated and, therefore, we do not give any details. Note, however, that for large γ

$$C \simeq 1 - \frac{1}{\gamma}\sqrt{\frac{2\alpha}{3}}$$

and hence, for $x > 0$,

$$\mathbf{P}_x\{\inf X_s > 0\} = \mathbf{P}_x\{L_\infty = 0\} = 1 - \left(1 + \sqrt{\frac{\alpha}{6}}x\right)^{-2}$$

4.4. *Total length of a killed critical branching Brownian tree*

Let $\hat{X}$ be a killed BBM such that particles hitting zero are immediately at that time removed from the process to a cemetery point Δ isolated from $\mathbf{R}$, where they are trapped. Consider the critical case $p_0 = p_2 = 1/2$ and assume that $\widehat{X}_0 = x > 0$. Let I_∞ be the functional introduced in (4.1). By Theorem 2.2 the function

$$\hat{u}(x) := \mathbf{E}_x(\exp(-\gamma I_\infty))$$

is the unique solution of the problem

$$\begin{cases} \frac{1}{2}\hat{u}'' - (\alpha+\gamma)\hat{u} + \alpha(\frac{1}{2}\hat{u}^2 + \frac{1}{2}) = 0, & x > 0, \\ \lim_{x\uparrow\infty} \hat{u}(x) = B, \quad \lim_{x\uparrow\infty} \hat{u}'(x) = 0, \\ \hat{u}(0) = 1, \\ 0 \le \hat{u} \le 1. \end{cases}$$

From this we obtain the same expression for $\hat{t} := \hat{u} - B$ as in (4.6) with $C = 1 - B$.

4.5. *Total length of a killed supercritical branching Brownian tree*

Let $\widehat{X}$ be as in Example 4.4 but now with $p_2 = 1$. Then

$$\hat{u}(x) := \mathbf{E}_x(\exp(-\gamma I_\infty))$$

is the unique solution of the problem

$$\begin{cases} \frac{1}{2}\hat{u}'' - (\alpha+\gamma)\hat{u} + \alpha\hat{u}^2 = 0, & x > 0, \\ \lim_{x\uparrow\infty} \hat{u}(x) = \lim_{x\uparrow\infty} \hat{u}'(x) = 0, \\ \hat{u}(0) = 1, \\ 0 \le \hat{u} \le 1. \end{cases}$$

Proceeding as in Example 4.2 we obtain

$$\hat{u}(x) = A\left(1 - \left(\frac{\sqrt{A-1} + \sqrt{A} + (\sqrt{A-1} - \sqrt{A})\exp(-\sqrt{\alpha+\gamma}x)}{\sqrt{A-1} + \sqrt{A} - (\sqrt{A-1} - \sqrt{A})\exp(-\sqrt{\alpha+\gamma}x)}\right)^2\right),$$

where $A = 3(\alpha+\gamma)/(2\alpha)$.

REFERENCES

Asmussen, S. and Hering, H. (1983). *Branching Processes.* Birkhäuser, Boston, Basel, Stuttgart.

Borodin, A.N. (1989). Distributions of functionals of Brownian local time.I. *Theory Probab. Appl.* **34**, 385–401.

Erdélyi, A., ed. (1954). *Tables of Integral Transforms,.* Vol. I. McGraw-Hill, New York, Toronto, London.

Ikeda, N., Nagasawa, M., and Watanabe, S. (1968a, 1968b, 1969). Branching Markov processes, I, II, III. *J. Math. Kyoto Univ.* **8**, 233–278; **8**, 365–410; **9**, 95–160.

Sevast'yanov, B.A. (1958). Branching stochastic processes for particles diffusing in a bounded domain with absorbing boundaries. *Theory Probab. Appl.* **3**, 111–126.

Watanabe, S. (1965). On the branching process for Brownian particles with an absorbing boundary. *J. Math. Kyoto Univ.* **4**, 385–398.

Frontiers in Pure and Applied Probab. 1, pp. 22 – 28
H. Niemi et al. (Eds.)
1993 TVP/VSP

CONVERGENCE RATES IN TRANSIENT PHENOMENA FOR BRANCHING PROCESSES

K. A. BOROVKOV
Steklov Mathematical Institute, Vavilov Street 42, 117966 Moscow GSP-1, Russia

ABSTRACT

We consider the limiting behaviour of nonhomogeneous Markov branching processes in discrete time when the offspring means are uniformly close to unity. In this set up, 1) the uniform convergence (over classes of processes with uniformly square-integrable offspring-numbers) of conditional laws to the exponential one is proved; 2) bounds for the rate of this convergence are given (these bounds are uniform over classes of processes with a bounded third moment); 3) asymptotic expansions are constructed which are uniform over classes of processes with uniformly cube-integrable offspring numbers.

We consider the limiting behaviour of conditional distributions of nonhomogeneous Markov branching processes in discrete time as time tends to infinity when the offspring means are close to unity uniformly over generations. The following three basic aspects are analysed here: the stability of transient phenomena under violation of time-homogenity, convergence rates in the integral limit theorem, and asymptotic expansions. Though all considerations are given for discrete-time processes, the results can be easily carried to the continuous-time case too by the standard embedding procedure.

First we prove the uniform (over suitable classes of processes) convergence of conditional distribution functions (d.f.'s) of the number of particles to the exponential law (the homogeneous case see in (Nagaev and Muhamedhanova, 1966a; 1968; Sevast'yanov, 1971; Fahady et al., 1971)). Then we establish uniform (over some classes of processes too) bounds for the rate of this convergence. Earlier only the case of one critical process has been investigated, see (Nagaev and Muhamedhanova, 1966b), the estimate there has the form $O(n^{-1}\log^2 n)$ when the third moment is finite. Asymptotic expansions are obtained here for Galton–Watson processes only; moreover, they are valid for the expectations of functions from a rather wide class that does not comprise, however, the indicators of intervals, and hence these expansions do not hold in the general case for d.f.'s. This part of paper answers a question posed in the author's paper (Borovkov, 1988): do the expansions derived there for Laplace transforms correspond to similar expansions for d.f.'s?

Now we introduce the basic notation. Let

$$\mu(j), \qquad j = 0, 1, \dots, n, \qquad \mu(0) = 1,$$

be a nonhomogeneous branching process,

$$f_j(z) = \mathbf{E}\Big(z^{\mu(j)} | \, \mu(j-1) = 1\Big)$$

be the generating function (g.f.) of the offspring distribution on the jth step. Set

$$A_j = f_j'(1), \qquad B_j = f_j''(1), \qquad C_j = f_j'''(1),$$

these are the left derivatives, as usual. Now let

$$\alpha = \alpha_n = \max_{j \le n} |A_j - 1|, \qquad A(j, m) = A_j A_{j+1} \dots A_m,$$

$$T(n) = \frac{1}{2} \sum_{j=1}^{n-1} B_j A(j+1, n)/A_j, \qquad S_m(x) = \frac{x^m - 1}{x - 1}, \qquad S_m(1) = m.$$

One can easily see that (Lemma 1 in Chapter 3 of (Sevast'yanov, 1971))

$$S_m(x) \ge \frac{1}{2} \min\Big(m, |x-1|^{-1}\Big).$$

Therefore, for $B_j \ge b > 0, \quad j = 1, \dots, n,$

$$T(n) \ge \frac{b}{2(1+\alpha)} S_n(1-\alpha) \ge \frac{b}{4(1+\alpha)} \min(n, \alpha^{-1}) \to \infty$$

as $n \to \infty$, $\alpha \to 0$. Clearly, the Galton–Watson process,

$$T(n) \ge \frac{B}{2A} S_n(A) \ge \frac{B}{4A} \min\Big(n, |A-1|^{-1}\Big).$$

We study the limiting behaviour of the conditional d.f.

$$F_n(x) = \mathbf{P}\Big\{\mu(n)/T(n) < x \big| \mu(n) > 0\Big\}$$

and the nonextinction probability

$$Q(n) = \mathbf{P}\Big\{\mu(n) > 0\Big\}.$$

Denote by

$$G_a(x) = \frac{1}{\Gamma(a)} \int_0^x y^{a-1} e^{-y} dy, \qquad x \ge 0,$$

the Gamma distribution with exponent $a > 0$, G_0 will denote the degenerate at zero distribution.

Now we introduce the following classes of functions. Let $\mathcal{F}$ be the class of all nonincreasing functions $\varphi: [0,\infty) \to (0,\infty)$ such that $\varphi(x) \to 0$ as $x \to \infty$. For an arbitrary function $\varphi \in \mathcal{F}$ and a natural number k, we introduce the class of g.f.'s

$$\mathcal{K}_{k,\varphi} = \Big\{f: \sum_{m\geq r} m^k p_m \leq \varphi(r), \quad r > 0\Big\}, \quad f(z) = \sum_{m\geq 0} z^m p_m.$$

Clearly, random variables with g.f.'s from the class $\mathcal{K}_{k,\varphi}$ are uniformly integrable to the kth power, and vice versa: for any class $\mathcal{K}$ of g.f.'s of random variables with uniformly integrable kth power there exists a $\varphi \in \mathcal{F}$ such that $\mathcal{K} \subseteq \mathcal{K}_{k,\varphi}$. We shall use this notation in the case $\varphi \equiv c = \text{const}$ too:

$$\mathcal{K}_{k,c} = \Big\{f: \sum_{m\geq 0} m^k p_m \leq c\Big\}.$$

Denote by $\mathcal{M}(\mathcal{K})$ the class of all branching processes of the type described above, such that all $f_j \in \mathcal{K}$, $j \geq 1$, for these processes, where $\mathcal{K}$ is a fixed class of g.f.'s. Now let $\mathcal{K}(b), b > 0$, be the following restriction of the class $\mathcal{K}$:

$$\mathcal{K}(b) = \mathcal{K} \cap \Big\{f: f''(1) \geq b\Big\}.$$

The notation $\mathcal{M}_0(\mathcal{K})$ will be used for the class of Galton–Watson processes, for which the g.f.'s f lie in $\mathcal{K}$.

Finally, $\mathcal{H}_\psi$, $\psi \in \mathcal{F}$, will denote the class of functions of bounded variation H on the real line, Fourier–Stiltjes transforms

$$h(t) = \int e^{itx} dH(x)$$

of which satisfy

$$\int_{|t|\geq r} \frac{|h(t)|}{1+|t|} dt \leq \psi(r), \qquad r \geq 0. \tag{1}$$

Our first main result can be stated as follows.

THEOREM 1. *For any $\varphi \in \mathcal{F}$ and $b > 0$,*

$$\Delta_n = \sup_x |F_n(x) - G(x)| \to 0, \qquad Q(n)T(n)/A(1,n) \to 1$$

uniformly in $\mathcal{M}(\mathcal{K}_{2,\varphi}(b))$ as $n \to \infty$, $\alpha \to 0$.

Now we consider the case of branching process starting with a large initial number of particles $\mu_N(0) = N$, and let

$$F_{N,n}(x) = \mathbf{P}\Big\{\mu_N(n)/T(n) < x\Big\}.$$

Denote by

$$\Pi_a[F] = e^{-a} \sum_{k \geq 0} a^k F^{*k}/k!$$

the compound Poisson law with parameter a and spectral measure F. The law $\Pi_\lambda[G_1]$ has the characteristic function $\exp(i\lambda t/(1-it))$; one can easily see that it has the atom $e^{-\lambda}$ at zero and its absolutely continuous component has the density

$$\lambda^{1/2} x^{-1/2} e^{-\lambda - x} I_1\left(\lambda^{1/2} x^{1/2}\right), \qquad x \geq 0,$$

where I_1 is the modified Bessel function of the first order.
The following theorem makes it possible to obtain results on convergence of $F_{N,n}$ from those for the conditional d.f.'s F_n including estimates of the rate of convergence.

THEOREM 2. *For* $\lambda = NQ(n)$,

$$\sup_x \left|F_{N,n}(x) - \Pi_\lambda[G_1](x)\right| \leq \lambda\left[\Delta_n + 2(1 - e^{-\lambda})Q(n)\right].$$

Bounds for the rate of convergence are obtained for the classes $\mathcal{M}(\mathcal{K}_{3,c}(b))$ with fixed $b > 0,\ c < \infty$. Below we single out the case, where

$$A_i = 1 + \beta(1 + \varepsilon_i), \qquad |\varepsilon_i| \leq \varepsilon \tag{2}$$

(if ε is small, one can get in this case more precise bounds).
Let $\bar{A} = \max_{1 \leq j \leq n+1} A(j,n),\ T = T(n)$.

THEOREM 3. *For any* $c < \infty, \quad b > 0$,

$$\Delta_n = O\left(\bar{A}T^{-1} \log^2 T\right)$$

uniformly over the class $\mathcal{M}(\mathcal{K}_{3,c}(b))$ *as* $n \to \infty,\ \alpha \to 0$. *In the case* (2) *one has*

$$\Delta_n = O\left(e^{\varkappa}\left(S^{-1} \log S + \left|\beta - \varepsilon|\beta|\right|\right) \log S\right),$$

where $S = S_n(1 + \beta - \varepsilon|\beta|),\ \varkappa = 4|\beta|\varepsilon n/(1 + \beta - \varepsilon|\beta|)$.

Note that in different circumstances each of these two estimates can prove to be preferable. In particular, for Galton–Watson processes we get the estimate

$$\Delta_n = O\left((S^{-1} \log S + |A - 1|) \log S\right), \qquad S = S_n(A), \tag{3}$$

uniformly in the class $\mathcal{M}_0(\mathcal{K}_{3,c}(b))$ as $n \to \infty,\ A \to 1$.
Making use of this result, one can get a simple nonuniform estimate for the difference $|F_n(x) - G_1(x)|$. To do this one should only use the following

LEMMA. *For any d.f. $F(x)$ on $[0,\infty)$,*

$$\left|F(x)-G_1(x)\right| \le \frac{4(\log^2\Delta+1)\Delta+\delta}{1+x^2}, \qquad x\ge 0,$$

where $\Delta=\sup_x\left|F(x)-G_1(x)\right|$, $\delta=\left|\int_0^\infty x^2d(F-G_1)\right|$.

As an immediate consequence of this lemma, we have the following estimate for Galton–Watson processes: for any $c<\infty$, $b>0$,

$$\left|F_n(x)-G_1(x)\right| = O\Big(\frac{\varepsilon\log^2\varepsilon}{1+x^2}\Big), \qquad x\ge 0, \qquad \varepsilon=|A-1|+S^{-1}\log S$$

uniformly in $\mathcal{M}_0(\mathcal{K}_{3,c}(b))$ as $n\to\infty$, $A\to 1$.

Remark 1. What we can say about the relation between the terms in the right-hand side of (3)? It is easy to see that $|A-1|=O(S^{-1})$ for $A\le 1+O(n^{-1})$, and hence the term $S^{-1}\log S$ is the main one. Moreover, the "critical" value of A here is $A=1+(1+h)n^{-1}\log\log n$. We have $|A-1|=o\Big(S^{-1}\log S\Big)$ for $h<0$, $S^{-1}\log S=o(|A-1|)$ for $h>0$.

Remark 2. Theorem 2 enables us to obtain the corresponding results for the processes with large initial number of particles from Theorem 3. For Galton–Watson processes, e.g., we have

$$Q(n)=O\Big(A^nS^{-1}\Big)=O\Big(|A-1|+S^{-1}\Big)$$

by Theorem 1, and hence (3) and Theorem 3 imply that

$$\sup_x\left|F_{N,n}(x)-\Pi_\lambda[G_1](x)\right| = O\bigg(\lambda\Big(S^{-1}\log S+|A-1|\Big)\log S\bigg)$$

uniformly in $\mathcal{M}_0$ $(\mathcal{K}_{3,c}(b))$.

Now let us turn back to asymptotic expansions. In (Borovkov, 1988) it was shown that in the case of one critical Galton–Watson process with finite third moment the following representation holds true for the Laplace transform φ_{0n} of the law

$$F_{0n}(x)=\mathbf{P}\Big\{Q(n)\mu(n)<x|\mu(n)>0\Big\}$$

which differs from F_n only by the normalizing factor:

$$\varphi_{0,n}(\lambda)=\frac{1}{1+\lambda}-\Big(\frac{\lambda}{1+\lambda}\Big)^2\gamma\frac{\log n}{n}+o\Big(\frac{\log n}{n}\Big), \qquad |\log\lambda|=o\Big(|\log n|\Big),$$

where $\gamma=1-2C/(3B^2)$. If this relation implies corresponding representation for d.f.'s, we would have that

$$F_{0n}(x)=\Big(1-2\gamma\frac{\log n}{n}\Big)\,G_1(x)+\gamma\frac{\log n}{n}\Big(G_0(x)+G_2(x)\Big)+o\Big(\frac{\log n}{n}\Big). \qquad (4)$$

But F_{0n} has, as is well-known, the jumps of order $O(n^{-1})$, and the approximant has (for $\gamma \neq 0$) a jump at zero of size $n^{-1}\log n$. Hence this equality can not hold uniformly in x. However, to this relation there correspond representations for expectations of functions of $\mu(n)$ from a rather wide class $\mathcal{H}_\psi$.
Of course, the class $\mathcal{H}_\psi$ does not contain step functions, cf. (1), but it can comprise functions that are rather close to them. For example, there exists a $\psi \in \mathcal{F}$ such that the classes

$$\left\{x^a h^{-a}\chi_{(0,h)}(x) + \chi_{[h,\infty)}(x)\colon\ a \geq a_0 > 0,\ h \geq h_0\right\},$$

$$\left\{B_{a,b}(x/h)\colon\ a \geq a_0,\ b \geq b_0,\ h \geq h_0\right\}$$

are contained in $\mathcal{H}_\psi$, where χ_A is the indicator of a set A, $B_{a,b}$ is the Beta d.f. with parameters a, b.
Set

$$F \circ H = \int_{-\infty}^{\infty} H(x)dF(x), \qquad H \in \mathcal{H}_\psi,$$

for a d.f. F, and introduce the d.f. M with the density $(x-1)e^{-x}\log x,\ x > 0$. Now we state our main result in this direction.

THEOREM 4. *For any fixed $\varphi, \psi \in \mathcal{F}$, $b > 0$, as $n \to \infty$, $A \to 1$,*

$$F_n \circ H = R \circ H + o\Big(|A-1| + S^{-1}\log S\Big), \qquad S = S_n(A),$$

uniformly in $\mathcal{M}_0(\mathcal{K}_{3,\varphi}(b))$ and $\mathcal{H}_\psi$, where

$$R(x) = \left[1 + \gamma\Big(S^{-1}\log S - (1-c_0)(A-1)\Big)\right]G_1(x)$$

$$-\gamma S^{-1}(\log S)G_0(x) + \gamma(2-c_0)(A-1)G_2 - \gamma(A-1)M(x),$$

$\gamma = 1 - 2C/(3B^2)$, $c_0 = 0,5772...$ *is the Euler's constant. In this case*

$$Q(n) = \frac{A-1+S^{-1}}{2B}\left[1 - (1-\gamma)(A-1) - \gamma\frac{\log S}{S} + o\Big(|A-1| + \frac{\log S}{S}\Big)\right]$$

uniformly in the classes under consideration.

A more detailed exposition with complete proofs is to appear in the "Theory of Probability and Its Applications".

REFERENCES

Borovkov, K.A. (1988). On a method of proving the limit theorems for branching processes. *Theory Probab. Appl.* **33**, 115–123.

Fahady, K.S., Quine, M.P., and Vere-Jones, D. (1971). Heavy traffic approximations for the Galton–Watson processes. *Adv. Appl. Probab.* **3**, 282–300.

Nagaev, S.V. and Muhamedhanova, R. (1966a). Transient phenomena in branching random processes in discrete time. In: *Limit Theorems and Statistical Inference.* Fan, Tashkent, pp. 83–89 (in Russian).

Nagaev, S.V. and Muhamedhanova, R. (1966b). Some limit theorems in the theory of branching processes. In: *Limit Theorems and Statistical Inference.* Fan, Tashkent, pp. 90–112 (in Russian).

Nagaev, S.V. and Muhamedhanova, R. (1968). Some comments on the published limit theorems in the theory of branching processes. In: *Probabilistic Models and Statistical Control.* Fan, Tashkent, pp. 46–49 (in Russian).

Sevast'yanov, B.A. (1971). *Branching Processes.* Nauka, Moscow (in Russian).

Frontiers in Pure and Applied Probab. 1, pp. 29 – 67
H. Niemi et al. (Eds.)
1993 TVP/VSP

THE WICK PRODUCT

H. GJESSING, H. HOLDEN, T. LINDSTRØM, B. ØKSENDAL, J. UBØE, and T.-S. ZHANG

Department of Mathematics, University of Bergen, Allégt. 55, N-5007 Bergen, Norway
Department of Mathematics, University of Trondheim (NTH), N-7034 Trondheim, Norway
Department of Mathematics, University of Oslo, Box 1053, Blindern, N-0316 Oslo, Norway
Department of Mathematics, University of Oslo, Box 1053, Blindern, N-0316 Oslo, Norway
Department of Mathematics, SSH, Skåregt. 103, N-5500 Haugesund, Norway
Department of Mathematics, University of Oslo, Box 1053, Blindern, N-0316 Oslo, Norway

ABSTRACT

We give a survey of the basic properties of the Wick product and its applications.

CONTENTS

Introduction 30
1. The Wick product in physics 30
2. The Wick product in stochastic analysis 33
2.1 *The white noise probability space* 33
2.2 *The Wiener–Ito chaos expansion* 35
2.3 *The Hida test function space* $(\mathcal{S})$ *and the Hida distribution space* $(\mathcal{S})^*$ 36
2.4 *The Wick product* 38
2.5 *Applications* 40
3. Discrete Wick calculus 47
4. Other characterizations of the Wick product 51
4.1 *The* $\mathcal{S}$*-transform and the Hermite transform* 51
4.2 *Connection with convolution* 54
4.3 *The backward heat equation* 55
5. Properties of the Wick product 58
5.1 *Algebraic properties* 58
5.2 *Analytic properties and estimates* 63
5.3 *Is the Wick product local?* 63
6. Some open problems 64
References 66

INTRODUCTION

The Wick product was introduced by G.C. Wick in 1950 in the context of quantum field theory. In 1965 T. Hida and N. Ikeda introduced a closely related concept in probability theory. The concept now plays a crucial role in both mathematical physics and stochastic analysis.

The purpose of this paper is to give an introduction to the Wick product and some of its basic properties. We want to point out some of the different (but related) approaches to the concept and we also hope to clarify the relation between them. Finally we discuss some of the recent applications of the Wick product in stochastic differential equations. No attempt has been made to be complete in any way, neither with respect to the history of the concept nor with respect to the selection of the associated topics. Indeed, there are so many different aspects of it — and open problems about them — that a complete survey seems impossible at the present time. But we hope this article will convince the reader about the importance of the Wick product and perhaps inspire him or her to consult the literature further and even pursue the concept by own research, for example, by solving some of the problems we list in the end!

1. THE WICK PRODUCT IN PHYSICS

In this section we give a brief discussion of the use of the Wick product in the theory of quantum fields. Quantum fields represent the synthesis of quantum mechanics with relativity.

The Wick product was introduced by Wick (Wick, 1950), who called it the "S-product", as a convenient notation to simplify certain complicated expressions in the study of Heissenberg's S-matrix.

We will here concentrate on a simple case, namely that of free Euclidean fields. In this formulation the fields are continued analytically to imaginary time. This changes the Minkowski metric into the technically simpler Euclidean metric and it changes the complex Schrödinger equation into the real heat equation (with a potential), which is more tractable from the point of view of stochastic analysis. The following presentation is based on Simon (Simon, 1974) and Glimm and Jaffe (Glimm and Jaffe, 1987).

Let $(\Omega, \mathcal{B}, \mathbf{P})$ be a probability space, and fix a random variable U. Consider a formal power series in U and define a formal derivation with respect to U by

$$\frac{\partial}{\partial U}\left(\sum_{n=0}^{\infty} a_n U^n\right) = \sum_{n=0}^{\infty}(n+1)a_{n+1}U^n. \tag{1.1}$$

The Wick order of U^k, $:U^k:$, can recursively be defined by

$$:U^0: = 1, \quad \frac{\partial}{\partial U} :U^n: = n :U^{n-1}:, \quad \mathbf{E}[:U^n:] = 0, \quad n = 1, 2, \ldots. \tag{1.2}$$

Hence,

$$:U: = U - \mathbf{E}(U),$$

$$: U^2 := U^2 - 2U\mathbf{E}(U) - \mathbf{E}(U^2) + 2(\mathbf{E}(U))^2,$$

$$: U^3 := U^3 - 3U^2\mathbf{E}(U) - 3U\mathbf{E}(U^2)$$

$$+6U\mathbf{E}(U)^2 - \mathbf{E}(U)^3 + 6\mathbf{E}(U)\mathbf{E}(U^2) - 6\mathbf{E}(U^3). \qquad (1.3)$$

Extend this definition by linearity to convergent power series to obtain, e.g.,

$$: e^{aU} := \mathbf{E}(e^{aU})^{-1} e^{aU}. \qquad (1.4)$$

We may extend this definition further to more than one random variable as follows: If $U_1, \ldots, U_n$ are random variables, then we define

$$: U_1^0 \ldots U_N^0 := 1, \qquad (1.5)$$

and

$$\frac{\partial}{\partial U_i} : U_1^{n_1} \ldots U_N^{n_N} := n_i : U_1^{n_1} \ldots U_i^{n_i - 1} \ldots U_N^{n_N} :,$$

$$\mathbf{E}(: U_1^{n_1} \cdots U_N^{n_N} :) = 0, \quad \text{not all} \quad n_i = 0. \qquad (1.6)$$

This gives, e.g.,

$$: UV := UV - U\mathbf{E}(V) - V\mathbf{E}(U) + 2\mathbf{E}(U)\mathbf{E}(V) - \mathbf{E}(UV). \qquad (1.7)$$

Sometimes $: UV :$ is called the Wick product of U and V. Here we will call it the *physical Wick product* to distinguish it from the Wick product we will define later.

The use of the physical Wick product requires extreme care, as most of the usual algebraic properties one is used to in fact no longer hold. We have, e.g.,

$$: UV : |_{V=U^2} = U^3 - U\mathbf{E}(U^2) - U^2\mathbf{E}(U) + 2\mathbf{E}(U)\mathbf{E}(U^2) - \mathbf{E}(U^3) \qquad (1.8)$$

which in general is different from $: U^3 :$ as given by (1.3)! Furthermore, using (1.7),

$$: U1 := U1 - U\mathbf{E}(1) - 1\mathbf{E}(U) + 2\mathbf{E}(U)\mathbf{E}(1) - \mathbf{E}(U1)$$

$$= U - U - \mathbf{E}(U) + 2\mathbf{E}(U) - \mathbf{E}(U) = 0$$

which clearly is different from $: U :$.

For a zero mean Gaussian random variable U with variance $\sigma^2 = \mathbf{E}[U^2]$, we find in particular

$$: U^n := \sigma^n h_n \left(\frac{U}{\sigma} \right), \qquad (1.9)$$

where h_n is the nth *Hermite polynomial*, defined by

$$h_n(x) = (-1)^n e^{x^2/2} \frac{d^n}{dx^n} \left(e^{-x^2/2} \right), \quad n = 0, 1, 2, \ldots, \qquad (1.10)$$

so that

$$h_0(x) = 1, \quad h_1(x) = x, \quad h_2(x) = x^2 - 1, \quad h_3(x) = x^3 - 3x, \\ h_4(x) = x^4 - 6x^2 + 3, \quad h_5(x) = x^5 - 10x^3 + 15x, \dots \tag{1.11}$$

We will now see how Wick products are used in the quantum field setting:
Let $\mathcal{E}$ be a positive, continuous, nondegenerate bilinear form on the Schwartz space $\mathcal{S} = \mathcal{S}(\mathbf{R}^d)$ of rapidly decreasing smooth functions on $\mathbf{R}^d$ (d being the dimension of space-time). As usual we let $\mathcal{S}' = \mathcal{S}'(\mathbf{R}^d)$ denote the space of tempered distributions on $\mathbf{R}^d$ ($\mathcal{S}'(\mathbf{R}^d)$ is the dual of $\mathcal{S}(\mathbf{R}^d)$). By the Bochner–Minlos theorem (Gelfand and Vilenkin, 1964), there exists a measure $\mu_{\mathcal{E}}$ on $\mathcal{S}'$ such that

$$T(\varphi) := \int_{\mathcal{S}'} e^{i\langle\omega,\varphi\rangle} d\mu_{\mathcal{E}}(\omega) = e^{-\mathcal{E}(\varphi,\varphi)/2}; \quad \varphi \in \mathcal{S}, \tag{1.12}$$

where $\langle\omega,\varphi\rangle = \omega(\varphi)$ is the action of $\omega \in \mathcal{S}'$ on $\varphi \in \mathcal{S}$.
Let $\mathbf{G} = L^2(\mathbf{R}^d, \mathcal{E})$. We can form the symmetric tensor product $\mathcal{G}_n$ of n copies of $\mathbf{G}$ and define the direct sum

$$\mathcal{G} = \bigoplus_{n=0}^{\infty} \mathcal{G}_n \tag{1.13}$$

which is called the *Fock space* associated with $\mathbf{G}$ and which is unitarily equivalent to $\mathbf{L}^2(\mathcal{S}'(\mathbf{R}^d), \mu_{\mathcal{E}})$. The space $\mathcal{G}_n$ is spanned by the functions

$$\omega \to : \langle\omega,\varphi_1\rangle \dots \langle\omega,\varphi_n\rangle : \quad (\varphi_i \in \mathcal{S})$$

and we have in fact that $:\langle\omega,\varphi_1\rangle \dots \langle\omega,\varphi_n\rangle:$ is the orthogonal projection of $\langle\omega,\varphi_1\rangle \dots \langle\omega,\varphi_n\rangle$ onto $\mathcal{G}_n$.
Quantum fields are described by probability measures on $\mathcal{S}'(\mathbf{R}^d)$ that satisfy the *Osterwalder–Schrader axioms*. Specifically,

(OS 0) $T(\varphi)$ should be *entire* (analytic) in the sense that the function

$$(z_1, \dots, z_n) \to T\left(\sum_{i=1}^{n} z_i\varphi_i\right), \quad z_i \in \mathbf{C}$$

is analytic in $\mathbf{C}^n$, for all n.

(OS 1) $|T(\varphi)| \leq \exp[c(\|\varphi\|_1 + \|\varphi\|_p^p)]$ for $p \in [1,2]$.

(OS 2) $T(\varphi)$ is invariant under Euclidean symmetries in $\mathbf{R}^d$.

(OS 3) $T(\varphi)$ should satisfy reflection positivity, i.e., the matrix

$$[T(\varphi_i - \theta\varphi_j)]_{i,j=1}^{n}$$

should be positive definite (θ is time reflection).

(OS 4) The fields should be ergodic in the sense that the time translation subgroup acts ergodically on the measure space $(\mathcal{S}'(\mathbf{R}^d), \mu_{\mathcal{E}})$.

Each axiom corresponds to a physical property that is not important in our context. In the case of a *free field* we choose

$$\mathcal{E}(\varphi,\psi) = (\varphi, (-\Delta_d + m^2)^{-1}\psi)_{\mathbf{L}^2(\mathbf{R}^d)},$$

where $\Delta_d = \frac{1}{2}\sum_{j=1}^{d} \partial^2/\partial x_j^2$ is the Laplacian in $\mathbf{R}^d$ and $m > 0$ represents the mass. One can then prove that the free field satisfies the Osterwalder–Schrader axioms above. One may consider the free field as infinitely many uncoupled harmonic oscillators. Often interacting fields are constructed by perturbing the free fields. In the interpretation of free fields the space $\mathcal{G}_n$ is called the space of n particle vectors and in particular $\mathcal{G}_0$ corresponds to the vacuum state.

2. THE WICK PRODUCT IN STOCHASTIC ANALYSIS

In stochastic analysis the Wick product was first introduced by T. Hida and N. Ikeda in 1965 (Hida and Ikeda, 1965). A systematic, general account of the traditions of both mathematical physics and probability theory regarding this subject was given by R.Z.Dobrushin and R.A.Minlos in 1977 (Dobrushin and Minlos, 1977). In 1989 P. A. Meyer and J. A. Yan (Meyer and Yan, 1989) defined the Wick product of two *Hida distributions* (or *white noise functionals*, see §2.3 below) in terms of their $\mathcal{S}$-*transforms* (see §4). Motivated by the interpretation and solution of certain stochastic (ordinary and partial) differential equations, the Wick product in the $\mathbf{L}^2(\mu)$- and the $\mathbf{L}^1(\mu)$-setting was introduced and applied in (Lindstrøm et al., 1991; 1992) and (Holden et al., 1991). (Here μ is the white noise probability measure defined in §2.1 below.)

Today the Wick product is important in the study of stochastic (ordinary and partial) differential equations. In general one can say that the use of this product corresponds to — and extends naturally — the use of Ito integrals. We now explain this in more detail.

2.1. *The white noise probability space*

Two fundamental concepts in stochastic analysis are *white noise* and *Brownian motion.* Let us, therefore, start by recalling some of the basic definitions and features of the *white noise probability space.* (Here we give only the most basic results. For a complete account we refer to (Hida et al., to appear).)

Referring to the construction in (1.12), let $\mu = \mu_I$ be the measure on $\mathcal{S}' = \mathcal{S}'(\mathbf{R}^d)$ corresponding to the bilinear form

$$\mathcal{E}_I(\varphi,\psi) = (\varphi,\psi)_{\mathbf{L}^2(\mathbf{R}^d)} = \int_{\mathbf{R}^d} \varphi\psi dx; \quad \varphi,\psi \in \mathcal{S}.$$

In other words, μ is defined by the property

$$\int_{\mathcal{S}'(\mathbf{R}^d)} e^{i\langle\omega,\varphi\rangle} d\mu(\omega) = e^{-\|\varphi\|^2/2}, \qquad \varphi \in \mathcal{S}(\mathbf{R}^d), \tag{2.1}$$

where $\|\varphi\|^2 = (\varphi, \varphi)_{\mathbf{L}^2(\mathbf{R}^d)}$.
Let $\mathcal{B}$ denote the Borel sets on $\mathcal{S}'$ (equipped with the weak star topology). Then the triple $(\mathcal{S}'(\mathbf{R}^d), \mathcal{B}, \mu)$ is called the *white noise probability space*.

Definition 2.1. The *white noise process* is a map

$$W\colon \mathcal{S} \times \mathcal{S}' \to \mathbf{R}$$

given by

$$W(\varphi, \omega) = W_\varphi(\omega) = \langle \omega, \varphi \rangle, \qquad \omega \in \mathcal{S}', \quad \varphi \in \mathcal{S}. \tag{2.2}$$

It is not difficult to prove that if $\varphi \in \mathbf{L}^2(\mathbf{R}^d)$ and we choose $\varphi_n \in \mathcal{S}$ such that $\varphi_n \to \varphi$ in $\mathbf{L}^2(\mathbf{R}^d)$, then

$$\langle \omega, \varphi \rangle := \lim_{n \to \infty} \langle \omega, \varphi_n \rangle \quad (\text{limit in } \mathbf{L}^2(\mu)) \tag{2.3}$$

exists and is independent of the choice of $\{\varphi_n\}$. In particular, if we define

$$\widetilde{B}_x(\omega) := \widetilde{B}_{x_1,\ldots,x_d}(\omega) := \langle \omega, \chi_{[0,x_1]\times\ldots\times[0,x_d]}(\cdot) \rangle, \tag{2.4}$$

then $\widetilde{B}_x$ has an *x-continuous version* B_x which then becomes a *d-parameter Brownian motion*.
The *d-parameter Wiener–Ito integral* of $\varphi \in \mathbf{L}^2(\mathbf{R}^d)$ is then defined by

$$\int_{\mathbf{R}^d} \varphi(y) dB_y(\omega) = \langle \omega, \varphi \rangle. \tag{2.5}$$

Combining (2.2) and (2.5) and using integration by parts for Wiener–Ito integrals, we see that the white noise may be treated as the distributional derivative of the Brownian motion:

$$W_{x_1,\ldots,x_d} = \frac{\partial^d B_{x_1,\ldots,x_d}}{\partial x_1 \partial x_2 \ldots \partial x_d}. \tag{2.6}$$

For more details see, e.g., (Lindstrøm et al., 1991) or (Holden et al., 1991).

Figure 1 shows computer simulations of the white noise process $W_{\varphi_x}(\omega)$, where $\varphi(y) = \chi_{[0,h]\times[0,h]}(y)$, $y \in \mathbf{R}^2$, and $\varphi_x(y) = \varphi(y-x)$ is the x-shift of φ $(x \in \mathbf{R}^2)$.

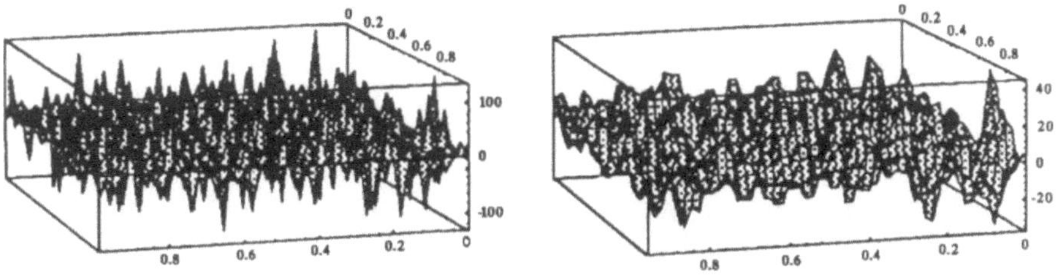

Fig. 1. Two sample paths of white noise ($h = \frac{1}{50}$, $h = \frac{1}{20}$).

2.2. *The Wiener–Ito chaos expansion*
Of special interest will now be the space $\mathbf{L}^2(\mathcal{S}'(\mathbf{R}^d), \mu)$ or $\mathbf{L}^2(\mu)$ for short. The celebrated *Wiener–Ito chaos expansion theorem* says that every $f \in \mathbf{L}^2(\mu)$ has the form

$$f(\omega) = \sum_{n=0}^{\infty} \int_{(\mathbf{R}^d)^n} f_n(u) dB_u^{\otimes n}, \tag{2.7}$$

where $f_n \in \hat{\mathbf{L}}^2(\mathbf{R}^{nd})$, i.e., f_n is *symmetric* in its nd variables (in the sense that $f_n(u_{\sigma_1}, \ldots, u_{\sigma_{nd}}) = f_n(u_1, \ldots, u_{nd})$ for all permutations σ of $(1, 2, \ldots, nd)$) and f_n is square integrable with respect to Lebesgue measure on $\mathbf{R}^{nd}$. The terms on the right-hand side of (2.7) are the *multiple Ito integrals*, defined in (Ito, 1951). (If $d = 1$, they are just iterated Ito integrals on the subspace $u_1 < u_2 < \ldots < u_n$ multiplied by $n!$.)
With f and f_n as in (2.7), we have

$$\|f\|^2_{\mathbf{L}^2(\mu)} = \sum_{n=0}^{\infty} n! \|f_n\|^2_{\mathbf{L}^2(\mathbf{R}^{nd})}. \tag{2.8}$$

There is an equivalent expansion of $f \in \mathbf{L}^2(\mu)$ in terms of the Hermite polynomials h_n defined in (1.10) above. We now explain this more closely:
For $n = 1, 2, \ldots$ let $\xi_n(x)$ be the *Hermite function* of order n, i.e.,

$$\xi_n(x) = \pi^{-1/4}((n-1)!)^{1/2} e^{-x^2/2} h_{n-1}(\sqrt{2}x), \quad x \in \mathbf{R}, \tag{2.9}$$

where h_{n-1} is defined in (1.10).
Then ξ_n is an eigenfunction with eigenvalue $2n$ of the operator

$$A = -\left(\frac{d}{dx}\right)^2 + x^2 + 1.$$

Moreover, $\{\xi_n\}_{n=1}^{\infty}$ forms an orthonormal basis for $\mathbf{L}^2(\mathbf{R})$. Therefore, the family $\{e_\alpha\}$ of tensor products

$$e_\alpha := e_{\alpha_1,\ldots,\alpha_d} := \xi_{\alpha_1} \otimes \ldots \otimes \xi_{\alpha_d} \tag{2.10}$$

(where α denotes the multi-index $(\alpha_1, \ldots, \alpha_d)$) forms an orthonormal basis for $\mathbf{L}^2(\mathbf{R}^d)$.
This is the basis we will use in the rest of this paper. Note that $e_\alpha \in \mathcal{S}(\mathbf{R}^d)$ for all $\alpha = (\alpha_1, \ldots, \alpha_d)$. Now assume that the family of all multi-indices $\beta = (\beta_1, \ldots, \beta_d)$ is given a fixed ordering

$$(\beta^{(1)}, \beta^{(2)}, \beta^{(3)}, \ldots, \beta^{(n)}, \ldots), \quad \text{where} \quad \beta^{(k)} = (\beta_1^{(k)}, \ldots, \beta_d^{(k)}),$$

and put $e_n = e_{\beta^{(n)}}$; $n = 1, 2, \ldots$.
Let $\alpha = (\alpha_1, \ldots, \alpha_m)$ be a multi-index. Then it was shown by Ito (Ito, 1951) that

$$\int\limits_{(\mathbf{R}^d)^n} e_1^{\hat{\otimes}\alpha_1} \hat{\otimes} e_2^{\hat{\otimes}\alpha_2} \hat{\otimes} \ldots \hat{\otimes} e_m^{\hat{\otimes}\alpha_m} dB^{\hat{\otimes}n} = \prod_{j=1}^{m} h_{\alpha_j}(\theta_j), \tag{2.11}$$

where $\theta_j(x) = \int\limits_{\mathbf{R}^d} e_j(x) dB_x(\omega)$, $n = |\alpha_1| + \cdots + |\alpha_m|$ and $\hat{\otimes}$ denotes the *symmetrized tensor product*, so that, e.g., $f\hat{\otimes}g(x_1, x_2) = [f(x_1)g(x_2) + f(x_2)g(x_1)]/2$ if $x_i \in \mathbf{R}$ and similarly for more than two variables.
If we define, for each multi-index $\alpha = (\alpha_1, \ldots, \alpha_m)$,

$$H_\alpha(\omega) = \prod_{j=1}^{m} h_{\alpha_j}(\theta_j), \tag{2.12}$$

then we see that (2.11) can be written as

$$\int\limits_{(\mathbf{R}^d)^n} e^{\hat{\otimes}\alpha} dB^{\otimes|\alpha|} = H_\alpha(\omega) \tag{2.13}$$

using the multi-index notation: $e^{\hat{\otimes}\alpha} = e_1^{\hat{\otimes}\alpha_1} \otimes \ldots \otimes e_m^{\hat{\otimes}\alpha_m}$ if $e = (e_1, e_2, \ldots)$. Since the family $\{e^{\hat{\otimes}\alpha}; |\alpha| = n\}$ forms an orthonormal basis for $\hat{\mathbf{L}}^2((\mathbf{R}^d)^n)$, we see by combining (2.7) and (2.13) that we also have the representation

$$f(\omega) = \sum_\alpha c_\alpha H_\alpha(\omega) \tag{2.14}$$

(the sum being taken over all multi-indices α of non-negative integers). Moreover, it can be proved that

$$\|f\|^2_{\mathbf{L}^2(\mu)} = \sum_\alpha \alpha! c_\alpha^2, \tag{2.15}$$

where $\alpha! = \alpha_1!\alpha_2!\ldots\alpha_m!$ if $\alpha = (\alpha_1, \ldots, \alpha_m)$.
We refer to (Hida et al., to appear) for more information.

2.3. *The Hida test function space $(\mathcal{S})$ and the Hida distribution space $(\mathcal{S})^*$*

There is a subspace of $\mathbf{L}^2(\mu)$ which in some sense corresponds to the Schwartz subspace $\mathcal{S}(\mathbf{R}^d)$ of $\mathbf{L}^2(\mathbf{R}^d)$. This space is called the *Hida test function space* and is denoted by $(\mathcal{S})$. Using the recent characterization due to one of us (see (Zhang, 1991)), a simple description of $(\mathcal{S})$ can be given as follows.

Definition 2.2. Let $f \in \mathbf{L}^2(\mu)$ have the chaos expansion

$$f(\omega) = \sum_\alpha c_\alpha H_\alpha(\omega). \tag{2.16}$$

Then f is a *Hida test function*, i.e., $f \in (\mathcal{S})$, if

$$A_f(k) := \sup_\alpha c_\alpha^2 \alpha!(2\mathcal{N})^{\alpha k} < \infty \quad \text{for all} \quad k < \infty, \tag{2.17}$$

where

$$(2\mathcal{N})^\alpha := \prod_{j=1}^m (2^d \beta_1^{(j)} \dots \beta_d^{(j)})^{\alpha_j} \quad \text{if} \quad \alpha = (\alpha_1, \dots, \alpha_m). \tag{2.18}$$

The topology of $(\mathcal{S})$ can be described by the metric given by

$$d(0, f) = \sum_{k=1}^\infty 2^{-k} \frac{A_f(k)}{1 + A_f(k)} \quad \text{for} \quad f \in (\mathcal{S}) \tag{2.19}$$

and

$$d(f_1, f_2) = d(0, f_1 - f_2) \quad \text{for} \quad f_1, f_2 \in (\mathcal{S}). \tag{2.20}$$

The *Hida distribution space* $(\mathcal{S})^*$ is just the dual of $(\mathcal{S})$. By (Zhang, 1991) we may describe this space as follows.

THEOREM 2.3. *A Hida distribution F is a formal series*

$$F = \sum_\alpha b_\alpha H_\alpha, \tag{2.21}$$

where

$$\sup_\alpha b_\alpha^2 \alpha!((2\mathcal{N})^{-\alpha})^q < \infty \quad \textit{for some} \quad q > 0. \tag{2.22}$$

If $F \in (\mathcal{S})^*$ is given by (2.21) and $f \in (\mathcal{S})$ is given by (2.14), the action of F on f is given by

$$\langle F, f \rangle = \sum_\alpha \alpha! b_\alpha c_\alpha. \tag{2.23}$$

Note that no assumptions are made regarding the convergence of the formal series in (2.21).

We can in a natural way regard $\mathbf{L}^2(\mu)$ as a subspace of $(\mathcal{S})^*$. In particular, if $X \in \mathbf{L}^2(\mu)$, then by (2.23) the action of X on $f \in (\mathcal{S})$ is given by

$$\langle X, f \rangle = \mathbf{E}[X \cdot f].$$

On the other hand, it is easy to see that $(\mathcal{S}) \subset \mathbf{L}^p(\mu)$ for all $p < \infty$ and the inclusion map is continuous. We conclude that

$$(\mathcal{S}) \subset \mathbf{L}^p(\mu) \subset \mathbf{L}^q(\mu) \subset (\mathcal{S})^* \quad \text{if} \quad 1 < q \le p < \infty. \tag{2.24}$$

However, it is not the case that $\mathbf{L}^1(\mu) \subset (\mathcal{S})^*$ (see (Holden et al., 1991)).

Example. The discussion below works for all d, but for simplicity let us assume that $d = 1$.
We have already introduced the white noise process $W_\varphi(\omega)$ as a map from $\mathcal{S} \times \mathcal{S}'$ into $\mathbf{R}$ (see (2.2)). It is also possible to define a *pointwise* version $W_t(\omega), t \in \mathbf{R}$, but then as an element of $(\mathcal{S})^*$, as follows. Define

$$W_t(\omega) = \sum_{k=1}^{\infty} e_k(t) H_{\varepsilon_k}(\omega) = \sum_{k=1}^{\infty} e_k(t) h(\theta_k), \tag{2.25}$$

where $\varepsilon_k = (0, \ldots, 1)$ with 1 on the kth place, $k = 1, 2, \ldots$. Then by (2.18) $(2\mathcal{N})^{\varepsilon_k} = 2k$ and, using the notation of (2.22) with $\alpha = \varepsilon_k$, we obtain

$$\sup_\alpha b_\alpha^2 \alpha! (2\mathcal{N})^{-\alpha q} = \sup_k e_k^2(t) \cdot 1 \cdot (2k)^{-q} < \infty$$

for some $q > 0$.
This shows that the *pointwise version* W_t *of the white noise* exists as an element of $(\mathcal{S})^*$. Since, for $\varphi \in \mathcal{S}(\mathbf{R}^d)$,

$$W_\varphi(\omega) = \langle \omega, \varphi \rangle = \sum_k (\varphi, e_k) \langle \omega, e_k \rangle = \sum_k (\varphi, e_k) H_{\varepsilon_k}(\omega),$$

we see that we may regard $W_t(\omega)$ as the limit in $(\mathcal{S})^*$ of $W_{\varphi_n}(\omega)$ as $\varphi_n \to \delta_t$, the point mass at t. Or, heuristically,

$$W_t(\omega) = \int_{\mathbf{R}} \delta_t(u) dB_u(\omega). \tag{2.26}$$

2.4. *The Wick product*

We are now ready to give the definition of the *Wick product* $F \diamond G$ of two Hida distributions F, G:

Definition 2.4. Let $F = \sum_\alpha a_\alpha H_\alpha$, $G = \sum_\beta b_\beta H_\beta$ be two elements of $(\mathcal{S})^*$. Then the *Wick product* of F and G is the element $F \diamond G$ in $(\mathcal{S})^*$ given by

$$F \diamond G = \sum_{\alpha,\beta} a_\beta b_\beta H_{\alpha+\beta} = \sum_\gamma c_\gamma H_\gamma, \tag{2.27}$$

where $c_\gamma = \sum\limits_{\alpha+\beta=\gamma} a_\alpha b_\beta$.

Remark. It follows from characterization (2.19) that the Wick product of two elements of $(\mathcal{S})^*$ again is an element of $(\mathcal{S})^*$. Similarly, from characterization (2.15) we see that

$$f, g \in (\mathcal{S}) \Rightarrow f \diamond g \in (\mathcal{S}).$$

If

$$f = \sum_{n=0}^{\infty} \int f_n dB^{\otimes n} \quad \text{and} \quad g = \sum_{m=0}^{\infty} \int g_m dB^{\otimes m} \tag{2.28}$$

are two functions in $\mathbf{L}^2(\mu)$, their Wick product can be expressed as

$$f \diamond g = \sum_{n,m=0}^{\infty} \int f_n \hat{\otimes} g_m dB^{\otimes(n+m)} \tag{2.29}$$

whenever the sum converges (in $\mathbf{L}^1(\mu)$).
Equivalently, if we use the Hermite expansions

$$f = \sum_{\alpha} a_\alpha H_\alpha \quad \text{and} \quad g = \sum_{\beta} b_\beta H_\beta, \tag{2.30}$$

then the Wick product is given by

$$f \diamond g = \sum_{\alpha,\beta} a_\alpha b_\beta H_{\alpha+\beta} = \sum_{\gamma} c_\gamma H_\gamma, \tag{2.31}$$

where $c_\gamma = \sum\limits_{\alpha+\beta=\gamma} a_\alpha b_\beta$.
In particular, if we choose $d = 1$ and f Gaussian, i.e.,

$$f(\omega) = \int_{\mathbf{R}} f_1(x) dB_x(\omega), \quad \text{where} \quad f_1 \in \mathbf{L}^2(\mathbf{R}) \quad \text{is deterministic}, \tag{2.32}$$

then one obtains from (2.31) and (1.9) that

$$f^{\diamond n} = \|f_1\|^n h_n \left(\frac{f}{\|f_1\|} \right) =: f^n : . \tag{2.33}$$

Thus the physical Wick power and the (stochastic analysis) Wick power coincide in this case. However, for other products they do not coincide. For example, if

$$U = \int \chi_{[0,t]}(s)\, dB_s = B_t \quad \text{and} \quad V = U^{\diamond 2},$$

we have by (2.24) and (2.28)

$$U \diamond V = U \diamond (U^{\diamond 2}) = U^{\diamond 3} = t^{3/2} h_3 \left(\frac{B_t}{\sqrt{t}} \right) = B_t^3 - 3tB_t$$

while, by (1.8),

$$: UV := B_t^3 - tB_t.$$

In fact, the physical Wick product $: UV :$ should not be called a *product*, because as an binary operation it is not even associative. On the other hand it is trivial to verify that the (stochastic analysis) Wick product $U \diamond V$ is associative, as well as commutative and distributive over addition. So, considered as a binary operation on $(\mathcal{S})$ or $(\mathcal{S})^*$ it forms a ring with unit 1 (the constant function).

2.5. *Applications*

In quantum physics the physical Wick product may be regarded as a renormalized product, introduced to avoid selfinteractions. In stochastic analysis the Wick product is natural because it is implicit in the Ito integral (and, more generally, in the Skorokhod integral if the integrand is not adapted). More precisely, if $Y_t(\omega)$ is a (suitable) stochastic process adapted to the filtration $\mathcal{F}_t$ generated by the (1-parameter) Brownian motion B_t, then we have

$$\int_{\mathbf{R}} Y_t(\omega) dB_t(\omega) = \int_{\mathbf{R}} Y_t \diamond W_t dt. \tag{2.34}$$

Here $W_t \in (\mathcal{S})^*$ is the pointwise white noise defined in (2.25) and the integral on the right is to be regarded as an integral in $(\mathcal{S})^*$. If one prefers to work with less singular objects one can reformulate (2.34) as follows:

$$\int_{\mathbf{R}} (\varphi * Y)_t \delta B_t = \int_{\mathbf{R}} Y_t \diamond W_{\varphi_t} dt\,; \quad \varphi \in \mathcal{S}(\mathbf{R}), \tag{2.35}$$

where $*$ denotes the convolution with respect to the Lebesgue measure on $\mathbf{R}$, i.e.,

$$(\varphi * Y)_t(\omega) = \int_{\mathbf{R}} \varphi(s) Y_{s-t}(\omega)\, ds \tag{2.36}$$

and

$$\varphi_t(s) = \varphi(s-t) \quad \text{is the } t\text{-shift of} \quad \varphi. \tag{2.37}$$

As before W_φ is the white noise "smeared out by φ" given by (2.2). The notation "δB_t" indicates that the integral should be interpreted as a *Skorokhod integral*, since the integrand in (2.35) need not be adapted even if Y_t is. In fact, (2.35) remains true for nonadapted processes Y_t. (The Skorokhod integral coincides with the Ito integral if the integrand is adapted. See, e.g., (Nualart and Zakai, 1986) for more information.) There are several versions of (2.34), see, e.g., (Lindstrøm et al., 1992) and (Ash and Potthoff, 1991). The "smeared out" version (2.35) is proved in (Øksendal and Zhang, 1992).

From (2.34) and (2.35) we conclude that if we interpret white noise differential equation

$$\frac{dX_t}{dt} = b(X_t) + \sigma(X_t) \cdot W_t\,, \quad X_0 = x \in \mathbf{R}^n, \tag{2.38}$$

as an Ito integral equation

$$X_t = x + \int_0^t b(X_s)ds + \int_0^t \sigma(X_s)dB_s, \tag{2.39}$$

we are really saying that the product on the right-hand side of (2.38) is interpreted as a Wick product:

$$\frac{dX_t}{dt} = b(X_t) + \sigma(X_t) \diamond W_t \; ; \quad X_0 = x. \tag{2.40}$$

Moreover, if all products in the equation are the Wick products, then the equation can be solved by *ordinary* (deterministic) calculus rules as long as all the products are taken in the Wick sense. This can facilitate the solution of stochastic differential equations considerably. We illustrate this by some examples.

Example 2.5. Consider the equation for, say, a population growth in a "white noise environment":

$$\frac{dX_t}{dt} = X_t W_t \; ; \quad X_0 \quad \text{given} \tag{2.41}$$

or, in the Ito differential notation,

$$dX_t = X_t dB_t \; ; \quad X_0 \quad \text{given.} \tag{2.42}$$

Let us first assume that $X_0 = x$ (constant). Rather than using the Ito calculus on (2.42) we use the Wick calculus on the Wick interpretation of (2.41), i.e.,

$$\frac{dX_t}{dt} = X_t \diamond W_t \; ; \quad X_0 = x. \tag{2.43}$$

Based on the analogy with deterministic calculus, we guess that the solution of (2.43) is

$$X_t = x \cdot \mathrm{Exp}\left(\int_0^t W_s \, ds\right) = x \cdot \mathrm{Exp} B_t, \tag{2.44}$$

where

$$\mathrm{Exp} B_t = \sum_{n=0}^{\infty} \frac{1}{n!} B_t^{\diamond n} \tag{2.45}$$

is the *Wick exponential* of B_t. (See Example 2.7 for a more general discussion of this.)
Indeed, we verify that (2.44) is a solution by verifying (using (2.45)) that

$$\frac{d}{dt}\left(\mathrm{Exp}\left(\int_0^t W_s \, ds\right)\right) = \mathrm{Exp}\left(\int_0^t W_s \, ds\right) \diamond W_t .$$

On the other hand, it is well known that the Ito calculus applied to (2.42) gives the solution

$$X_t = x \cdot \exp\left(B_t - \frac{1}{2}t\right). \tag{2.46}$$

The two solutions (2.44) and (2.46) are indeed the same, as can be seen by combining (2.33) with the following well known equation for the Hermite polynomials:

$$\sum_{n=0}^{\infty} h_n(x)\frac{t^n}{n!} = \exp\left(tx - \frac{1}{2}t^2\right). \tag{2.47}$$

Now suppose that the initial value X_0 is not a constant x but a random variable which is *anticipating*, i.e., depending on $\mathcal{F}_t$ for some $t > 0$. Then X_s will not be adapted and the Ito differential in (2.42) is no longer well-defined. However, the Wick product interpretation (2.43) makes sense as before and the Wick calculus now leads to the solution

$$X_t = X_0 \diamond \operatorname{Exp} B_t. \tag{2.48}$$

(If $X_0 = x$ (constant), then (2.48) reduces to (2.44), since the Wick product reduces to the ordinary product if one of the factors are deterministic.) Making use of (2.35), we see that the process X_t given by (2.48) solves the *Skorokhod integral equation*

$$X_t = X_0 + \int_0^t X_s \delta B_s\,; \quad t \geq 0. \tag{2.49}$$

This is an extra advantage with the Wick calculus: It works equally well for adapted and nonadapted processes.
It is interesting to compare the Wick solution (2.48) with a solution obtained by completely different methods by Buckdahn (Buckdahn, 1991). Specializing his general solution formula to our case, we get the solution

$$X_t(\omega) = X_0(A_t\omega) \cdot \exp\left(B_t - \frac{1}{2}t\right), \tag{2.50}$$

where A_t is the shift operator on $C[0,\infty)$ given by

$$(A_t\omega)(s) = \omega(s) - t \wedge s\,; \qquad s \geq 0, \quad \omega \in C[0,\infty).$$

By uniqueness of equation (2.41), we know that two solutions (2.48) and (2.50) must be the same, but this is not easy to see directly!

Example 2.6 (*The stochastic Volterra equation*). The classical deterministic Volterra equation (of the second kind) has the form

$$X_t = Y_t + \int_0^t \gamma(t,s)X_s\,ds\,; \quad t \geq 0, \tag{2.51}$$

where $\gamma(t,s)$, Y_t are given functions. If the system is randomly perturbed or if there is insufficient information about the function $\gamma(t,s)$, then a natural mathematical formulation would be to put

$$\gamma(t,s) = b(t,s) + \sigma(t,s)W_s \tag{2.52}$$

which would lead to the *stochastic Volterra equation*

$$X_t = Y_t + \int_0^t b(t,s)X_s\,ds + \int_0^t \sigma(t,s)X_s \cdot W_s\,ds. \tag{2.53}$$

Again there is a question what we should mean by the product $\sigma(t,s)X_s \cdot W_s$ in the right-hand side of (2.53). In the general case we allow Y_t to be a random variable, possibly anticipating. And then — following the discussion in the previous example — it is natural to interpret (2.53) in the Skorokhod sense, i.e.,

$$X_t = Y_t + \int_0^t b(t,s)X_s\,ds + \int_0^t \sigma(t,s)X_s\delta B_s, \tag{2.54}$$

since we cannot expect X_t to be adapted in this case.
It is proved in (Øksendal and Zhang, 1992) that under suitable conditions the solution X_t of (2.54) is given by

$$X_t = Y_t + \int_0^t H(t,s) \diamond Y_s\,ds, \tag{2.55}$$

where

$$H(t,s) = H(t,s,\omega) = \sum_{n=1}^{\infty} K_n(t,s);$$

with

$$K_1(t,s) = K_1(t,s,\omega) = b(t,s) + \sigma(t,s)W_s$$

and inductively

$$K_{n+1}(t,s) = \int_0^t K_1(t,u) \diamond K_n(u,s)\,du.$$

In particular, if $Y_s(\omega) = Y(\omega)$ (i.e., constant with respect to s), we get a striking, and perhaps surprising, connection between this general solution X_t and the solution x_t in the special case when $Y_t \equiv 1$, i.e.,

$$x_t = 1 + \int_0^t b(t,s)x_s\,ds + \int_0^t \sigma(t,s)x_s\,dB_s.$$

By (2.55), we see that the connection is

$$X_t = Y \diamond x_t.$$

For details we refer to (Øksendal and Zhang, 1992).

Example 2.7 *(Fluid flow in porous media).* The pressure equation for the flow of an incompressible fluid in a porous medium is

$$\begin{cases} \operatorname{div}(K \cdot \nabla p) = -f & \text{in} \quad D, \\ p = 0 & \text{on} \quad \partial D, \end{cases} \tag{2.56}$$

where (at a fixed instant of time t) D is a given domain in $\mathbf{R}^d$ $(d = 3)$, f is the given source rate for the fluid in D and $K = K(x)$ is the *permeability* of the medium at the point x. In a typical porous rock the values of $K(x)$ are changing rapidly with x and it is hopeless to find an analytic description of this function. Therefore, it is natural to use a stochastic approach, where we represent the function $K(x)$ by a *positive noise* $K(x, \omega)$. Measurements of porous rock samples indicate that such a positive noise should have — at least approximately — the following properties:

(i) $x \neq y \Rightarrow K(x, \cdot)$ and $K(y, \cdot)$ are independent,

(ii) $\{K(x, \cdot)\}_{x \in \mathbf{R}^d}$ is a stationary process, i.e., for all $x_1, \ldots, x_n \in \mathbf{R}^d$, the distribution of $(K(x_1 + h, \cdot), K(x_2 + h, \cdot), \ldots, K(x_n + h, \cdot))$ is independent of $h \in \mathbf{R}^d$,

(iii) $K(x, \cdot)$ has a lognormal distribution, for all $x \in \mathbf{R}^d$.

There does not exist an ordinary stochastic process with these properties. However, if we consider *distribution valued* (or generalized) stochastic processes $K(\varphi, \omega)$ and interpret (i) to mean

(i)′ $\operatorname{supp} \varphi \cap \operatorname{supp} \psi = \emptyset \Rightarrow K(\varphi, \cdot)$ and $K(\varphi, \cdot)$ are independent,

then a good candidate for such a positive noise is the *Wick exponential of white noise*. In other words, we put

$$K(x, \omega) := \operatorname{Exp} W_{\varphi_x}(\omega) := \sum_{n=0}^{\infty} \frac{1}{n!} W_{\varphi_x}^{\diamond n}(\omega), \tag{2.57}$$

where as before $\varphi_x(y) = \varphi(y - x)$ is the x-shift of the test function $\varphi \in \mathcal{S}(\mathbf{R}^3)$. Combining (2.11) and (2.47), we see that

$$\operatorname{Exp} W_\varphi = \exp\left(W_\varphi - \frac{1}{2}\|\varphi\|^2\right); \qquad \varphi \in \mathcal{S}(\mathbf{R}^d).$$

Moreover, the series (2.57) converges in $\mathbf{L}^p(\mu)$ for all $p < \infty$. This follows from the Carlen–Kree estimate (Carlen and Kree, 1991) (see Theorem 5.10).

Computer simulations of the 2-parameter (i.e., $d = 2$) positive noise $K(x, \omega)$ are shown on Figure 2.

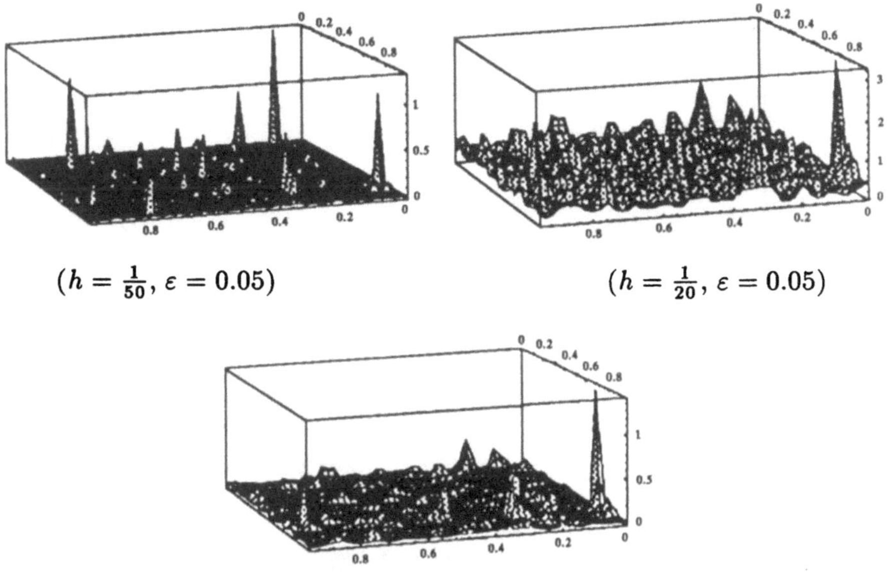

$(h = \frac{1}{50},\ \varepsilon = 0.05)$ $(h = \frac{1}{20},\ \varepsilon = 0.05)$

$(h = \frac{1}{20},\ \varepsilon = 0.1)$

Fig. 2. Three sample paths of positive noise $\mathrm{Exp}[\varepsilon W_\varphi]$

At this point it is necessary to make precise what we mean by a *solution* p of (2.56). It has been known for a long time (see (Walsh, 1986)) that in higher dimensions stochastic partial differential equations only have solutions in the (Schwartz) distribution sense, i.e., solutions which are distribution valued stochastic processes. So we should look for solutions $p = p(\varphi, \omega)$, $\varphi \in \mathcal{S}(\mathbf{R}^d)$, $\omega \in \mathcal{S}(\mathbf{R}^d)$. If we fix a test function φ and consider its shifts $\varphi_x(\cdot)$, $x \in \mathbf{R}^d$, we get the process

$$X(\varphi, x, \omega) = p(\varphi_x, \omega). \tag{2.58}$$

Now note that taking distributional derivatives of p with respect to φ is (by integration by parts) the same as taking derivatives of X with respect to x. For example, if $d = 1$, $F \in \mathcal{S}'(\mathbf{R})$, and D denotes the differentiation operator, we have

$$\langle DF, \varphi_x(\cdot)\rangle = -\langle F, D\varphi_x(\cdot)\rangle = -\langle F, \frac{\partial}{\partial y}\varphi(y - x)\rangle = \langle F, \frac{d}{dx}\varphi_x(y)\rangle.$$

This leads to the following definition.

Definition 2.8. Let $p \geq 1$. An $\mathbf{L}^p$ *functional process* is a function

$$X\colon \mathcal{S}(\mathbf{R}^d) \times \mathbf{R}^d \times \mathcal{S}'(\mathbf{R}^d) \to \mathbf{R}$$

such that

(i) $x \to X(\varphi, x, \omega)$ is (Borel) measurable for all $\varphi \in \mathcal{S}$, $\omega \in \mathcal{S}'$,

and

(ii) $\omega \to X(\varphi, x, \omega)$ belongs to $\mathbf{L}^p(\mu)$ for all $\varphi \in \mathcal{S}$, $x \in \mathbf{R}^d$.

The heuristic interpretation of X is that $X(\varphi, x, \omega)$ is the result we get if we measure X using the test function (or "window") φ shifted to the point x and in the "experiment" ω. Note that the positive noise K given by (2.57) is an $\mathbf{L}^2$ functional process (in fact $\mathbf{L}^p$ for all $p < \infty$) according to this definition.
If K in (2.56) is represented by (2.57), the pressure p is represented by an $\mathbf{L}^p$ functional process $X(\varphi, x, \omega)$ and the product "$K \cdot \nabla p$" is interpreted as the Wick product $K \diamond \nabla X$, then equation (2.56) becomes

$$\begin{cases} \operatorname{div}(K(\varphi, x) \diamond \nabla X(\varphi, x)) = -(f * \varphi)(x)\,; & x \in D, \\ X(\varphi, x) = 0\,; & x \in \partial D, \end{cases} \tag{2.59}$$

where $(f * \varphi)(x) = \int_{\mathbf{R}^d} f(y)\varphi(y - x)dy$ is the convolution and the derivatives (div, grad) are taken with respect to x.
This equation seems suitable for the Wick calculus approach. However, in this case we encounter a new difficulty. It turns out that even in the 1-dimensional case ($d = 1$) there does not exist an $\mathbf{L}^p$ functional process solution of (2.59) for $p > 1$. In fact, this equation does not even have solution in $(\mathcal{S})^*$ (see §4). So in order to handle such equations it is necessary to extend the definition of the Wick product beyond $\mathbf{L}^2(\mu)$ — and even beyond $(\mathcal{S})^*$. In (Holden et al., 1991) the Wick product is (partially) extended to $\mathbf{L}^1(\mu)$ and this is sufficient to solve, for example, the stochastic Schrödinger equation with a positive noise potential. The extension goes as follows.

Definition 2.9. Let $X, Y \in \mathbf{L}^1(\mu)$ and suppose there exist $X_n, Y_n \in \mathbf{L}^2(\mu)$ such that

$$X_n \to X \quad \text{in} \quad \mathbf{L}^1(\mu), \quad Y_n \to Y \quad \text{in} \quad \mathbf{L}^1(\mu) \tag{2.60}$$

and

$$X_n \diamond Y_n \quad \text{converges in} \quad \mathbf{L}^1(\mu) \quad \text{in} \quad Z, \quad \text{say.} \tag{2.61}$$

Then we define $X \diamond Y = Z$.
The definition does not depend on the sequences $\{X_n\}, \{Y_n\}$. In fact, if (2.60)–(2.61) hold, then

$$\mathcal{F}[Z](\varphi) = e^{\|\varphi\|^2/2}\mathcal{F}[X](\varphi)\mathcal{F}[Y](\varphi)\,; \quad \varphi \in \mathcal{S}(\mathbf{R}^d), \tag{2.62}$$

where

$$\mathcal{F}[Z](\varphi) = \int e^{i\langle\omega,\varphi\rangle} Z(\omega)\, d\mu(\omega) \tag{2.63}$$

is the *Fourier transform* of Z, and similarly with X and Y. Note that if we choose $\varphi = 0$ in (2.62), we get the following result of independent interest.

COROLLARY 2.10. *Suppose* $X, Y \in \mathbf{L}^1(\mu)$ *and that* $X \diamond Y \in \mathbf{L}^1(\mu)$ *exists. Then*

$$\mathbf{E}[X \diamond Y] = \mathbf{E}[X] \cdot \mathbf{E}[Y].$$

For a further discussion of equation (2.59) we refer to (Lindstrøm et al., 1991b).

3. DISCRETE WICK CALCULUS

It is possible to develop a discrete version of the Wick calculus which relates to the Bernoulli random walks more or less the same way as the continuous theory relates to the Brownian motion. The discrete Wick calculus has close connection to Meyer's "toy Fock space" version of quantum probability (Meyer, 1986; 1987) and it has much of the same purpose — to serve as a laboratory where one can observe the basic algebraic and probabilistic ideas without being disturbed by the heavy analytical machinery needed in the continuous theory. In this section we shall sketch briefly the basic ideas of the discrete Wick calculus and indicate the connections to the continuous theory; more detailed information can be found in (Hølden et al., 1992).

Fix a finite set T, and let

$$\Omega = \{\omega \mid \omega\colon T \to \{-1,1\}\}$$

be the space of all Bernoulli trials over T. By the *toy Fock space* over T we shall simply mean the space $\mathbf{L}^2(\Omega, \mathbf{P})$, where $\mathbf{P}$ is the uniform probability on Ω; i.e., the set of all functions $X\colon \Omega \to \mathbf{C}$ with the norm

$$\|X\| = \left(\sum |X(\omega)|^2 \mathbf{P}(\omega)\right)^{1/2}.$$

For each $A \subset T$, define a function $\chi_A\colon \Omega \to \mathbf{C}$ by

$$\chi_A(\omega) = \begin{cases} \prod_{a \in A} \omega(a) & \text{if } A \neq \emptyset, \\ 1 & \text{if } A = \emptyset. \end{cases}$$

Clearly, $\{\chi_A\colon A \subset T\}$ is an orthonormal set in $\mathbf{L}^2(\Omega, \mathbf{P})$, and since its cardinality equals the dimension of $\mathbf{L}^2(\Omega, \mathbf{P})$, it must be a basis. Hence any function $X \in \mathbf{L}^2(\Omega, \mathbf{P})$ can be written uniquely as a sum

$$X = \sum_{A \subset T} X(A) \chi_A,$$

where each $X(A)$ is a complex number. We shall refer to this as the *Walsh decomposition* of X, and — as we shall see later — it is a close relative of the Wiener–Ito chaos.

We are now ready to introduce the discrete Wick product on $\mathbf{L}^2(\Omega, \mathbf{P})$. Intuitively, the idea is that we want an algebraic operation $\diamond$ such that

$$\chi_A \diamond \chi_B = \begin{cases} \chi_{A \cup B} & \text{if } A \cap B = \emptyset, \\ 0 & \text{otherwise,} \end{cases}$$

and if we extend by ordinary algebraic rules, we are led to the following definition:

Definition 3.1. If $X = \sum X(A)\chi_A$ and $Y = \sum Y(B)\chi_B$ are two elements of $\mathbf{L}^2(\Omega, \mathbf{P})$, their *Wick product* $Z = X \diamond Y$ is defined by $Z = \sum Z(C)\chi_C$, where

$$Z(C) = \sum\{X(A)Y(B) |\ A \cup B = C, A \cap B = \emptyset\}.$$

Remark 3.2. If U is the ordinary product of X and Y, then the Walsh components of U are

$$U(C) = \sum\{X(A)Y(B) |\ A \Delta B = C\},$$

and hence in one respect the Wick product is simpler than the ordinary product; in order to compute $Z(C)$ we only need to know the Walsh components $X(A)$ and $Y(B)$ for *subsets* of C, while $U(C)$ depends on all Walsh components of X and Y. Let us make a useful but trivial observation.

LEMMA 3.3. $(\mathbf{L}^2(\Omega, \mathbf{P}), +, \diamond)$ *is a commutative ring with unit element* $\chi_\emptyset = 1$.

With a well-defined Wick product it is easy to define Wick polynomials. If $p(z_1, \ldots, z_k) = \sum_{|\alpha|=0}^{m} c_\alpha z^\alpha$ is a complex polynomial in k variables (where $\alpha = (\alpha_1, \alpha_2, \ldots, \alpha_k)$ runs over multi-indices), and $X_1, \ldots, X_k \in \mathbf{L}^2(\Omega, \mathbf{P})$, we define

$$p^\diamond(X_1, \ldots, X_k) = \sum_{|\alpha|=0}^{m} c_\alpha X_1^{\diamond\alpha_1} \diamond \ldots \diamond X_k^{\diamond\alpha_k},$$

where, of course, $X^{\diamond n} = X \diamond \ldots \diamond X$ (n times) for every positive integer n, and $X^{\diamond 0} = 1$. Power series can be treated similarly as long as we are a little careful with convergence problems (see (Holden et al., 1992) for the details).
To state the main result of the discrete Wick calculus, we need the following notation. If $\alpha = (\alpha_1, \ldots, \alpha_k)$ is a multi-index and A is a subset of T, then $\mathcal{P}_\alpha(A)$ is the family of all sequences of sets

$$\langle\{A_1^{(1)}, A_2^{(1)}, \ldots, A_{\alpha_1}^{(1)}\}, \{A_1^{(2)}, \ldots, A_{\alpha_2}^{(2)}\}, \ldots, \{A_1^{(k)}, \ldots, A_{\alpha_k}^{(k)}\}\rangle,$$

such that the sets $A_j^{(i)}$ are non-empty and $\bigcup_{i,j} A_j^{(i)} = A$. If $\alpha_m = 0$, we just let $\{A_1^{(m)}, \ldots, A_{\alpha_m}^{(m)}\}$ be the empty set. We shall write $\langle A_j^{(i)}\rangle$ as an abbreviation of $\langle\{A_1^{(1)}, \ldots, A_{\alpha_1}^{(1)}\}, \ldots, \{A_1^{(k)}, \ldots, A_{\alpha_k}^{(k)}\}\rangle$.

THEOREM 3.4. *Assume that* $F\colon \mathbf{C}^N \times \mathbf{C}^k \to \mathbf{C}^N$ *is an analytic function, and let* $Y_1, \ldots, Y_k\colon \Omega \to \mathbf{C}$ *be random variables with expectations* $y_1, \ldots, y_k$, *respectively. Assume also that the Jacobian matrix*

$$JF(x_1, \ldots, x_N; y_1, \ldots, y_k) = \begin{pmatrix} \frac{\partial F_1}{\partial x_1} & \frac{\partial F_1}{\partial x_2} & \cdots & \frac{\partial F_1}{\partial x_N} \\ \vdots & & & \\ \frac{\partial F_N}{\partial x_1} & \frac{\partial F_N}{\partial x_2} & \cdots & \frac{\partial F_N}{\partial x_N} \end{pmatrix}$$

is nonsingular. Then there is a one-to-one correspondence between solutions $x_1, \ldots, x_N$ *of the deterministic equation*

$$F(x_1, \ldots, x_N, y_1, \ldots, y_k) = 0$$

and solutions $X_1, \ldots, X_N$ of the stochastic equation

$$F^{\diamond}(X_1, \ldots, X_N, Y_1, \ldots, Y_k) = 0.$$

Given a solution $x = (x_1, \ldots, x_N)$ of the deterministic equation, the corresponding solution $X = (X_1, \ldots, X_N)$ of the stochastic equation is given by the following hierarchy of equations:

$$X(\varnothing) = x,$$

and if $X(B)$ has been determined for all proper subsets B of A, then $X(A)$ is given by

$$X(A) = -JF(x_1, \ldots, x_N; y_1, \ldots, y_k)^{-1} \Big[\sum_{l=1}^{k} \frac{\partial F}{\partial y_l}(x_1, \ldots, y_k) Y_l(A)$$

$$+ \sum_{|\alpha|>1} \sum_{(\{A_j^i\}) \in \mathcal{P}_\alpha(A)} D_\alpha F(x_1, \ldots, y_k) X_1(A_1^{(1)}) \ldots Y_k(A_{\alpha_{N+k}}^{(N+k)}) \Big].$$

The proof of this theorem can be found in (Holden et al., 1992). Although the statement may seem quite complicated and confusing in the present setting, the main message is simple; if we can solve the nonlinear, deterministic equation

$$F(x_1, \ldots, x_N, y_1, \ldots, y_k) = 0,$$

we can also solve the nonlinear, stochastic equation

$$F^{\diamond}(X_1, \ldots, X_N, Y_1, \ldots, Y_k) = 0,$$

by solving an additional hierarchy of very simple linear equations (but, it should be admitted, this hierarchy is very large). Note that the form of our equation is more than general enough to cover all kinds of finite difference equations (a few examples are worked out in (Holden et al., 1992)).

Let us end this section by pointing out the close formal relationship between the discrete and the continuous Wick calculus. For simplicity, we shall only compare the discrete theory with the one-dimensional version of the continuous theory. Assume that our underlying space T is a timeline

$$T = \{-T_0, -T_0 + \Delta t, \ldots, T_0 - \Delta t, T_0\},$$

where we think of Δt as a small time-increment. Let $B\colon \Omega \times T \to \mathbf{R}$ be the random walk

$$B(\omega, t) = \sum_{s<t} \omega(s)\sqrt{\Delta t},$$

and note that $\Delta B(\omega,t) \equiv B(\omega,t+\Delta t) - B(\omega,t) = \omega(t)\sqrt{\Delta t}$.
Given an $X \in \mathbf{L}^2(\Omega,\mathbf{P})$, we now define a function $X_n\colon T^n \to \mathbf{C}$ for each $n \in \mathbf{N}$ by

$$X_n(t_1,\ldots,t_n) = \begin{cases} \frac{1}{(\Delta t)^{n/2} n!} X(\{t_1,\ldots,t_n\}) & \text{if } t_1,\ldots,t_n \text{ are distinct,} \\ 0 & \text{otherwise,} \end{cases}$$

where $X(\{t_1,\ldots,t_n\})$ is the Walsh component of X. Note that X_n is symmetric by definition, and that by rewriting the Walsh decomposition, we get

$$\begin{aligned} X &= \sum_n \sum_{t_1<t_2<\ldots<t_n} n! X_n(t_1,\ldots,t_n)\Delta t^{n/2} \chi_{\{t_1,\ldots,t_n\}} \\ &= \sum_n \sum_{(t_1,\ldots,t_n)\in T^n} X_n(t_1,\ldots,t_n)\Delta B(t_1)\ldots\Delta B(t_n) \end{aligned}$$

which is the discrete Wiener–Ito decomposition.
Returning to the Wick product, we now try to compute $(X \diamond Y)_n(t_1,\ldots,t_n)$. We get

$$(X \diamond Y)_n(t_1,\ldots,t_n) = \frac{1}{(\Delta t)^{n/2} n!}(X \diamond Y)\{t_1,\ldots,t_n\}$$

$$= \frac{1}{(\Delta t)^{n/2} n!} \sum_{k=0}^{n} \sum_{t_{i_1},\ldots,t_{i_k}} X\{t_{i_1},\ldots,t_{i_k}\} \cdot Y\{t_{j_1},\ldots,t_{j_{n-k}}\}$$

where $t_{i_1},\ldots,t_{i_k}$ is a k-element subset of $\{t_1,\ldots,t_n\}$ and $t_{j_1},\ldots,t_{j_{n-k}}$ are the remaining elements in $\{t_1,\ldots,t_n\}$. Continuing, we get

$$(X \diamond Y)_n(t_1,\ldots,t_n)$$

$$= \frac{1}{(\Delta t)^{n/2} n!} \sum_{k=0}^{n} \sum_{t_{i_1}\ldots t_{i_k}} (\Delta t)^{k/2} k! X_k(t_{i_1},\ldots,t_{i_k})(\Delta t)^{\frac{n-k}{2}}(n-k)! Y_{n-k}(t_{j_1},\ldots,t_{j_{n-k}})$$

$$= \sum_{k=0}^{n} \frac{1}{\binom{n}{k}} \sum_{t_{i_1}\ldots t_{i_k}} X_k(t_{i_1},\ldots,t_{i_k}) Y_{n-k}(t_{j_1},\ldots,t_{j_{n-k}}) = \sum_{k=0}^{n} (X_k \hat{\otimes} Y_{n-k})(t_1,\ldots,t_n),$$

where the last step uses that since X_k and Y_{n-k} already are symmetric, it suffices to symmetrize over all possible ways of distributing the t's between X_k and Y_{n-k}. It follows that

$$X \diamond Y = \sum_{n,m} \sum_{(t_1,\ldots,t_{n+m})} (X_n \hat{\otimes} Y_m)(t_1,\ldots,t_{n+m})\Delta B(t_1)\ldots\Delta B(t_{n+m})$$

and hence the discrete and the continuous Wick products take on exactly the same form.

4. OTHER CHARACTERIZATIONS OF THE WICK PRODUCT

4.1. *The $\mathcal{S}$-transform and the Hermite transform*
If $F \in (\mathcal{S})^*$, then the *$\mathcal{T}$-transform* of $F, \mathcal{T}F$, is a map from $\mathcal{S}(\mathbf{R}^d)$ into the complex plane $\mathbf{C}$ defined by

$$\mathcal{T}F(\varphi) = \langle F, \exp(i\langle \cdot, \varphi\rangle)\rangle, \tag{4.1}$$

(It can be proved that the function $\omega \to \exp(i\langle\omega,\varphi\rangle); \omega \in \mathcal{S}'(\mathbf{R}^d)$ belongs to $(\mathcal{S})$, so (4.1) is well-defined.)
Note that if $F \in \mathbf{L}^2(\mu)$, then

$$\mathcal{T}F(\varphi) = \int_{\mathcal{S}'(\mathbf{R}^d)} \exp(i\langle\omega,\varphi\rangle)F(\omega)d\mu(\omega) \tag{4.2}$$

so $\mathcal{T}$ coincides with the Fourier transform $\mathcal{F}$ in this case.
The *$\mathcal{S}$-transform* of $F, \mathcal{S}F$, is a map from $\mathcal{S}(\mathbf{R}^d)$ into $\mathbf{C}$ defined by

$$\mathcal{S}F(\varphi) = e^{-\|\varphi\|^2/2}\mathcal{T}F(-i\varphi), \tag{4.3}$$

where $\|\varphi\|^2 = \int_{\mathbf{R}^d} |\varphi|^2 dx$ as before.
In particular, if $F \in \mathbf{L}^2(\mu)$, we get

$$\mathcal{S}F(\varphi) = e^{-\|\varphi\|^2/2} \int_{\mathbf{R}^d} \exp(\langle\omega,\varphi\rangle)F(\omega)\, d\mu(\omega). \tag{4.4}$$

The *Hermite transform* of $F, \mathcal{H}F$ or $\tilde{F}$, is a map from the space $\mathbf{C}_0^{\mathbf{N}}$ of all finite sequences of complex numbers into $\mathbf{C}$ defined by

$$\mathcal{H}F(z_1,z_2,\ldots) = \tilde{F}(z_1,z_2,\ldots) = \mathcal{S}F(z_1e_1 + z_2e_2 + \ldots)\ ;\ (z_1,z_2,\ldots) \in \mathbf{C}_0^{\mathbf{N}}, \tag{4.5}$$

where as before $\{e_n\}$ is a given orthonormal basis of $\mathbf{L}^2(\mathbf{R}^d)$. Equivalently (see (Lindstrøm et al., 1991a), Theorem 5.7), if $F \in (\mathcal{S})^*$ has the expansion

$$F(\omega) = \sum_\alpha c_\alpha H_\alpha(\omega)$$

then, using the multi-index notation $z^\alpha = z_1^{\alpha_1} z_2^{\alpha_2} \ldots$ if $z = (z_1, z_2, \ldots)$ and $\alpha = (\alpha_1, \alpha_2, \ldots)$, we have

$$\mathcal{H}F(z) = \sum_\alpha c_\alpha z^\alpha\ ; \quad z \in \mathbf{C}_0^{\mathbf{N}}. \tag{4.6}$$

It can be shown that this series converges and represents an analytic function of $z \in \mathbf{C}_0^{\mathbf{N}}$, for all $F \in (\mathcal{S})^*$. (See (Hida et al., to appear).)
The $\mathcal{S}$-transform was first introduced in (Kubo and Takenaka, 1980). The $\mathcal{H}$-transform and its inverse (see below) was first used in (Lindstrøm et al., 1991a; 1992) (on $\mathbf{L}^2(\mu)$) as a tool to solve stochastic differential equations.

One important property of the $\mathcal{H}$-transform is the following, which follows directly from (4.6) and the definition of the Wick product.

THEOREM 4.1. a) *Let $F, G \in (\mathcal{S})^*$. Then*

$$\mathcal{H}(F \diamond G)(z) = \mathcal{H}F(z) \cdot \mathcal{H}G(z)\ ; \quad z \in \mathbf{C}_0^{\mathbf{N}},$$

where the product on the right-hand side is the usual complex product.
b) *Alternatively, using the $\mathcal{S}$-transforms, we get*

$$\mathcal{S}(F \diamond G)(\varphi) = \mathcal{S}F(\varphi) \cdot \mathcal{S}G(\varphi)\ ; \quad \varphi \in \mathcal{S}(\mathbf{R}^d).$$

Example 4.2. We can use the Hermite transform to prove that there does not exist an $\mathbf{L}^p$ function process with $p > 1$ which solves the 1-dimensional pressure equation (2.59) for fluid flow in a porous medium:
Suppose $X(\varphi, x, \omega)$ is an $\mathbf{L}^p$ functional process satisfying the equation

$$\begin{cases} (K(x) \diamond X'(x))' = -a\ ; & x \in (0,1), \\ X(0) = X(1) = 0, \end{cases} \tag{4.7}$$

where $a = \int_{\mathbf{R}} \varphi(y)dy$, $\varphi \in \mathcal{S}(\mathbf{R})$ is fixed. As before $K(x,\omega) = \operatorname{Exp} W_{\varphi_x}(\omega)$ (see (2.57)).
Applying the Hermite transform to this equation, we get

$$\begin{cases} (\tilde{K} \cdot \tilde{X}'(x))' = -a\ ; & x \in (0,1), \\ \tilde{X}(0) = \tilde{X}(1) = 0. \end{cases}$$

The solution of this equation is

$$\tilde{X}(x) = \int_0^x (-at + A) \exp(-\tilde{W}_{\varphi_t})\, dt,$$

where A is the constant determined by the requirement $X(1) = 0$, i.e.,

$$A = A(z) = a \int_0^1 t \exp(-\tilde{W}_{\varphi_t}) dt \cdot \left(\int_0^1 \exp(-\tilde{W}_{\varphi_t}) dt \right)^{-1}.$$

However, as a function of $z = (z_1, z_2, \ldots)$ we see that $A(z)$ is not *analytic* because it has poles at the zeroes of the function

$$B(z) = \int_0^1 \exp(-\tilde{W}_{\varphi_t}(z)) dt.$$

Thus $\tilde{X}(z)$ is not analytic. Therefore, X is not even in $(\mathcal{S})^*$ and, therefore, not in $\mathbf{L}^p(\mu)$ for $p > 1$.

This shows the need for including the $\mathbf{L}^1$ functional processes in the general study of stochastic differential equations, as pointed out in §2.5.
We end this subsection by giving an explicit form of the *inverse of the $\mathcal{H}$-transform:* Suppose $X(\omega) = \sum_\alpha c_\alpha H_\alpha(\omega) \in \mathbf{L}^2(\mu)$ has the Hermite transform

$$\tilde{X}(z) = \sum_\alpha c_\alpha z^\alpha \; ; \quad z = (z_1, z_2, \ldots) \in \mathbf{C}_0^{\mathbf{N}}.$$

Let λ be the probability measure on $\mathbf{R}^{\mathbf{N}}$ defined by

$$\int_{\mathbf{R}^{\mathbf{N}}} f(y_1, \ldots, y_n) d\lambda(y) = (2\pi)^{-n/2} \int_{\mathbf{R}^n} f(y) e^{-|y|^2/2} dy$$

if f is a bounded measurable function of $y \in \mathbf{R}^{\mathbf{N}}$ depending only on the first n coordinates $y_1, \ldots, y_n$ of y. In other words, λ is the infinite product of the normalized Gaussian measure in $\mathbf{R}$. Then we obtain X from $\tilde{X}(z)$ by

$$X(\omega) = \lim_{n,k \to \infty} \int_{\mathbf{R}^{\mathbf{N}}} \tilde{X}^{(n,k)}(\theta + iy) d\lambda(y) \quad (\text{limit in } \mathbf{L}^2(\mu)), \tag{4.8}$$

where $\theta + iy = (\theta_1 + iy_1, \theta_2 + iy_2, \ldots)$ (with $\theta_k = \int_{\mathbf{R}^d} e_k dB$ as in (2.11)) and

$$\tilde{X}^{(n,k)}(z) = \sum_{\alpha \in J_{n,k}} c_\alpha z^\alpha \; ; \quad J_{n,k} = \{\alpha;\, |\alpha| \leq n \text{ and } \alpha_j = 0 \quad \text{for } j > k\}$$

is the doubly truncated series for $\tilde{X}(z)$.
To prove (4.8), we combine (2.11) with the following well known formula for the Hermite polynomials:

$$h_n(x) = \int_{\mathbf{R}} (x + iy)^n e^{-y^2/2} \frac{dy}{\sqrt{2\pi}}.$$

Example. If we choose, e.g., $\varphi = e$, then $K_\varphi = \operatorname{Exp} W_\varphi = \exp(W_\varphi - 1/2)$ has the $\mathcal{H}$-transform $\tilde{K}_\varphi(z) = \exp \tilde{W}_\varphi(z) = \exp z_1$. We can now verify directly that

$$\int_{\mathbf{R}^{\mathbf{N}}} \tilde{K}_\varphi(\theta + iy) d\lambda(y) = \int_{\mathbf{R}^{\mathbf{N}}} \exp(\theta_1 + iy_1) d\lambda(y) = \frac{1}{\sqrt{2\pi}} \int_{\mathbf{R}} \exp(\theta_1 + iy_1 - \frac{1}{2}|y_1|^2) dy_1$$

$$= \exp(\theta_1 - \frac{1}{2}) = \operatorname{Exp} W_\varphi = K_\varphi,$$

by using the well-known formula

$$\int_{\mathbf{R}} e^{i\alpha x - \varepsilon x^2} dx = \sqrt{\frac{\pi}{\varepsilon}} \exp(-\frac{\alpha^2}{4\varepsilon}) \; ; \qquad \varepsilon > 0, \quad \alpha \in \mathbf{R}.$$

4.2. *Connection with convolution*
If $f \in (\mathcal{S})$ and $\eta \in \mathcal{S}'(\mathbf{R}^d)$, we define the η-shift of f, $\tau_\eta f$, as follows:

$$\tau_\eta f(\omega) = f(\omega + \eta), \quad \omega \in \mathcal{S}'(\mathbf{R}^d).$$

The following results are known (Hida et al., to appear)

(i) For each $\eta \in \mathcal{S}'(\mathbf{R}^d)$ the map $f \to \tau_\eta f$ is continuous from $(\mathcal{S})$ into $(\mathcal{S})$.

(ii) For each $f \in (\mathcal{S})$ the map $\eta \to \tau_\eta f(\cdot)$ is continuous from $\mathcal{S}'(\mathbf{R}^d)$ into $(\mathcal{S})$.

Moreover, from (Zhang, 1992) we have

THEOREM 4.3. *For $f \in (\mathcal{S})$, $F \in (\mathcal{S})^*$, define*

$$\Gamma(f, F) \colon \mathcal{S}'(\mathbf{R}^d) \to \mathbf{R}$$

by

$$\Gamma(f, F)(\eta) = \langle F, \tau_\eta f \rangle \; ; \quad \eta \in \mathcal{S}'(\mathbf{R}^d).$$

Then Γ is a continuous map from $(\mathcal{S}) \times (\mathcal{S})^*$ into $(\mathcal{S})$. In particular, the function

$$\Gamma(f, F)(\cdot) = \langle F, \tau.f \rangle \quad \text{is in} \quad (\mathcal{S}).$$

This makes it possible to give the following definition.

Definition 4.4. Let $F, G \in (\mathcal{S})^*$. Then we define the *convolution* of F and G, $F * G$, by

$$\langle F * G, f \rangle = \langle F, \langle G, \tau.f \rangle \rangle = \langle G, \langle F, \tau.f \rangle \rangle \; ; \quad f \in (\mathcal{S}). \tag{4.9}$$

Remark. The idcntity in (4.9) can be proved by using the $\mathcal{S}$-transform: Define $Z(f) = \langle F, \langle G, \tau.f \rangle \rangle$; $f \in (\mathcal{S})$. Then, for $\varphi \in \mathcal{S}(\mathbf{R}^d)$,

$$\mathcal{S}Z(\varphi) = \langle Z, \exp\langle \cdot, \varphi \rangle \rangle e^{-\|\varphi\|^2/2} = \langle F, \langle G, \tau.(\exp\langle \cdot, \varphi \rangle) \rangle \rangle e^{-\|\varphi\|^2/2}$$

$$= \langle F, \langle G, \exp\langle \cdot, \varphi \rangle \rangle \exp\langle \cdot, \varphi \rangle \rangle \cdot e^{-\|\varphi\|^2/2}$$

$$= \langle F, \exp\langle \cdot, \varphi \rangle \rangle \cdot \langle G, \exp\langle \cdot, \varphi \rangle \rangle \cdot e^{-\|\varphi\|^2/2} = \mathcal{S}F(\varphi) \cdot \mathcal{S}G(\varphi) \cdot e^{\|\varphi\|^2/2}$$

which shows that we may interchange F and G in (4.9).
Note that this calculation shows that

$$\mathcal{S}(F * G)(\varphi) = \mathcal{S}F(\varphi) \cdot \mathcal{S}G(\varphi) \cdot \exp\left(\frac{1}{2}\|\varphi\|^2\right) \tag{4.10}$$

and, hence, if also $H \in (\mathcal{S})^*$,

$$\mathcal{S}(F * G * H)(\varphi) = \mathcal{S}F(\varphi) \cdot \mathcal{S}G(\varphi) \cdot \mathcal{S}H(\varphi) \cdot \exp(\|\varphi\|^2). \tag{4.11}$$

According to the characterization theorem (Theorem 4.40 in (Hida et al., to appear)) there exist J, $K \in (\mathcal{S})^*$ such that, for $\varphi \in \mathcal{S}(\mathbf{R}^d)$,

$$\mathcal{S}J(\varphi) = \exp\left(\frac{1}{2}\|\varphi\|^2\right) \quad \text{and} \quad \mathcal{S}K(\varphi) = \exp(-\|\varphi\|^2). \tag{4.12}$$

So if we combine Theorem 4.1 b) with (4.10) and (4.11), respectively, we get the following result (Kuo, 1991).

COROLLARY 4.5. *Let F, $G \in (\mathcal{S})^*$. Then*

$$F * G = F \diamond G \diamond J \tag{4.13}$$

and

$$F \diamond G = F * G * K. \tag{4.14}$$

Remark. In (Kuo, 1991) (4.13) is used a *definition* of the convolution $*$ and the identity (4.14) is proved from this. Thus, Corollary 4.5 shows that our definition (4.9) is equivalent to the definition in (Kuo, 1991).

4.3. *The backward heat equation*

There is a simple and striking representation of the Wick product in terms of the backward (or, more precisely, the ill-posed) heat equation problem:
Given a function $f(x)$, $x \in \mathbf{R}$, find $F(x,t)$ such that

$$\begin{cases} \frac{\partial}{\partial t}F(x,t) + \frac{1}{2}\frac{\partial^2}{\partial x^2}F(x,t) = 0 & \text{in} \quad D = \{(x,t); t > 0\}, \\ \lim\limits_{t\to 0+} F(x,t) = f(x); & x \in \mathbf{R}. \end{cases} \tag{4.15}$$

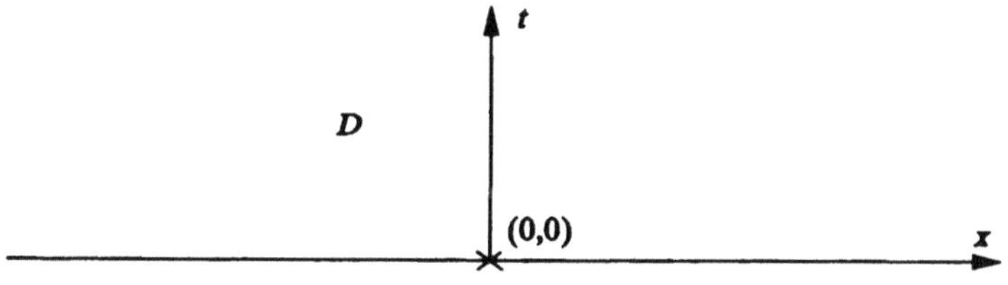

Fig. 3

Remark. Note that if the domain $D = \{(x,t); t > 0\}$ is changed to $D' = \{(x,t); t < 0\}$, then the solution of the corresponding problem (4.15) is given (under weak conditions on f) by

$$F(x,t) = \mathbf{E}^{(x,t)}[f(B_\tau)],$$

where $\mathbf{E}^{(x,t)}$ is the expectation with respect to the law $\mathbf{P}^{(x,t)}$ of the space-time Brownian motion $X_s = (B_s^x, t+s)$, $s \geq 0$, starting at (x,t), and τ is the first exit time from D' of X_s. However, in the case considered in (4.15) one can only expect a solution to exist for more special choices of f.

The connection between the Wick products and problem (4.15) seems to be a part of the folklore among experts in quantum field theory, although it seems hard to find it in the literature. Recently a more thorough investigation has been done (Norberg, 1992). We will here only sketch the main ideas and refer to (Norberg, 1992) for proofs and more details.

First we establish that if $f(x) = x^n$, then the solution of (4.15) is given by

$$F(x,t) = H_n(x,t) := t^{n/2} h_n\left(\frac{x}{\sqrt{t}}\right), \tag{4.16}$$

where $\{h_n\}$ are the Hermite polynomials defined in (1.10). $H_n(x,\sigma^2)$ is sometimes called the nth *Hermite polynomial with parameter* $\sigma > 0$. To prove that $H_n(x,t)$ solves (4.15), we first note that by (2.47) we have

$$\sum_{n=0}^{\infty} \frac{u^n}{n!} H_n(x,t) = \exp\left(ux - \frac{1}{2}u^2 t\right); \quad u \in \mathbf{R}. \tag{4.17}$$

From (4.17) we deduce that

$$\frac{\partial}{\partial x} H_n(x,t) = n H_{n-1}(x,t) \tag{4.18}$$

and

$$\frac{\partial}{\partial t} H_n(x,t) = -\frac{n(n-1)}{2} H_{n-2}(x,t). \tag{4.19}$$

It follows that $F(x,t) = H_n(x,t)$ satisfies the backward heat equation. Moreover, we see that

$$\lim_{t \to 0} H_n(x,t) = x^n. \tag{4.20}$$

We conclude that to the given boundary value $f(x) = x^n$ there corresponds the solution $F(x,t) = H_n(x,t)$.

More generally, we are interested in the correspondence $\mathcal{K}$ given by

$$f \to F = \mathcal{K} f$$

between boundary values f and the corresponding solutions F of (4.15). The correspondence is clearly linear, so taking limits of sums of the polynomials above we obtain the following list of examples:

$f(x)$	$F(x,t) = \mathcal{K}f(x,t)$
x^n	$H_n(x,t)$
$e^{cx} = \sum\limits_{n=0}^{\infty} \frac{1}{n!}(cx)^n$	$\sum\limits_{n=0}^{\infty} \frac{c^n}{n!} H_n(x,t) = \exp\left(cx - \frac{1}{2}c^2 t\right)$
$\cos x = \frac{1}{2}(e^{ix} + e^{-ix})$	$\frac{1}{2}\left(\exp(ix + \frac{1}{2}t) + \exp(-ix + \frac{1}{2}t)\right) = e^{t/2}\cos x$
$\sin x$	$e^{t/2}\sin x$
$\sum\limits_{n=0}^{\infty} a_n x^n$ (with suitable convergence conditions)	$\sum\limits_{n=0}^{\infty} a_n H_n(x,t)$

How does this correspondence behave under products? The table above indicates the answer:
If $f(x) = \sum\limits_{n=0}^{\infty} a_n x^n$, $g(x) = \sum\limits_{m=0}^{\infty} b_m x^m$, then $fg(x) = \sum\limits_{k=0}^{\infty} c_k x^k$, where $c_k = \sum\limits_{n+m=k} a_n b_m$. Therefore, if

$$\mathcal{K}f = F = \sum_{n=0}^{\infty} a_n H_n(x,t), \quad \mathcal{K}g = G = \sum_{m=0}^{\infty} b_m H_m(x,t),$$

then

$$\mathcal{K}(fg) = \sum_{k=0}^{\infty} c_k H_k(x,t),$$

which is strikingly similar to formula (2.31) for the Wick product based on the Hermite expansions in the white noise probability space. In view of this the following definition is natural.

Definition 4.6. Suppose $F = \mathcal{K}f, G = \mathcal{K}g$ are solutions of (4.15) with boundary values f, g. Then the *backward heat equation* (b.h.e.) *Wick product* of F and G, $F\tilde{\diamond}G$, is defined by

$$F\tilde{\diamond}G = \mathcal{K}(f \cdot g),$$

i.e., $F\tilde{\diamond}G$ is the solution of (4.15) with boundary values $f \cdot g$.
In this setting the b.h.e. Wick product is purely deterministic: It is just a product on a subset of the set of smooth functions on D. To see the explicit connection to the Wick product in stochastic analysis, we use the following result, which may be deduced from (2.11):

$$n! \int\limits_{0 \le u_1 \le \dots \le u_n < t} \dots \left(\int \left(\int dB_{u_1} \right) dB_{u_2} \right) \dots dB_{u_n} = t^{n/2} h_n \left(\frac{B_t}{\sqrt{t}} \right). \qquad (4.21)$$

By (2.29) we can rewrite this as

$$B_t^{\diamond n} = H_n(B_t, t) = x^{\hat{\diamond} n}|_{x=B_t}. \tag{4.22}$$

More generally, if X_t is a (real) continuous semimartingale we define the *(semimartingale) Wick powers* of X_t by

$$X_t^{\hat{\diamond} n} = H_n(X_t, [X]_t), \tag{4.23}$$

where $[X]_t = \lim_{\substack{\Delta t_j \to 0 \\ t_j \leq t}} \sum_j (X_{t_{j+1}} - X_{t_j})^2$ is the *quadratic variation process* of X_t.
(Since this is not necessarily the same as the stochastic analysis Wick power applied to the random variable $X_t(\cdot)$ we use the notation $\hat{\diamond}$ here.) This gives the following definition of the (semimartingale) Wick product $(X\hat{\diamond}Y)_t$ of two semimartingales X_t, Y_t:

$$\begin{aligned}(X\hat{\diamond}Y)_t &= \frac{1}{4}\left\{(X_t+Y_t)^{\hat{\diamond}2} - (X_t - Y_t)^{\hat{\diamond}2}\right\} \\ &= \frac{1}{4}\left\{H_2(X_t + Y_t[X+Y]_t) - H_2(X_t - Y_t, [X-Y]_t)\right\} \\ &= \frac{1}{4}\left\{(X_t+Y_t)^2 - [X+Y]_t - ((X_t-Y_t)^2 - [X-Y]_t)\right\} \\ &= X_tY_t - \frac{1}{4}\left\{[X+Y]_t - [X-Y]_t\right\} = X_tY_t - [XY]_t,\end{aligned} \tag{4.24}$$

where $[XY]_t$ is the *quadratic covariation* of X and Y.
Finally, we mention that the exponential in the semimartingale Wick sense gets the form

$$\hat{\text{E}}\text{xp}(X_t) := \sum_{n=0}^{\infty} \frac{1}{n!} X_t^{\diamond n} = \exp\left(X_t - \frac{1}{2}[X]_t\right). \tag{4.25}$$

Fore more information we refer to (Norberg, 1992).

5. PROPERTIES OF THE WICK PRODUCT

The discussion in §2 shows that the stochastic calculus based on Ito integrals (and Ito's differentiation rules) can be reformulated as the Wick calculus where the usual products are replaced by the Wick products and ordinary differentiation rules apply. This represents an obvious technical advantage and in addition it generalizes directly to the anticipative case (with Ito integrals replaced by Skorokhod integrals). In view of this it is important to investigate further the properties of the Wick product.

5.1. *Algebraic properties*

Fix $Y \in (\mathcal{S})$ and consider the operator

$$X \to X \diamond Y\,; \quad X \in \mathbf{L}^2(\mu).$$

This is, for a generic choice of Y, an unbounded densely defined linear operator (see, e.g., (Meyer, 1987)). As such the Wick product is not well behaved and must be treated with care. On the other hand it exhibits a number of nice algebraic properties, a few of which are listed below. Recall that if $F,\ G \in (\mathcal{S})^*$, then $F \diamond G \in (\mathcal{S})^*$ always. Similarly, $f,\ g \in (\mathcal{S}) \Rightarrow f \diamond g \in (\mathcal{S})$. If $X,\ Y \in \mathbf{L}^1(\mu)$ however, we do not know if $X \diamond Y \in \mathbf{L}^1(\mu)$ always exists.

PROPOSITION 5.1.

(i) *The Wick product is commutative, distributive, and associative on* $(\mathcal{S})^*$ *and — when defined — on* $\mathbf{L}^1(\mu)$.

(ii) *If* $X, Y \in (\mathcal{S})^*$ *(or* $X,\ Y \in \mathbf{L}^1(\mu)$ *and* $X \diamond Y \in \mathbf{L}^1(\mu)$ *exists) and* $X \diamond Y = 0$, *then either* $X = 0$ *or* $Y = 0$.

(iii) *If* $X, Y \in \mathbf{L}^1(\mu)$ *and* $X \diamond Y$ *is defined in* $\mathbf{L}^1(\mu)$, *then*

$$\mathcal{F}[X \diamond Y](\varphi) = e^{\|\varphi\|^2/2}\mathcal{F}[X](\varphi) \cdot \mathcal{F}[Y](\varphi), \quad \forall \varphi \in \mathcal{S}(\mathbf{R}^d).$$

(See formula (2.62).)

(iv) *If* $X,\ Y \in \mathbf{L}^1(\mu)$ *and* $X \diamond Y \in \mathbf{L}^1(\mu)$ *is defined, then*

$$\mathbf{E}[X \diamond Y] = \mathbf{E}[X] \cdot \mathbf{E}[Y] \qquad (\textit{special case of } \text{(iii)}).$$

(v) $F,\ G \in (\mathcal{S})^* \Rightarrow \mathcal{H}(F \diamond G)(z) = \mathcal{H}F(z) \cdot \mathcal{H}G(z)\ ; \quad z \in \mathbf{C}_0^{\mathbf{N}}$ *and* $\mathcal{S}(F \diamond G)(\varphi) = \mathcal{S}F(\varphi) \cdot \mathcal{S}G(\varphi)\ ; \quad \varphi \in \mathcal{S}(\mathbf{R}^d)$ *(Theorem* 4.1).

(vi) $F,\ G \in (\mathcal{S})^* \Rightarrow F * G = F \diamond G \diamond J$ *and* $F \diamond G = F * G * K$ *(Corollary* 4.5).

Proof. The only statement that is not proved already is (ii). This was first proved in (Gjessing, 1992). In the $(\mathcal{S})^*$ setting we note that a short proof can be given using the Hermite transform.
If $X, Y \in (\mathcal{S})^*$ and $X \diamond Y = 0$, then

$$0 = \mathcal{H}(X \diamond Y) = \mathcal{H}X(z) \cdot \mathcal{H}Y(z); \quad z \in \mathbf{C}_0^{\mathbf{N}}. \tag{5.1}$$

Fix M and consider the analytic functions

$$\tilde{X}^{(M)}(z_1, \ldots, z_M) = \mathcal{H}X(z_1, \ldots, z_M, 0, \ldots)\ ; \quad (z_1, \ldots, z_M) \in \mathbf{C}^M,$$
$$\tilde{Y}^{(M)}(z_1, \ldots, z_M) = \mathcal{H}Y(z_1, \ldots, z_M, 0, \ldots)\ ; \quad (z_1, \ldots, z_M) \in \mathbf{C}^M.$$

By (5.1) we have $\tilde{X}^{(M)}\tilde{Y}^{(M)} \equiv 0$ on $\mathbf{C}^M$ and this is only possible if either $\tilde{X}^{(M)} \equiv 0$ or $\tilde{Y}^{(M)} \equiv 0$. Letting $M \to \infty$, we see that either $\tilde{X}^{(M)} \equiv 0$ for infinitely many M or $\tilde{Y}^{(M)} \equiv 0$ for infinitely many M. This gives that either $\mathcal{H}X \equiv 0$ or $\mathcal{H}Y \equiv 0$ and the conclusion follows.
We refer to (Gjessing, 1992) for a proof in the $\mathbf{L}^1(\mu)$ case.

Remarks. Property (iv) has the consequence that the average value $\mathbf{E}[X(\varphi, x, \cdot)]$ of the solution $X(\varphi, x, \omega)$ of a stochastic differential equation involving the Wick products coincides with the solution of the corresponding deterministic equation where all the noise terms are replaced by their expected value. For example, the expected value $u(\varphi, x) = \mathbf{E}[X(\varphi, x, \cdot)]$ of a solution X of equation (2.59) must satisfy the equation

$$\Delta_x u(\varphi, x) = -(f * \varphi)(x); \quad x \in D \subset \mathbf{R}^d,$$

(where $\Delta_x = \sum_{i=1}^{d} \partial^2/\partial x_i^2$ is the Laplace operator), since

$$\mathbf{E}[K(\varphi, x)] = \mathbf{E}[\mathrm{Exp}W_{\varphi_x}] = 1.$$

It is natural to build analytic functions of white noise using the Taylor expansions with powers replaced by the Wick powers. One example is $\mathrm{Exp}W_\varphi$, which was introduced in Example 2.7, formula (2.57). More generally, given $X \in \mathbf{L}^1(\mu)$, we put

$$\mathrm{Exp}[X] := \sum_{n=0}^{\infty} \frac{1}{n!} X^{\diamond n}, \tag{5.2}$$

(when all the Wick powers of X exist and the series converges in $\mathbf{L}^1(\mu)$). Here is a list of examples and their properties:

PROPOSITION 5.2.

(i) $\mathrm{Exp}[W_\varphi] := \sum_{n=0}^{\infty} \frac{1}{n!} W_\varphi^{\diamond n} = \exp(W_\varphi - \frac{1}{2}\|\varphi\|^2)$.

(ii) $\mathrm{Cos}[W_\varphi] := \sum_{n=0}^{\infty} \frac{(-1)^n}{(2n)!} W_\varphi^{\diamond 2n} = \exp(\frac{1}{2}\|\varphi\|^2) \cdot \cos W_\varphi$.

(iii) $\mathrm{Sin}[W_\varphi] := \sum_{n=0}^{\infty} \frac{(-1)^n}{(2n+1)!} W_\varphi^{\diamond(2n+1)} = \exp(\frac{1}{2}\|\varphi\|^2) \cdot \sin W_\varphi$.

(iv) $\mathrm{Exp}[X] \diamond \mathrm{Exp}[Y] = \mathrm{Exp}[X + Y]$ (*when defined*).

(v) $\mathrm{Exp}[-\mathrm{Exp}[W_\varphi]] = \sum_{n=0}^{\infty} \frac{(-1)^n}{n!} \exp(nW_\varphi - \frac{n^2}{2}\|\varphi\|^2)$.

Remarks. We have already proved (i) (see Example 2.7). Then (ii) and (iii) can be proved by using that

$$\mathrm{Cos}[W_\varphi] = \frac{1}{2}(\mathrm{Exp}[iW_\varphi] + \mathrm{Exp}[-iW_\varphi])$$

and

$$\mathrm{Sin}[W_\varphi] = \frac{1}{2i}(\mathrm{Exp}[iW_\varphi] - \mathrm{Exp}[-iW_\varphi])$$

(note that $iW_\varphi = W_{i\varphi}$).

To prove (iv) we repeat the algebraic proof that $\exp x \cdot \exp y = \exp(x+y)$, $x, y \in \mathbf{R}$, but with the Wick products replacing ordinary products.

We establish (v) by applying the Exp-definition twice and noting that we have convergence in $\mathbf{L}^1(\mu)$. Formula (v) shows that *in general* $\mathrm{Exp}[X]$ *need not be positive a.s. — in fact not even bounded from below.*

For example, if $y^2 = x/2$, then it is easily seen that

$$f(x,y) := \sum_{n=0}^{\infty} \frac{(-1)^n}{n!} \exp(nx - n^2y^2) < -M,$$

for $x > 2\log(3+M)$.
We now discuss positivity in more detail.

Definition 5.3. If $F \in (\mathcal{S})^*$, we say that F is *positive* (and write $F \geq 0$) if

$$\langle F, f\rangle \geq 0 \quad \text{for all} \quad f \in (\mathcal{S}), \quad f \geq 0. \tag{5.3}$$

In particular, if $X \in \mathbf{L}^2(\mu)$, then $X \geq 0$ iff

$$\langle X, f\rangle = \mathbf{E}[X \cdot f] \geq 0 \quad \text{for all} \quad f \in (\mathcal{S}), \quad f \geq 0,$$

(see (2.23) and below). This occurs iff

$$X(\omega) \geq 0 \quad \text{for} \quad \text{a.a. } \omega.$$

In view of this the following definition is natural:

Definition 5.4. An $\mathbf{L}^p$ functional process $X(\varphi, x, \omega)$ is called *positive* if

$$X(\varphi, x, \omega) \geq 0 \quad \text{a.s.}, \qquad \forall \varphi \in \mathcal{S}(\mathbf{R}^d), \quad \forall x \in \mathbf{R}^d.$$

It is known (Kuo, 1991) that, in general, positivity is *not* preserved by the Wick products. However, we do have the following result ((Lindstrøm et al., 1991a), Corollary 7.5).

THEOREM 5.5. *Let*

$$X(\varphi, x, \omega) = \sum_{\alpha} a_\alpha(\varphi, x) H_\alpha(\omega)$$

and

$$Y(\varphi, x, \omega) = \sum_{\beta} b_\beta(\varphi, x) H_\beta(\omega)$$

be $\mathbf{L}^2$ functional processes which are homogeneous in the sense that

$$a_\alpha(c\varphi, x) = c^{|\alpha|} a_\alpha(\varphi, x), \tag{5.4}$$

$$b_\beta(c\varphi, x) = c^{|\beta|} b_\beta(\varphi, x) \tag{5.5}$$

for all multi-indices α, β and all $c \in \mathbf{R}$, $\varphi \in \mathcal{S}(\mathbf{R}^d)$. Suppose X, Y are positive (as functional processes). Then

$$X \diamond Y$$

is also a positive functional process (when defined).

In general, one might say that the Wick multiplication relates to the Hermite polynomials in the same way as ordinary multiplication relates to the polynomials

x^n. It is, therefore, of interest to investigate the algebraic properties of the Hermite polynomials. We mention one useful result, a direct proof of which can be found in (Holden et al., 1991):

PROPOSITION 5.6. *If* $a = (a_1, \ldots, a_N)$ *with* $\sum_{j=1}^{N} a_j^2 = 1$, *then*

$$h_n\Big(\sum_{j=1}^{N} a_j x_j\Big) = \sum_{\alpha} \frac{n!}{\alpha!} a^{\alpha} h_{\alpha}(x), \tag{5.6}$$

the summation being taken over all $\alpha = (\alpha_1, \ldots, \alpha_N)$ *such that* $|\alpha| = n$ ($\alpha! = \alpha_1! \ldots \alpha_N!$ *and* $h_\alpha(x) = \prod_{j=1}^{N} h_{\alpha_j}(x_j), x = (x_1, \ldots, x_N)$).

Remark. It is of interest to note that the Wick calculus we have already established can be used to prove results about the Hermite polynomials. For example, we can now give a short proof of Proposition 5.6 as follows.
Choose $c = (c_1, \ldots, c_N) \in \mathbf{R}^N$ and put $\varphi = \sum_{j=1}^{N} c_j e_j$. Then, by (2.33), we have

$$\|\varphi\|^2 h_n\left(\frac{W_\varphi}{\|\varphi\|}\right) = W_\varphi^{\diamond n} = \left(\sum_{j=1}^{N} c_j W_{e_j}\right)^{\diamond n} = \sum_{\alpha} \frac{n!}{\alpha!} c^{\alpha} W_{(\alpha)}^{\diamond \alpha}, \tag{5.7}$$

where we have used the fact that the multinomial formula looks the same for the Wick products as for ordinary products, and we have adopted the notation

$$W_{(\alpha)}^{\diamond \alpha} = W_{e_{\alpha_1}}^{\diamond \alpha_1} \diamond \ldots \diamond W_{e_{\alpha_N}}^{\diamond \alpha_N}. \tag{5.8}$$

By (2.29) and (2.11), we have

$$W_{e_{\alpha_1}}^{\diamond \alpha_1} \diamond \ldots \diamond W_{e_{\alpha_N}}^{\diamond \alpha_N} = \prod_{j=1}^{N} h_{\alpha_j}(\theta_j). \tag{5.9}$$

Combining (5.7)–(5.9), we conclude that

$$\|\varphi\|^2 h_n\left(\frac{W_\varphi}{\|\varphi\|}\right) = \sum_{\alpha} \frac{n!}{\alpha!} c^{\alpha} h_{\alpha}(\theta), \tag{5.10}$$

where $\theta = (\theta_1, \ldots, \theta_N)$. In particular, if $\sum_{j=1}^{N} c_j^2 = 1$ we get

$$h_n\left(\sum_{j=1}^{N} c_j \theta_j\right) = \sum_{\alpha} \frac{n!}{\alpha!} c^{\alpha} h_{\alpha}(\theta). \tag{5.11}$$

Since this holds a.s. and $\theta_1, \ldots, \theta_N$ are independent normal random variables it must hold for all possible values of $x_j = \theta_j(\omega)$ and (5.6) follows.

5.2. *Analytic properties and estimates*
Unfortunately, very little is known regarding $\mathbf{L}^p$-estimates and continuity of the Wick product. Here is a basic connection between X and $\tilde{X}$ ((Holden et al., 1991), Theorem 4.2).

THEOREM 5.7. *Let $X \in \mathbf{L}^2(\mu)$ and let $\tilde{X}^{(n,k)}(z)$ be the truncated Hermite transform of X (see §4.1). Then*

$$\mathbf{E}[|X|^p] \leq \liminf_{n,k\to\infty} \int\int |\tilde{X}^{(n,k)}(\xi + i\eta)|^p d\lambda(\xi) d\lambda(\eta)$$

for all $p \geq 1$.

COROLLARY 5.8. *Let $X_n \in \mathbf{L}^2(\mu)$ for all n and assume that $\{\tilde{X}_n\}_{n=1}^\infty$ is a convergent sequence in $\mathbf{L}^p(\lambda \times \lambda)$ for some $p \geq 1$. Then $\{X_n\}$ converges in $\mathbf{L}^p(\mu)$.*

COROLLARY 5.9. *Suppose $X \in \mathbf{L}^2(\mu)$ satisfies the condition*

$$\sum \frac{1}{n!} \|\tilde{X}^n\|_{\mathbf{L}^p(\lambda\times\lambda)} < \infty$$

for some $p \geq 1$. Then

$$\mathrm{Exp}(X) \in \mathbf{L}^p(\mu).$$

Unfortunately, Theorem 5.7 and its Corollaries are too crude to be really useful. In many important cases $\tilde{X}$ is not even in $\mathbf{L}^1(\lambda \times \lambda)$. Some stronger results can be found in (Gjessing, 1992).
For the Wick powers of white noise there exists a sharp estimate due to Carlen and Kree (Carlen and Kree, 1991) (already mentioned in Example 2.7). Formulated in our setting the result is the following:

THEOREM 5.10. *Let $\varphi \in \mathcal{S}(\mathbf{R}^d)$ and $p \geq 1$. There exist constants $A_{p,n}, B_{p,n}$ such that*

$$n! A_{p,n} \|\varphi\|^n \leq \mathbf{E}[|W_\varphi^{\diamond n}|^p]^{1/p} \leq n! B_{p,n} \|\varphi\|^n$$

and

$$\log A_{p,n} \sim \log B_{p,n} \sim -\frac{n}{2} \log n \quad \text{as} \quad n \to \infty.$$

5.3. *Is the Wick product local?*
Unlike the ordinary (pointwise) product $(X \cdot Y)(\omega) = X(\omega) \cdot Y(\omega)$ of two random variables X, Y on $\mathcal{S}'$, the value of the Wick product $X \diamond Y$ at $\omega_0 \in \mathcal{S}'$ is in general not determined by $X(\omega_0)$ and $Y(\omega_0)$ alone but by (possibly) all the values $\{X(\omega);\ \omega \in \mathcal{S}'\}$ and $\{Y(\omega);\ \omega \in \mathcal{S}'\}$. This is not so surprising in view of the definition

$$(X \diamond Y)(\omega) = \sum_{\gamma} \left(\sum_{\alpha+\beta=\gamma} a_\alpha b_\beta \right) H_\gamma(\omega)$$

(if $X = \sum_\alpha a_\alpha H_\alpha, Y = \sum_\beta b_\beta H_\beta$) or in view of its connection to convolution (Corollary 4.5)

$$X \diamond Y = F * G * K.$$

In fact, in general the Wick product is not even *local*, in the following sense.

Definition 5.11. Let $\mathcal{M} \subset L^1(\mu)$. We say that the Wick product is *local* on $\mathcal{M}$ if for all $X, Y \in \mathcal{M}$, all $\omega_0 \in \mathcal{S}'$ and for all (nonempty) open neighborhoods U of ω_0 (in the weak-star topology on $\mathcal{S}'$) the value of $(X \diamond Y)(\omega_0)$ is determined by the values of X and Y on U alone.

PROPOSITION 5.12. *Let $A \neq \emptyset$ be an open subset of $\mathcal{S}'$ such that $B = (\overline{A})^c \neq \emptyset$ and $\mu(\partial A) = 0$, where ∂A is the boundary of A and $\overline{A}$ is the closure of A. Suppose $\mathcal{M}$ contains X_1, X_2, Y_1, Y_2 such that*

$$X_1 = X_2 \quad \text{a.s. on} \quad A, \quad Y_1 = Y_2 \quad \text{a.s. on} \quad B$$

and

$$\mu(X_1 \neq X_2) > 0, \quad \mu(Y_1 \neq Y_2) > 0.$$

Then the Wick product is not local on $\mathcal{M}$.

Proof. Note that

$$(X_1 - X_2) \diamond (Y_1 - Y_2) = \begin{cases} 0 \diamond (Y_1 - Y_2) & \text{on} \quad A, \\ (X_1 - X_2) \diamond 0 & \text{on} \quad B. \end{cases}$$

Hence, if $\diamond$ is local on $\mathcal{M}$, then $(X_1 - X_2) \diamond (Y_1 - Y_2) = 0$ on $A \cup B$ and hence a.s. on $\mathcal{S}'$. So, from Proposition 5.1 (ii), we conclude that either $X_1 = X_2$ a.s. or $Y_1 = Y_2$ a.s., which contradicts our assumptions.
In view of Proposition 5.12 we conclude that usually the Wick product is not local unless each of the factors X, Y themselves are locally determined. For more details we refer to (Gjessing, 1992).

6. SOME OPEN PROBLEMS

There are many interesting and natural problems related to the Wick product. We list some of them below. Some of the problems may be hard to solve and some easy — or even known. We would appreciate hearing from those who have information about any of them.

PROBLEM 1. *Do there exist $p, q \geq 1$ such that*

$$X, Y \in \mathbf{L}^p(\mu) \Rightarrow X \diamond Y \quad \text{is defined and lies in } \mathbf{L}^q(\mu)?$$

(Probably not, but it would be useful to have a counterexample.)

PROBLEM 2. *Extend the definition of* $\diamond$ *from* $\mathbf{L}^1(\mu)$ *to a class* $\mathcal{W} \supset \mathbf{L}^1(\mu)$ *of measurable functions (on* $\mathcal{S}'$*) in such a way that* $X \diamond Y \in \mathcal{W}$ *for all* $X,\ Y \in \mathcal{W}$.

PROBLEM 3. *For what* X *does there exist* Y *such that* $X \diamond Y = 1$?

PROBLEM 4. *If* $X, Y \in \mathbf{L}^1(\mu)$ *are independent, is*

$$X \diamond Y = X \cdot Y?$$

Remark. The conclusion can be shown to hold if, e.g.,

$$X = \sum_n \int f_n dB^{\otimes n} \in \mathbf{L}^2(\mu) \quad \text{and} \quad Y = \sum_m \int g_m dB^{\otimes m} \in \mathbf{L}^2(\mu)$$

are *strongly independent* in the sense that for all n, m the coordinate projections of supp $f_n \subset \mathbf{R}^n$ and supp $g_m \subset \mathbf{R}^m$ are disjoint.

PROBLEM 5. *Extend the Hermite transform to* $\mathbf{L}^1(\mu)$.

Remark. Example 4.2 indicates that for such an extension of $\mathcal{H}$ to $\mathbf{L}^1(\mu)$ we can no longer expect that $\mathcal{H}X(z)$ is always analytic on $\mathbf{C}_0^{\mathbf{N}}$: For some $X \in \mathbf{L}^1(\mu)$, the function $\mathcal{H}X(z)$ may have poles.

PROBLEM 6. *For* $X \in \mathbf{L}^1(\mu)$ *and* $f\colon \mathbf{C} \to \mathbf{C}$ *analytic there is a natural Wick version of* $f(X)$*, denoted by* $f^\diamond(X)$*, defined as follows:*

$$f^\diamond(X) = \sum_{n=0}^{\infty} a_n X^{\diamond n} \quad \textit{if} \quad f(z) = \sum_{n=0}^{\infty} a_n z^n$$

(assuming that all the Wick powers exist and that the series on the left converges). More generally, suppose $f\colon \mathbf{R} \to \mathbf{R}$ *is a given function (not necessarily real analytic), when can one make sense of a Wick version* $f^\diamond(X)$?

Remark. This question is discussed in connection with stochastic differential equations in (Lindstrøm et al., 1992).

PROBLEM 7. *If* $X, Y \in \mathbf{L}^2(\mu)$ *and* (X, Y) *is identical in law with another pair* (X', Y') *with* $X', Y' \in \mathbf{L}^2(\mu)$*, is* $X \diamond Y$ *identical in law with* $X' \diamond Y'$ *(when defined)?*

Recently one of us (H. Gjessing) has proved that the answers to Problems 4 and 7 are *no*. Morever, regarding Problem 1 he has proved that there exist $x, Y \in \mathbf{L}^\infty(\mu)$ such that $X \diamond Y \notin \mathbf{L}^p(\mu)$, for all $p > 1$ (see (Gjessing, 1992)).

Acknowledgements

This work is supported by VISTA, a research cooperation between The Norwegian Academy of Science and Letters and Den Norske Stats Oljeselskap A.S. (Statoil).

REFERENCES

Ash, J. and Potthoff, J. (1991). Ito's lemma without nonanticipatory conditions. *Probab. Theory Rel. Fields.* **88**, 17–46.

Buckdahn, R. (1991). Linear Skorokhod stochastic differential equations. *Probab. Theory Rel. Fields.* **90**, 223–240.

Carlen, E. and Kree, P. (1991). L^p-Estimates on Iterated Stochastic Integrals. *Ann. Probab.* **19**, 354–368.

Dobrushin, R.L. and Minlos, R.A. (1977). Polynomials in linear random functions. *Russian Math. Surveys.* **32**, 71–127.

Gelfand, I.M. and Vilenkin, N.Y. (1964). *Generalized Functions, Vol. 4: Applications of Harmonic Analysis.* Academic Press, New York.

Gjessing, H. (1992). *Some Properties of the Wick Product and the Hermite Transform.* Manuscript.

Glimm, J. and Jaffe, A. (1987). *Quantum Physics.* (Second ed.) Springer-Verlag, New York, Berlin.

Hida, T. (1980). *Brownian Motion.* Springer-Verlag, New York, Berlin.

Hida, T. and Ikeda, N. (1965). Analysis on Hilbert space with reproducing kernel arising from multiple Wiener integral. In: *Proceedings of the Fifth Berkeley Symposium on Mathematical Statistics and Probability* II, Part 1. California Univ. Press, Los Angeles, pp. 117–143.

Hida, T., Kuo, H.-H., Potthoff, J., and Streit, L. *White Noise Analysis* (to appear).

Holden, H., Lindstrøm, T., Øksendal, B., Ubøe, J., and Zhang, T.-S. (1991). *Stochastic Boundary Value Problems. A White Noise Functional Approach.* Preprint. Oslo Univ. Press, Oslo.

Holden, H., Lindstrøm, T., Øksendal, B., and Ubøe, J. (1992). *Discrete Wick Calculus and Stochastic Functional Equations.* Preprint. Oslo Univ. Press, Oslo (to appear in *Potential Analysis*).

Ito, K. (1951). Multiple Wiener integral. *J. Math. Soc. Japan.* **3**, 157–169.

Kubo, I. and Takenaka, S. (1980). Calculus on Gaussian white noise I. *Proc. Japan Acad.* **56**, 376–380.

Kuo, H.-H. (1991). *Convolution and Fourier Transform of Hida Distributions.* Manuscript.

Lindstrøm, T., Øksendal, B., and Ubøe, J. (1991 a). Stochastic differential equations involving positive noise. In: *Stochastic Analysis,* Ed. by M. Barlow and N. Bingham. Cambridge Univ. Press, Cambridge, pp. 261–303.

Lindstrøm, T., Øksendal, B., and Ubøe, J. (1991 b). *Stochastic Modelling of Fluid Flow in Porous Media.* Preprint. Oslo Univ. Press, Oslo.

Lindstrøm, T., Øksendal, B., and Ubøe, J. (1992). Wick multiplication and Ito–Skorohod stochastic differential equations. In: *Ideas and Methods in Mathematics Analysis.* Ed. by S. Albeverio et al., Cambridge Univ. Press, Cambridge (to appear).

Meyer, P.A. (1986). A finite approximation to Boson Fock space. In: *Stochastic Processes in Classical and Quantum Systems.* Ed. by S. Albeverio et al., Springer-Verlag, New York, Berlin, pp. 405–410.

Meyer, P.A. (1987). Fock space and probability theory. In: *Stochastic Processes — Mathematics and Physics.* Ed. by S. Albeverio et al., Springer-Verlag, New York, Berlin, pp. 160–170.

Meyer, P.A. and Yan, J.A. (1989). Distributions sur l'espace de Wiener (suite). *Sém. de Probabilités* XXIII, Ed. by P.A. Meyer and M. Yor, Springer-Verlag, New York, Berlin.

Norberg, T. (1992). *A Note on Solutions of the Backward Heat Equation.* Manuscript.

Nualart, D. and Pardoux, E. (1988). Stochastic calculus with anticipating integrands. *Probab. Theory Rel. Fields.* **78**, 535–581.

Nualart, D. and Zakai, M. (1986). Generalized stochastic integrals and the Malliavin calculus. *Probab. Theory Rel. Fields.* **73**, 255–280.

Øksendal, B. and Zhang, T.-S. (1992). *The Stochastic Volterra Equation.* Manuscript.

Potthoff, J. (1991). *White Noise Methods for Stochastic Partial Differential Equations.* Manuscript.

Simon, B. (1974). *The $P(\varphi)_2$ Euclidean (Quantum) Field Theory.* Princeton Univ. Press, Princeton.

Walsh, J.B. (1986). An introduction to stochastic partial differential equations. In: *École d'Été de Probalités de Saint-Flour* XIV–1984. Ed by R. Carmona, H. Kesten and J.B. Walsh. Springer-Verlag, New York, London, pp. 265–437.

Wick, G.C. (1950). The evaluation of the collinear matrix. *Phys. Rev.* **80**, 268–272.

Zhang, T.-S. (1991). *Characterizations of White Noise Test Functions and Hida Distributions.* Preprint. Oslo Univ. Press, Oslo (to appear in *Stochastics*).

Zhang, T.-S. (1992). *On Convolution of Hida Distributions.* Preprint. Oslo Univ. Press, Oslo.

Frontiers in Pure and Applied Probab. 1, pp. 68 – 85
H. Niemi et al. (Eds.)
1993 TVP/VSP

MODELLING VARMAX-PROCESSES BY EXTENDED SAMPLE AUTOCORRELATION AND LINEAR REGRESSION TECHNIQUES

R. HÖGLUND and R. ÖSTERMARK

Åbo Akademi University, Department of Statistics, Fänriksgatan 3B, 20500 Åbo, Finland

Åbo Akademi University, Department of Business Administration, Henriksgatan 7, 20500 Åbo, Finland

ABSTRACT

The paper examines alternative time domain techniques for modelling VARMAX-processes. In particular, we focus on available methods for determining model order and/or estimating parameter values using generalized least squares, recursive residuals, extended sample autocorrelations and minimal order and rank conditions of ARMA-processes.

1. INTRODUCTION

In this paper we present five alternative algorithms for estimating the parameters of the VARMAX Model. The first algorithm is a direct extension of SPLIID's algorithm (Spliid, 1983), the second uses GLS to generate improved estimates, the third is based on the RWNAR procedure, the fourth utilizes the Extended Autocorrelation Approach of Tsay and Tiao (Tsay and Tiao, 1984) and the fifth applies the Corner Method (see (Beguin et al., 1981)).
The algorithms evolve in the following principal steps:

STAGE 1. *Determining the model order.*

Step 0. Estimating the orders p, q, and r. This step is divided in two parts, the first generating the tuple (s, r), where s represents the sum $q + p$. In the second part, the optimal order (p^*, q^*) will be determined in the vicinity of the reference point s. The outcome of Step 0 then is the order vector (p^*, q^*, r^*) and initial estimates of the parameter vector δ to be used in Steps 1–3.

STAGE 2. *Iterative estimation of model parameter values.*

Step 1. Compute model residuals of current iteration, the first time using parameter estimates of Step 0.

Step 2. Compute new parameter estimates.

Step 3. Check convergence and return to Step 1 if needed.

In the seminal paper Spliid (Spliid, 1983) did not consider methods for determining the orders, which he assumed to be known a priori. In the present study, we estimate the orders p, q, and r besides the parameter vector δ.

2. THE MODEL AND NOTATIONS

Let $\mathbf{y}_t$ be a k-dimensional time series $\mathbf{y}_t = (y_{1t}, y_{2t}, ..., y_{kt})'$ and let $\mathbf{x}_t$ be an m-dimensional time series $x_t = (x_{1t}, x_{2t}, ..., x_{mt})'$.
Then a VARMAX model could be expressed as

$$\Phi(B)\mathbf{y}_t = \beta(B)\mathbf{x}_t + \theta(B)\mathbf{a}_t, \tag{1}$$

where $\Phi(B) = \mathbf{I} - \Phi_1 B - ... - \Phi_p B^p$ and $\theta(B) = \mathbf{I} - \theta_1 B - ... - \theta_q B^q$ are $k \times k$ matrix polynomials and $\beta(B) = \beta_0 + \beta_1 B + ... + \beta_{r-1} B^{r-1}$ is a $k \times m$ matrix polynomial. B is the usual backward shift operator, $B^j \mathbf{y}_t = \mathbf{y}_{t-j}$, $\mathbf{a}_t$ is a k-dimensional white noise innovation vector with covariance matrix Σ.
The VARMAX model (1) could be also expressed as a vector TFN (transfer function noise) model

$$\mathbf{y}_t = \frac{\omega(B)}{\delta(B)}\mathbf{x}_t + \mathbf{n}_t, \quad \mathbf{n}_t = \Phi^{-1}(B)\theta(B)\mathbf{a}_t$$

$$\Rightarrow \Phi(B)\mathbf{y}_t = \Phi(B)\frac{\omega(B)}{\delta(B)}\mathbf{x}_t + \theta(B)\mathbf{a}_t,$$

which is model (1) with $\beta(B) = \Phi(B)\omega(B)\delta(B)^{-1}\,\mathbf{x}_t$.
In order to estimate model (1), we define the following entities:

$$\mathbf{Y}_{t,p} = (B\mathbf{y}_t, B^2\mathbf{y}_t, ..., B^p\mathbf{y}_t); \quad \mathbf{A}_{t,q} = (B\mathbf{a}_t, B^2\mathbf{a}_t, ..., B^q\mathbf{a}_t);$$

$$\mathbf{X}_{t,r} = (\mathbf{x}_t, B\mathbf{x}_t, ..., B^{r-1}\mathbf{x}_t); \quad \mathbf{U}_t = (-\mathbf{A}_{t,q}, \mathbf{Y}_{t,p}, \mathbf{X}_{t,r}) \quad \text{and}$$

$$\delta = (\theta_1, \theta_2, ..., \theta_q, \varphi_1, \varphi_2, ..., \varphi_p, \beta_0, \beta_1, ..., \beta_{r-1})'.$$

Then we can write model (1) as

$$\mathbf{y}_t = \mathbf{U}_t\delta' + \mathbf{a}_t. \tag{2}$$

Algorithm 1. *A Revised* SPLIID *Algorithm for* VARMAX *Model Estimation*

In this algorithm the model orders are estimated by an extended Hannan–Rissanen (Hannan and Rissanen, 1982) procedure (cf.(Granger and Newbold, 1986)) as follows.

STAGE 1. *Determining the Model Order.*

Step 0. Initiating the iterative procedure and estimating p (lag-order of $\mathbf{Y}_t$), q (lag-order of $\mathbf{A}_t$) and r (lag-order of $\mathbf{X}_t$).

Part 1. Determining the (s,r)-order. For each combination of $s = 0,1,\ldots$ and $r = 1,2,\ldots$, we

(i) build the matrices, $\mathbf{Y}_{t,s}$, $\mathbf{X}_{t,r}$, and $\mathbf{W}_t = (\mathbf{Y}_{t,s}, \mathbf{X}_{t,r}) \in \mathbf{R}^{1\times kp+rm}$,

(ii) estimate the model $\mathbf{y}_t = \mathbf{W}_t\gamma' + \mathbf{e}_t$, e.g., with OLS, which gives

$$\hat{\gamma} = (\mathbf{W}'\mathbf{W})^{-1}\mathbf{W}'\mathbf{y} \in \mathbf{R}^{kp+rm}, \quad \hat{\mathbf{e}}_t = \mathbf{y}_t - \mathbf{W}_t\hat{\gamma}, \quad \widehat{\Sigma}_e = \mathrm{cov}(\hat{\mathbf{e}}_t), \tag{3}$$

(iii) calculate the Akaike information criterion

$$\mathrm{AIC} = \log|\widehat{\Sigma}_e| + \frac{2(s+r)(k+m)^2}{n}. \tag{4}$$

We choose the combination (s,r) which minimizes AIC. This gives an estimate of the true order r of the $\mathbf{X}$ matrix. In calculating the OLS-estimates $\hat{\gamma}$, we can use a method of adding/deleting variables (see, e.g., (Seber, 1976, pp. 65–71, pp. 338–342)).

Now, let us call the residuals, $\hat{\mathbf{e}}_t$, obtained for the chosen (s,r) order $\hat{\mathbf{a}}_t(0)$. These residuals are then used to estimate the orders p and q and a new parameter vector δ in the following way.

Part 2. Determining the (p,q)-order. We note that the residuals $\hat{\mathbf{a}}_t(0)$ of Step 0 remain unchanged in all regressions of Part 2. For each combination of $p = 0,1,\ldots$ and $q = 0,1,\ldots$ around the reference point $s = p+q$ we

(i) build the matrices $\mathbf{Y}_{t,p}$ as above,

$$\hat{\mathbf{A}}_{t,q}(0) = \big(B\hat{\mathbf{a}}_t(0),\, B^2\hat{\mathbf{a}}_t(0), \ldots, B^q\hat{\mathbf{a}}_t(0)\big), \text{ and } \hat{\mathbf{U}}_t(0) = \big(-\hat{\mathbf{A}}_{t,q}(0),\, \mathbf{Y}_{t,p},\, \mathbf{X}_{t,r}\big),$$

(ii) estimate the model $\mathbf{y}_t = \hat{\mathbf{U}}_t(0)\delta(1) + \mathbf{e}_t$ with OLS, which gives

$$\hat{\delta}(1) = \big(\hat{\mathbf{U}}'(0)\hat{\mathbf{U}}(0)\big)^{-1}\hat{\mathbf{U}}(0)\mathbf{y}_t, \quad \hat{\mathbf{e}}_t = \mathbf{y}_t - \hat{\mathbf{U}}_t(0)\hat{\delta}(1), \quad \widehat{\Sigma}_{\mathbf{e}} = \mathrm{cov}(\hat{\mathbf{e}}_t), \tag{5}$$

(iii) estimate the Schwartz–Rissanen criterion

$$\mathrm{SR} = \log|\widehat{\Sigma}_e| + (p+q+r)(k+m)^2\frac{\log n}{n}. \tag{6}$$

We then choose that combination (p^*,q^*) which minimizes SR and set $\hat{\mathbf{a}}_t(1) = \hat{\mathbf{e}}_t(p^*,q^*)$. Now, let $j=1$ and proceed to Step 2 of Stage 2. We note that for $j=1$ the parameter values are inherited from Step 0.

STAGE 2. *Iterative Estimation of Model Parameter Values.*

Step 1. Calculating the residuals (j is the iteration number)

$$\hat{\mathbf{a}}_t(j) = \mathbf{y}_t - \sum_{i=1}^{p}\hat{\varphi}_i(j)\mathbf{y}_{t-i} + \sum_{i=1}^{q}\hat{\theta}_i(j)\hat{\mathbf{a}}_{t-i}(j) - \sum_{i=0}^{r-1}\hat{\beta}_i(j)\mathbf{x}_{t-i}; \quad t = 1,2,\ldots,n, \tag{7}$$

when $\hat{\delta}(i) = (\hat{\theta}_1(i), ..., \hat{\theta}_q(i), \hat{\varphi}_1(i), ..., \hat{\varphi}_p(i), \hat{\beta}_0(i), ..., \hat{\beta}_{r-1}(i))$.

Step 2. Calculating new parameter estimates.

Let $\hat{\mathbf{A}}_{t,q}(j) = (B\hat{\mathbf{a}}_t(j), ..., B^q\hat{\mathbf{a}}_t(j))$ and $\hat{\mathbf{U}}_t(j) = (-\hat{\mathbf{A}}_{t,q}(j), \mathbf{Y}_{t,p}, \mathbf{X}_{t,r})$. Then estimate

$$\mathbf{y}_t = \hat{\mathbf{U}}_t(j)\delta(j+1) + \mathbf{a}_t(j+1) \tag{8}$$

using OLS.

Step 3. Check for convergence.

If $|\hat{\delta}(j+1) - \hat{\delta}(j)| > \varepsilon$; $\varepsilon > 0$ is some convergence criterion, then let $j = j+1$ and go to Step 1.

Else let $\hat{\delta} = \hat{\delta}(j+1)$, $\hat{\delta} = (\hat{\theta}_1, ..., \hat{\theta}_q, \hat{\varphi}_1, ..., \hat{\varphi}_p, \hat{\beta}_0, ..., \hat{\beta}_{r-1})$ and stop.

Algorithm 2. *Using* GLS *to compute new parameter estimates*

In Step 2(8) Spliid (Spliid, 1983) used OLS to estimate the parameter matrix $\hat{\delta}(j+1)$, but since the regressors are usually correlated we could perhaps improve the precision of the estimates, e.g., by using Ridge regression or GLS. Here we employ a GLS procedure, originally presented by Koreisha and Pukkila (Koreisha and Pukkila, 1990; Pukkila et al., 1990) in identification of univariate ARMA models. The idea of Koreisha and Pukkila (Koreisha and Pukkila, 1990) is to state the ARMA-model in such a form that the autocorrelated residuals can be conveniently extracted. In the present paper the approach is extended to VARMAX-processes. We assume that the innovations $\mathbf{a}_t$ in model (1) are equal to the estimated $\hat{\mathbf{a}}_t$ plus a random error vector η_t, thus

$$\mathbf{R}^{k\times 1} \ni \mathbf{a}_t = \hat{\mathbf{a}}_t + \eta_t \tag{9}$$

which we substitute into model (1) to give

$$\Phi(B)\mathbf{y}_t = \beta(B)\mathbf{x}_t + \theta(B)(\hat{\mathbf{a}}_t + \eta_t) \Rightarrow$$

$$\mathbf{y}_t - \varphi_1\mathbf{y}_{t-1} - ... - \varphi_p\mathbf{y}_{t-p} = \beta_0\mathbf{x}_t + \beta_1\mathbf{x}_{t-1} + ... + \beta_{r-1}\mathbf{x}_{t-r+1} + \hat{\mathbf{a}}_t - \theta_1\hat{\mathbf{a}}_{t-1} - ... - \theta_q\hat{\mathbf{a}}_{t-q}$$

$$+\theta(B)\eta_t \Rightarrow \mathbf{y}_t - \hat{\mathbf{a}}_t = \sum_{i=1}^{p}\varphi_1\mathbf{y}_{t-1} + \sum_{i=0}^{r-1}\beta_i\mathbf{x}_{t-i} - \sum_{i=1}^{q}\theta_i\hat{\mathbf{a}}_{t-i} + \theta(B)\eta_t. \tag{10a}$$

This is a regression model with moving average residuals $\theta(B)\eta_t$, of order q. Now, the estimator of the previous equation at iteration j,

$$\mathbf{y}_t - \hat{\mathbf{a}}_t(j) = \sum_{i=1}^{p}\hat{\varphi}_1\mathbf{y}_{t-1} + \sum_{i=0}^{r-1}\hat{\beta}_i\mathbf{x}_{t-i} - \sum_{i=1}^{q}\hat{\theta}_i\hat{\mathbf{a}}_{t-i}(j) + \hat{\theta}(B)\eta_t, \tag{10b}$$

provides consistent estimates of θ (Pukkila et al., 1990) and we can construct an estimated covariance matrix, $\widehat{\Sigma}_\eta$, associated with the error vector $\hat{\theta}(B)\eta_t$, which allows for estimation of the parameters using GLS.

Step 2 above is then modified as follows:

Let $\hat{\mathbf{A}}_{t,q}(j) = (B\hat{\mathbf{a}}_t(j), ..., B^q\hat{\mathbf{a}}_t(j))$, $\hat{\mathbf{U}}_t(j) = (-\hat{\mathbf{A}}_{t,q}(j), \mathbf{Y}_{t,p}, \mathbf{X}_{t,r})$ as before, $\hat{\mathbf{z}}_t(j) = \mathbf{y}_t - \hat{\mathbf{a}}_t(j)$ and $\varepsilon_t(j) = \hat{\theta}(B)\eta_t(j)$.
Then estimate $\hat{\mathbf{z}}_t(j) = \hat{\mathbf{U}}_t(j)\delta(j+1) + \varepsilon_t(j+1)$, using GLS with $\widehat{\Sigma} = \widehat{\Sigma}_{\varepsilon(j)}$ based on $\hat{\varepsilon}_t(j+1) = \hat{\mathbf{z}}_t(j) - \hat{\mathbf{U}}_t(j)\hat{\delta}_t(j+1)$, where

$$\hat{\delta}(j+1) = \left[\hat{\mathbf{U}}'(j)\widehat{\Sigma}^{-1}_{\varepsilon(j)}\hat{\mathbf{U}}'(j)\right]^{-1}\hat{\mathbf{U}}'(j)\widehat{\Sigma}^{-1}_{\varepsilon(j)}\hat{z}(j). \quad (11)$$

Naturally, this procedure could also be implemented in Step 0 and we could use the RWNAR procedure of Pukkila (Pukkila et al., 1990) to identify the ARMA (p, q) orders.

Algorithm 3. Using the RWNAR *procedure for determining model order*

The AIC and SR criteria in (iii) of Part 1 and Part 2 of Step 0 could respectively be used in a residual white noise-AR (RWNAR) procedure as follows:
If $\hat{\mathbf{e}}_t$ in (3) or (5) is a white noise vector, it can be considered as a VAR(0) process. A nonwhite noise process can be represented as a VAR $(p^*), p^* > 0$. We calculate AIC or SR for VAR(0) and VAR(1) of $\hat{\mathbf{e}}_t$, and select the orders (s, r), and (p, q), respectively, for those combinations for which the selection criterion for the VAR(0) model is less than for the VAR(1) model. When the order (p^*, q^*, r^*) is selected let $j = 1$ and proceed to Step 2.

Algorithm 4. The Extended Sample Autocorrelation Approach

Yet another possibility in identifying the (p, q) order (Step 0, Part 2) is to use the extended autocorrelation introduced by Tsay and Tiao (Tsay and Tiao, 1984). This procedure consists in principle of iterated regressions of increasing order autoregressions, and estimation of the autocorrelation of the residuals. For the univariate case suppose an ARMA(p, q) process

$$y_t = \sum_{i=1}^{p} \varphi_i y_{t-i} + a_t - \sum_{i=1}^{q} \theta_i a_{t-i}. \quad (12)$$

Now we

(0) estimate an AR (p)-model $y_t = \sum_{i=1}^{p} \varphi^{(0)}_{i(p)} y_{t-i} + e^{(0)}_{t(p)}$; $e^{(0)}_{t(p)} = a_t - \sum_{i=1}^{q} \theta_i a_{t-i} \Rightarrow$ estimates $\hat{\varphi}^{(0)}_{i(p)}$ and $\hat{e}^{(0)}_{t(p)} = y_t - \sum_{i=1}^{p} \hat{\varphi}^{(0)}_{i(p)} y_{t-i}$.

If $q > 0$, then $e^{(0)}_{t(p)}$ is not a white noise and $\hat{e}^{(0)}_{t(p)}$ contains some information about the process y_t. (The value of q is determined below.)

(1) Define the first iterated AR(p) regression as $y_t = \sum_{i=1}^{p} \varphi^{(1)}_{i(p)} y_{t-i} + \beta^{(1)}_{1(p)} \hat{e}^{(0)}_{t-1(p)} + e^{(1)}_{t(p)} \Rightarrow$ estimates $\hat{\varphi}^{(1)}_{i(p)}, \hat{\beta}^{(1)}_{1(p)} \wedge \hat{e}^{(1)}_{t(p)} = y_t - \sum \hat{\varphi}^{(1)}_{i(p)} y_{t-1} - \hat{\beta}^{(1)}_{1(p)} \hat{e}^{(0)}_{t-1(p)}$.

If $q > 1$, then $e^{(1)}_{t(p)}$ is not a white noise and $e^{(1)}_{t(p)}$ contains some information about the process y_t.

(2) Then define the second iterated AR(p) regression as $y_t = \sum_{i=1}^{p} \varphi^{(2)}_{i(p)} y_{t-i} + \beta^{(2)}_{1(p)} \hat{e}^{(1)}_{t-1} + \beta^{(2)}_{2(p)} \hat{e}^{(0)}_{t-2(p)} + e^{(2)}_{t(p)} \Rightarrow \hat{\varphi}^{(2)}_{i(p)}, \hat{\beta}^{(2)}_{1(p)}, \hat{\beta}^{(2)}_{2(p)}$, and $\hat{e}^{(2)}_{t(p)}$.

If $q > 2$, then $\hat{e}^{(2)}_{t(p)}$ contains information about y_2 and the iteration continues.
In practice, the order (p, q) is unknown and has to be estimated. For this we use the above procedure of iterated AR-regressions. Define the jth iterated AR(u) regression of a time series as $y_t = \sum_{i=1}^{u} \varphi^{(j)}_{i(u)} y_{t-i} + \sum_{i=1}^{j} \beta^{(j)}_{i(u)} \hat{e}^{(j-i)}_{t-i(u)} + e^{(j)}_{t(u)}$; $t = u + j - 1, ..., n$; $j = 0, ...$; $u = 1, 2, ... \Rightarrow \hat{\varphi}^{(j)}_{i(u)}, \hat{\beta}^{(j)}_{i(u)}$, where

$$\hat{e}^{(l)}_{t(u)} = y_t - \sum_{i=1}^{u} \hat{\varphi}^{(l)}_{i(u)} y_{t-i} + \sum_{i=1}^{l} \hat{\beta}^{(l)}_{i(u)} \hat{e}^{(l-1)}_{t-i(u)} \tag{13}$$

is the estimated residual of the jth iterated AR(u) regression. We note that the coefficients $\beta^{(q)}_{j(p)}$ of an ARMA (p, q)-process correspond to the moving average coefficients of model (1) as follows: $\theta_j = \beta^{(q)}_{j(p)}$, $j = 1, 2, ..., q$.
We also note that in the ESACF-procedure, we still have to fix the search interval $[0, q_{\max}]$ to be used with each AR(p)-model investigated.
The iterated AR estimates $\hat{\varphi}^{(j)}_{i(u)}$ satisfy the recursion

$$\hat{\varphi}^{(j)}_{i(u)} = \hat{\varphi}^{(j-1)}_{i(u+1)} - \frac{\hat{\varphi}^{(j-1)}_{i-1(u)} \hat{\varphi}^{(j-1)}_{u+1(u+1)}}{\hat{\varphi}^{(j-1)}_{u(u)}}, \tag{14}$$

where $\hat{\varphi}^{(j-1)}_{0(u)} = -1$; $i = 1, ..., u$; $u \geq 1 \wedge j \geq 1$ (Tsay and Tiao, 1984). To estimate the orders (p, q) we use the iterated regression estimates and define the extended sample autocorrelation function in the following manner:

$$r_m(u) = r_m(\hat{W}^{(m)}_{ut}) = \frac{\sum_{i=1}^{N-m} (\hat{W}^{(m)}_{ui} - \overline{\hat{W}}^{(m)})(\hat{W}^{(m)}_{u,i+m} - \overline{\hat{W}}^{(m)})}{\sum_{i=1}^{N} (\hat{W}^{(m)}_{ui} - \overline{\hat{W}}^{(m)})^2}, \tag{15}$$

where u =AR-order of iterated regression model, m =MA–order, $\hat{W}^{(m)}_{u,t} = y_t - \sum_{j=1}^{u} \hat{\varphi}^{(m)}_{j(u)} y_{t-i}$.

To check whether $r_m(u)$ is effectively zero beyond m, we use Bartlett's approximation (Bartlett, 1946) to the variance of the estimated autocorrelation function of a stationary normal process (cf. (Östermark and Höglund, 1988)):

$$V[r_m(u)] \simeq \frac{1}{N}[1 + 2\sum_{v=1}^{q} r^2(v)], \quad m > q. \tag{16}$$

Tsay and Tiao (Tsay and Tiao, 1984) proved that, for an ARMA(p, q),

$$r_m(u)\begin{cases} = 0, & m > q, \quad u = p, \\ \neq 0, & m = q, \quad u = p. \end{cases}$$

Thus, the pth ESACF of an ARMA(p, q) process has the same cutting off property as the ordinary SACF of a pure MA(q) process. For known p this cutting off property is useful in identifying q. In practice, however, problems arise when both the fitted AR-order u exceeds p and the iterated MA-order m exceeds q. The effect of overfitting $u - p > 0$ autoregressive lags will lead to an increase in the order of the MA-polynomial of the transformed series $W_{u,t}^{(m)}$ and the number of additional terms is given by $\min\{u - p, m - q\}$. In particular, for an ARMA(p, q)-model

$$r_m(u) = \begin{cases} c(u - p, m - q), & 0 \leq m - q \leq u - p, \\ 0, & m - q > u - p \leq 0, \end{cases}$$

where $c(\cdot)$ is some nonzero constant or a continuous random variable bounded by -1 and $+1$.

The ESACF could then be arranged in a two-way table as follows

	MA				
AR	0	1	2	3	...
0	$r_{1(0)}$	$r_{2(0)}$	$r_{3(0)}$	$r_{4(0)}$	...
1	$r_{1(1)}$	$r_{2(1)}$	$r_{3(1)}$	$r_{4(1)}$	...
2	$r_{1(2)}$	$r_{2(2)}$	$r_{3(2)}$	$r_{4(2)}$	...
3	$r_{1(3)}$	$r_{2(3)}$	$r_{3(3)}$	$r_{4(3)}$	...
⋮	⋮	⋮	⋮	⋮	...

Then this table will contain zero and nonzero elements. The zero values for an ARMA(p, q) process form a triangle with the boundaries given by $u = p$ and $u - m = q$. Then the row and column coordinates of the vertex correspond to the AR and MA order, respectively. For $u = 2$, we see that $r_m(u) = 0$ for $m \geq 2$, for $u = 3$, $r_m(u) = 0$ for $m \geq 3$, and so on, so the ESACF pattern will be

		MA(m)					
u	AR	0(1)	1(2)	2(3)	3(4)	4(5)	...
0	0	X	X	X	X	X	...
1	1	X	X	X	X	X	...
2	2	X	0	0	0	0	...
3	3	X	X	0	0	0	...
4	4	X	X	X	0	0	...
⋮	⋮	⋮	⋮	⋮	⋮	⋮	⋮

X = nonzero value 0 = zero value

This procedure can now readily be generalized to VARMA and hence to VARMAX processes in the following way.

In Step 0, Part 1, we have estimated (s,r) of which s is chosen as the appropriate order for the $\beta(B)$ polynom. Part 2 is completely rewritten. We use the ESACF approach to estimate the order (p,q), by first eliminating the $\mathbf{X}$-effect, that is by using the estimates of β from Part 1 we construct a new variate

$$\mathbf{z}_t = \mathbf{y}_t - \sum_{i=0}^{r-1} \hat{\beta}_i^{(0)} \mathbf{x}_{t-i}, \quad \text{such that} \quad \mathbf{z}_t \approx \sum_{i=1}^{p} \varphi_i \mathbf{y}_{t-i} + \mathbf{a}_t - \sum_{i=1}^{q} \theta_i \mathbf{a}_{t-i}.$$

Then we calculate the SACF for the $\mathbf{z}_t$ series $\mathbf{r}_{j(0)}$ for $j = 0,1,\dots$. Given the AR-order $u = 1,2,\dots$ we do the iterated regressions for $j = 0,1,\dots$ and calculate the ESACF $\mathbf{r}_{j(u)}$ (cf. Spliid, 1983):

$$\mathbf{z}_t = \sum_{i=1}^{u} \varphi_{i(u)}^{(j)} \mathbf{y}_{t-i} + \sum_{i=1}^{j} \beta_{i(u)}^{(j)} \hat{\mathbf{e}}_{t-i(u)}^{(j-i)} + \mathbf{e}_{t(u)}^{(j)} \quad \Rightarrow \hat{\varphi}_{i(u)}^{(j)},\ \hat{\beta}_{i(u)}^{(j)},\ \hat{\mathbf{e}}_{t(u)}^{(j)},$$

$$\mathbf{W}_{u,t}^{(j)} = \mathbf{z}_t - \sum \hat{\varphi}_{i(u)}^{(j)} \mathbf{y}_{t-i}, \quad \mathbf{r}_j(\mathbf{W}_{u,t}^{(j)}) = r_{j(u)}; \quad u = 1,2,\dots; \quad j = 0,1,\dots .$$

The zero pattern of $\mathbf{r}_{j(u)}$; $u = 0,1,2,\dots$, $j = 0,1,2,\dots$, can then be used to identify the orders (p,q).

Algorithm 5. *The Corner Method*

The (p,q) order in Part 2 of Step 0 can also be identified by the Corner Method of Beguin (Beguin et al., 1981). They showed that the minimal order and rank conditions that characterize an ARMA(p,q) process are equivalent to the conditions:

$$\Delta(i,j) = 0 \quad (i \geq q+1, j \geq p+1), \quad \Delta(i,p) \neq 0 \quad (i \geq q), \quad \Delta(q,j) \neq 0 \quad (j \geq p),$$

where $\Delta(i,j)$ is the determinant of the matrix

$$D(i,j) = \begin{bmatrix} \rho_i & \rho_{i-1} & \cdots & \rho_{i-j+1} \\ \rho_{i+1} & \rho_i & \cdots & \rho_{i-j+2} \\ \vdots & \vdots & \ddots & \vdots \\ \rho_{i+j-1} & \rho_{i+j-2} & \cdots & \rho_i \end{bmatrix} \qquad (i \geq 0,\ j \geq 1).$$

The determinants can be ordered as in (Koreisha and Yoshimoto, 1991) in an array to deduce the order of the observed process:

	j						
i	1	2	...	p	$p+1$	...	K
1	$\hat{\Delta}(1,1)$	$\hat{\Delta}(1,2)$	...	$\hat{\Delta}(1,p)$	$\hat{\Delta}(1,p+1)$	...	$\hat{\Delta}(1,K)$
2	$\hat{\Delta}(2,1)$	$\hat{\Delta}(2,2)$	...	$\hat{\Delta}(2,p)$			
$\vdots$	$\vdots$	$\vdots$		$\vdots$			
q	$\hat{\Delta}(q,1)$	$\hat{\Delta}(q,2)$	...	$\hat{\Delta}(q,p)$			
$q+1$							
$\vdots$	$\vdots$					**0**	
K	$\hat{\Delta}(K,1)$						

The southeast corner of the above matrix with elements $i > q,\ j > p$ is zero, thus indicating that the northwest corner $i = q,\ j = p$ is the order of the ARMA (p,q)-process. It is also possible to test the significance of the assertion that $\hat{\Delta}(i,j) = 0,\ i > Q,\ j > q$; e.g., for a subset of $\hat{\Delta}(i,j)$. Under the null hypothesis that the process is ARMA(p,q), i.e., a specific subset of n determinants in the zero area $i > q,\ j > p$ is zero, the statistic $Q_n = n(\hat{\Delta}_1,...,\hat{\Delta}_n)(\hat{\mathbf{A}}\hat{\mathbf{C}}\hat{\mathbf{A}}^T)^{-1}(\hat{\Delta}_1,...,\hat{\Delta}_n)^T$, where element (i,j) of $\hat{\mathbf{C}}$ is $\sum_{h=-\infty}^{\infty} r_h r_{h-i+1} + r_{h+1} r_{h-1} - 2 r_h r_j r_{h-i} - 2 r_i r_j r_{h-j} + 2 r_i r_j r_h^2$ and element (i,j) of $\mathbf{A}$ is $\partial\hat{\Delta}_i/\partial\hat{\Delta}_j$, converges in distribution to $\chi^2(n)$.
Petrucelli and Davis (Petrucelli and Davies, 1984) showed, however, that the test of significance is unstable, since the inverse matrix $[\hat{\mathbf{A}}\hat{\mathbf{C}}\hat{\mathbf{A}}^T]^{-1}$ may be singular. Hence, the level of significance of the elements in the zero area is not tested statistically in this study. Instead, the array of determinants is used as a supporting tool in making judgements of the true order only.

3. PARALLEL PROCESSING OF TIME SERIES ALGORITHMS

A key issue in efficient time series modelling is the time absorbance of the computations. Although several parts of the identification and estimation stages can be automated and left for the computer to provide credible model contenders, the computing time is usually significant even with high performance computers (c.f., e.g., (Höglund and Östermark, 1991; Östermark and Höglund, 1988; Östermark, 1989)). The presentation in §2 suggests that trial and error still is a consistuent element of most iterative time series algorithms. In vector time series modelling the problems accentuate.

To overcome the time absorbance problem the performance of the algorithms may be significantly improved by multiprocessing (cf.(Fox et al., 1988)). Efficient utilization of multiprocessing requires that the problem be decomposable into computational sub-blocks that communicate with each other in a balanced way relative to productive computation. The idea is to spread the total amount of computation on at least two processors that work in parallel. The more independent the computational subproblems, the faster the multiprocessing system will be compared to the sequential algorithm processed on a single processor. For the time being, efforts to parallelize existing sequential time series algorithms or to design completely parallel time series systems from the beginning are reported scarcely.

In vector time series modelling at least the following main development paths towards user-friendly time series algorithms seem possible. Firstly, the computers of a multiprocessor system could be loaded with the same algorithm but with different parameter settings of the parametric search space in different processors. Secondly, completely different time series algorithms could be loaded on the processors to work on the same data set. Thirdly, meaningful sub-blocks of the same time series algorithm could be distributed to the multiprocessor family. In the first two frameworks the processors would be completely independent, save for possible resharing occuring in data base access. In the third framework, the processors are dependent on each other to the extent that the output of one processor serves as input for another. One of the most fruitful starting points for multiprocessor design of time series algorithms within the last framework probably will be the efficient utilization of parallel schemes for matrix inversion. As to the algorithms discussed in the present paper, parallel inversion of Cholesky factored symmetric positive definite matrices provides one a basis for designing high performance time series systems.

4. CONCLUSION

In this paper we have.discussed five algorithms for modelling VARMAX-processes. The frame of reference was provided by a simple OLS-method for parameter estimation, originally put forward by Spliid (Spliid, 1983). Since his approach did not consider the important topics of order determination, multicollinearity, heteroscedasticity and autocorrelated residuals, we wished to examine various methods to improve the performance of Spliid's algorithm, with explicit recognition of some of these problems.

At the present state of research, the above issues have been intensively studied in the univariate case, whereas vector-valued processes have gained less attention. We suggest ways to extend some well-known and powerful techniques for determining model order and dealing with multicollinearity, autocorrelation and heteroscedasticity in the univariate case to vector-valued time series.

Our intention is to implement these methods on a computer for comparative tests in our future research.

Acknowledgements
We thank MBA Jaana Aaltonen for her assistance in writing this paper. Financial support of the Academy of Finland is gratefully acknowledged.

APPENDIX

EXTRACTING SCALAR COMPONENT MODELS FOR VECTOR PROCESSES

In this appendix we present a somewhat different approach by Tiao and Tsay (Tiao and Tsay, 1989) to modelling vector valued time series. The idea is to reduce the dimension of the series by a suitable transformation.

1. *Background and Notation*
Assume that we have the following vector ARMA process:

$$\varphi(B)\mathbf{z}_t = \Theta(B)\mathbf{a}_t, \quad \text{s.t.}$$

$$(\mathbf{I} - \varphi_1 B - \ldots - \varphi_p B^p)\mathbf{z}_t = (\mathbf{I} - \Theta_1 B - \ldots - \Theta_q B^q)\mathbf{a}_t, \quad \varphi_i, \Theta_j \in \mathbf{R}^{k\times k}.$$

Premultiply by suitable preconditioner $\mathbf{T} \in \mathbf{R}^{k\times k}$ to be defined below : $\mathbf{T}\varphi(B)\mathbf{z}_t = \mathbf{T}\Theta(B)\mathbf{a}_t$

$$\Rightarrow (\mathbf{T} - \mathbf{T}\varphi_1 B - \ldots - \mathbf{T}\varphi_p B^p)\mathbf{z}_t = (\mathbf{T} - \mathbf{T}\Theta_1 B - \ldots - \mathbf{T}\Theta_q b^q)\mathbf{a}_t$$

$$\Rightarrow \mathbf{T}\mathbf{z}_t + (-\mathbf{T}\varphi_1 - \ldots - \mathbf{T}\varphi_p B^{p-1})B\mathbf{z}_t = \mathbf{T}\mathbf{a}_t + (-\mathbf{T}\Theta_1 - \ldots - \mathbf{T}\Theta_1 B^{q-1})B\mathbf{a}_t \quad (1)$$

$$\Rightarrow \mathbf{y}_t + \mathbf{G}\mathbf{z}_{t-1} = \mathbf{b}_t + \mathbf{H}\mathbf{a}_{t-1}.$$

Let us rewrite (1) with each row of $\mathbf{T}$ explicated:

$$\mathbf{T} = \left[\mathbf{v}_0^{(1)^T}\ \mathbf{v}_0^{(2)^T} \ldots \mathbf{v}_0^{(k)^T}\right]'$$

$$\Rightarrow \mathbf{y}_t + \begin{bmatrix} -\mathbf{v}_0^{(1)^T} \\ -\mathbf{v}_0^{(2)^T} \\ \vdots \\ -\mathbf{v}_0^{(k)^T} \end{bmatrix} (\varphi_1 B + \ldots + \varphi_p B^{p-1})\mathbf{z}_{t-1} = \begin{bmatrix} \mathbf{v}_0^{(1)^T} \\ \mathbf{v}_0^{(2)^T} \\ \vdots \\ \mathbf{v}_0^{(k)^T} \end{bmatrix} \mathbf{a}_t$$

$$+\begin{bmatrix} -\mathbf{v}_0^{(1)^T} \\ -\mathbf{v}_0^{(2)^T} \\ \vdots \\ -\mathbf{v}_0^{(k)^T} \end{bmatrix} (\theta_1 B - \ldots - \theta_q B^q)$$

$$\Rightarrow \begin{bmatrix} y_{1t} \\ y_{2t} \\ \vdots \\ y_{kt} \end{bmatrix} + \begin{bmatrix} \sum_{l=1}^{p_1} -\mathbf{v}_0^{(1)^T} \varphi_l \mathbf{z}_{t-l} \\ \sum_{l=1}^{p_2} -\mathbf{v}_0^{(2)^T} \varphi_l \mathbf{z}_{t-l} \\ \vdots \\ \sum_{l=1}^{p_k} -\mathbf{v}_0^{(k)^T} \varphi_l \mathbf{z}_{t-l} \end{bmatrix} = \begin{bmatrix} \mathbf{v}_0^{(1)^T} \\ \mathbf{v}_0^{(2)^T} \\ \vdots \\ \mathbf{v}_0^{(k)^T} \end{bmatrix} \mathbf{a}_t + \begin{bmatrix} -\sum_{l=1}^{q_1} \mathbf{v}_0^{(1)^T} \theta_l \mathbf{a}_{t-l} \\ -\sum_{l=1}^{q_2} \mathbf{v}_0^{(2)^T} \theta_l \mathbf{a}_{t-l} \\ \vdots \\ -\sum_{l=1}^{q_k} \mathbf{v}_0^{(k)^T} \theta_l \mathbf{a}_{t-l} \end{bmatrix}.$$

Let $\mathbf{v}_l^{(i)^T} = -\mathbf{v}_0^{(i)^T} \varphi_i$, $\mathbf{h}_l^{(i)^T} = -\mathbf{v}_0^{(i)^T} \theta_l$

$$\Rightarrow \begin{bmatrix} y_{1t} \\ y_{2t} \\ \vdots \\ y_{kt} \end{bmatrix} + \begin{bmatrix} \sum_{l=1}^{p_1} \mathbf{v}_l^{(1)^T} \mathbf{z}_{t-l} \\ \sum_{l=1}^{p_2} \mathbf{v}_l^{(2)^T} \mathbf{z}_{t-l} \\ \vdots \\ \sum_{l=1}^{p_k} \mathbf{v}_l^{(k)^T} \mathbf{z}_{t-l} \end{bmatrix} = \begin{bmatrix} \mathbf{v}_0^{(1)^T} \\ \mathbf{v}_0^{(2)^T} \\ \vdots \\ \mathbf{v}_0^{(k)^T} \end{bmatrix} \mathbf{a}_t + \begin{bmatrix} \sum_{l=1}^{q_1} \mathbf{h}_l^{(1)^T} \mathbf{a}_{t-l} \\ \sum_{l=1}^{q_2} \mathbf{h}_l^{(2)^T} \mathbf{a}_{t-l} \\ \vdots \\ \sum_{l=1}^{q_k} \mathbf{h}_l^{(k)^T} \mathbf{a}_{t-l} \end{bmatrix}. \tag{2}$$

Denote the model structure of y_{it} as SCM (p_i, q_i). Then, from (2),

$$y_{it} + \sum_{l=1}^{p_i} \mathbf{v}_l^{(i)^T} \mathbf{z}_{t-l} = \mathbf{v}_0^{(i)^T} \mathbf{a}_t + \sum_{l=1}^{q_i} \mathbf{h}_l^{(i)^T} \mathbf{a}_{t-l}. \tag{3}$$

Now, let $r = \max_i(p_i)$, $s = \max_i(q_i)$ and restate (2) as follows:

$$\begin{bmatrix} y_{1t} \\ y_{2t} \\ \vdots \\ y_{kt} \end{bmatrix} + \begin{bmatrix} \mathbf{v}_1^{(1)^T} \\ \mathbf{v}_1^{(2)^T} \\ \vdots \\ \mathbf{v}_1^{(k)^T} \end{bmatrix} \mathbf{z}_{t-1} + \begin{bmatrix} \mathbf{v}_2^{(1)^T} \\ \mathbf{v}_2^{(2)^T} \\ \vdots \\ \mathbf{v}_2^{(k)^T} \end{bmatrix} \mathbf{z}_{t-2} + \ldots + \begin{bmatrix} \mathbf{v}_r^{(1)^T} \\ \mathbf{v}_r^{(2)^T} \\ \vdots \\ \mathbf{v}_r^{(k)^T} \end{bmatrix} \mathbf{z}_{t-r}$$

$$= \mathbf{T}\mathbf{a}_t + \begin{bmatrix} \mathbf{h}_1^{(1)^T} \\ \mathbf{h}_1^{(2)^T} \\ \vdots \\ \mathbf{h}_1^{(k)^T} \end{bmatrix} \mathbf{a}_{t-1} + \begin{bmatrix} \mathbf{h}_2^{(1)^T} \\ \mathbf{h}_2^{(2)^T} \\ \vdots \\ \mathbf{h}_2^{(k)^T} \end{bmatrix} \mathbf{a}_{t-2} + \ldots + \begin{bmatrix} \mathbf{h}_s^{(1)^T} \\ \mathbf{h}_s^{(2)^T} \\ \vdots \\ \mathbf{h}_s^{(k)^T} \end{bmatrix} \mathbf{a}_{t-s}. \tag{4}$$

We note that $\mathbf{v}_l^{(i)} = \mathbf{0}$ for $p_i < l \leq r$ and $\mathbf{h}_l^{(i)} = \mathbf{0}$ for $q_i < l \leq s$.
Let $\mathbf{G}_l = (\mathbf{v}_l^{(1)^T}\ \mathbf{v}_l^{(2)^T} \ldots \mathbf{v}_l^{(k)^T})'$, $\mathbf{H}_l = (\mathbf{h}_l^{(1)^T}\ \mathbf{h}_l^{(2)^T} \ldots \mathbf{h}_l^{(k)^T})'$.

Then (4) can be expressed succinctly as

$$\mathbf{y}_t + \sum_{l=1}^{r} \mathbf{G}_l \mathbf{z}_{t-l} = \mathbf{T}\mathbf{a}_t + \sum_{l=1}^{s} \mathbf{H}_l \mathbf{a}_{t-l} \tag{5}$$

$$\Rightarrow \mathbf{y}_t + \sum_{l=1}^{r} \mathbf{G}_l \mathbf{T}^{-1}\mathbf{T}\mathbf{z}_{t-l} = \mathbf{T}\mathbf{a}_t + \sum_{l=1}^{s} \mathbf{H}_l \mathbf{T}^{-1}\mathbf{T}\mathbf{a}_{t-l}.$$

Define $\varphi_l^* = -\mathbf{G}_l\mathbf{T}^{-1}$, $\theta^* = -\mathbf{H}_l\mathbf{T}^{-1}$. Then $\mathbf{y}_t = \mathbf{T}\mathbf{z}_t$, $\mathbf{b}_t = \mathbf{T}\mathbf{a}_t$

$$\Rightarrow \mathbf{y}_t - \sum_{l=1}^{r} \varphi_l^* \mathbf{y}_{t-l} = \mathbf{b}_t - \sum_{l=1}^{s} \theta^* \mathbf{b}_{t-l}. \tag{6}$$

By construction, the number of nonzero rows for the ith component of $\mathbf{y}_t$ in (6) is exactly $(p_i + q_i)$.
Our objective is to find $\mathbf{T}$ such that $\sum_{i=1}^{k}(p_i + q_i)$ is minimized, or: Minimize $\sum_{i=1}^{k}(p_i + q_i)$ such that $\mathbf{y}_t = \mathbf{T}\mathbf{z}_t$ is of minimal order and the k components $\{y_{it}\}$ are linearly independent.
We note that the transformation $\mathbf{T}$ reduces $\varphi(B)\mathbf{z}_t = \theta(B)\mathbf{a}_t$ to a parsimonious representation to the extent *linearly possible.*
Let $\mathbf{Y}_{m,t} = (\mathbf{z}_t^T, \mathbf{z}_{t-1}^T, \ldots, \mathbf{z}_{t-m}^T)^T \in \mathbf{R}^{k(m+1)\times 1} \Rightarrow \mathbf{Y}_{h,t-j-1} = (\mathbf{z}_{t-j-1}^T, \mathbf{z}_{t-j-2}^T, \ldots, \mathbf{z}_{t-j-1-h}^T)^T \in \mathbf{R}^{k(h+1)\times 1}$. Then $\Gamma(m,h,j) = \mathbf{E}[\mathbf{Y}_{h,t-j-1}\mathbf{Y}_{m,t}^T] = \mathbf{E}[(\mathbf{z}_{t-j-1}^T, \mathbf{z}_{t-j-2}^T, \ldots, \mathbf{z}_{t-j-1-h}^T)^T(\mathbf{z}_t^T, \mathbf{z}_{t-1}^T, \ldots, \mathbf{z}_{t-m}^T)]$

$$= \begin{bmatrix} \Gamma_{j+1} & \Gamma_j & \cdots & \Gamma_{j+1-m} \\ \Gamma_{j+2} & \Gamma_{j+1} & \cdots & \Gamma_{j+2-m} \\ \vdots & \vdots & \ddots & \vdots \\ \Gamma_{j+h+1} & \Gamma_{j+h} & \cdots & \Gamma_{j+h+1-m} \end{bmatrix} \in \mathbf{R}^{k(h+1)\times k(m+1)}.$$

In order to alleviate interpretations of Theorem 1 of (Tiao and Tsay, 1989), we present a few variations of $\Gamma(\cdot)$ below.

$$\Gamma(p_i,h,j) = \begin{bmatrix} \Gamma_{j+1} & \Gamma_j & \cdots & \Gamma_{j+1-p_i} \\ \Gamma_{j+2} & \Gamma_{j+1} & \cdots & \Gamma_{j+2-p_i} \\ \vdots & \vdots & \ddots & \vdots \\ \Gamma_{j+h+1} & \Gamma_{j+h} & \cdots & \Gamma_{j+h+1-p_i} \end{bmatrix} \in \mathbf{R}^{k(1+h)\times k(1+p_i)},$$

$$\Gamma(p_i+1,h,q_i) =$$

$$\begin{bmatrix} \Gamma_{q_i+1} & \Gamma_{q_i} & \cdots & \Gamma_{q_i+1-p_i} & \Gamma_{q_i-p_i} \\ \Gamma_{q_i+2} & \Gamma_{q_i+1} & \cdots & \Gamma_{q_i+2-p_i} & \Gamma_{q_i+1-p_i} \\ \vdots & \vdots & \ddots & \vdots & \vdots \\ \Gamma_{q_i+1+h} & \Gamma_{q_i+h} & \cdots & \Gamma_{q_i+1+h-p_i} & \Gamma_{q_i+h-p_i} \end{bmatrix} \in \mathbf{R}^{k(1+h)\times k(p_i+2)},$$

$$\Gamma(p_i+1,h,q_i+1) =$$

$$\begin{bmatrix} \Gamma_{q_i+2} & \Gamma_{q_i+1} & \cdots & \Gamma_{q_i+2-p_i} & \Gamma_{q_i+1-p_i} \\ \Gamma_{q_i+3} & \Gamma_{q_i+2} & \cdots & \Gamma_{q_i+3-p_i} & \Gamma_{q_i+2-p_i} \\ \vdots & \vdots & \ddots & \vdots & \vdots \\ \Gamma_{q_i+2+h} & \Gamma_{q_i+1+h} & \cdots & \Gamma_{q_i+h+2-p_i} & \Gamma_{q_i+h+1-p_i} \end{bmatrix} \in \mathbf{R}^{k(1+h)\times k(p_i+2)}.$$

By (20) and (23) in (Tiao and Tsay, 1989, p. 166–167) we see that $\Gamma(p_i, h, j)$ $\mathbf{v}^{(i)} = \mathbf{0}$ for $h \geq p_i$ and $j \geq q_i$,

$$\left.\begin{aligned} \Gamma(p_i+1, h, q_i)\begin{pmatrix}\mathbf{v}^{(i)}\\ 0\end{pmatrix} &= \mathbf{0} \\ \Gamma(p_i+1, h, q_i+1)\begin{pmatrix}\mathbf{v}^{(i)}\\ 0\end{pmatrix} &= \mathbf{0} \\ \Gamma(p_i+1, h, q_i+1)\begin{pmatrix}0\\ \mathbf{v}^{(i)}\end{pmatrix} &= \mathbf{0} \end{aligned}\right\} \text{for } h \geq 1+p_i \ (\Gamma(\cdot)\text{may have full column rank}).$$

The condition $h \geq p_i$ above must hold since we wish to check rank properties from the *column rank* of $\Gamma(\cdot)$ (cf. (Tiao and Tsay, 1989, p. 166, where $r \geq s$ by assumption)). We now present Theorem 1 of Tiao and Tsay (Tiao and Tsay, 1989).

THEOREM 1(a). *$\Gamma(p_i, h, j)\mathbf{v}^{(i)} = 0$ for $h \geq p_i$ and $j \geq q_i$.*

This result holds directly by (20) and (23), (24).

THEOREM 1(b). *$\Gamma(m, h, q_i)\mathbf{v}^{(i)}(m, p_i, 1) = 0$ for $h \geq m > p_i$, where*

$$\Gamma(m, h, q_i) = \begin{bmatrix} \Gamma_{q_i+1} & \Gamma_{q_i} & \cdots & \Gamma_{q_i+1-m} \\ \Gamma_{q_i+2} & & & \\ \vdots & & & \\ \Gamma_{q_i+1+h} & & & \end{bmatrix} \in \mathbf{R}^{k(h+1)\times k(m+1)},$$

$\mathbf{v}^{(i)}(m, p_i, 1) = (\mathbf{v}^{(i)^T}, \mathbf{0}^T)$, where $\mathbf{v}^{(i)} \in \mathbf{R}^{k(p_i+1)}$, $\mathbf{0} \in \mathbf{R}^{k(m-p_i)} \Rightarrow \mathbf{v}^{(i)}(m, p_i, 1) \in \mathbf{R}^{k(p_i+1)+k(m-p_i)} \equiv \mathbf{R}^{k(m+1)}$.

THEOREM 1(c). *$\Gamma(m, h, j)\mathbf{v}^{(i)}(m, p_i, l) = \mathbf{0}$, $l = 1, \ldots, s$ for $h \geq m > p_i$, $j > q_i$ and $s = \min(m - p_i + 1, j - q_i + 1)$.*

$$\Gamma(m, h, j) =$$

$$\left[\begin{array}{cccc|ccc} \Gamma_{j+1} & \Gamma_j & \cdots & \Gamma_{j+1-p_i} & \Gamma_{j-p_i} & \cdots & \Gamma_{j+1-m} \\ \Gamma_{j+2} & & & & & & \\ \vdots & & m > p_i & & & & \\ \Gamma_{j+1+m} & & & & & & \\ \hline \Gamma_{j+2+m} & & & & & & \\ \vdots & & & h \geq m > p_i & & & \\ \Gamma_{j+1+h} & & & & & & \end{array}\right] \in \mathbf{R}^{k(1+h)\times k(1+m)},$$

$\mathbf{v}(m, p_i, l) = (0, ..., 0, \mathbf{v}^{(i)^T}, 0, ..., 0) \in \mathbf{R}^{k(l-1+p_i+1+m+1-p_i-l)} = \mathbf{R}^{k(m+1)}$, *such that* $k(l-1)$ *zeros precede* $\mathbf{v}^{(i)} \in \mathbf{R}^{k(p_i+1)}$ *and* $k(m+1-p_i-l)$ *zeros succeed* $\mathbf{v}^{(i)}$. *As* l *increases, the left* **0**-*vector of* $\mathbf{v}^{(i)}(m, p_i, l)$ *is prolonged and the right one shortened.*

Since $h \geq m$, the number of rows in $\Gamma(m, h, j)$ is $k(1+h) = r \geq c = k(1+m)$. We then know that the rank properties of $\Gamma(m, h, j)$ are such that $r(\Gamma(m, h, j)) \leq c \leq r \Rightarrow r(\Gamma(m, h, j)) = c - s = c - \min(m - p_i + 1, j - q_i + 1) = \min(k(1+m) - (m - p_i + 1),\ k(1+m) - (j - q_i + 1))$.
Assume that, for example, $l = 1$. Then under the assumption above, $\Gamma(m, h, j = 1) \times \mathbf{v}^{(i)}(m, p_i, 1) = 0$ corresponds to the **NW** submatrix above. For $l = 2$, we move one submatrix column (k elements) to the right.
Let

$$\Gamma(m, h, j) = \begin{bmatrix} \Gamma_{j+1} & \Gamma_j & \dots & \Gamma_{j+1-m} \\ \Gamma_{j+2} & & & \\ \vdots & & & \\ \Gamma_{j+1+h} & & & \end{bmatrix} \in \mathbf{R}^{k(1+h)\times k(1+m)},$$

$\mathbf{v}^{(i)}(m, p_i, l) = (0, ..., 0, \mathbf{v}^{(i)^T}, 0, ..., 0)^T \in \mathbf{R}^{k(m+1)}$, be such that $k(l-1)$ zeros precede $\mathbf{v}^{(i)} \in \mathbf{R}^{k(p_i+1)}$ and $k(m+1-p_i-l)$ zeros succeed $\mathbf{v}^{(i)}$.
According to Tiao and Tsay (Tiao and Tsay, 1989) $\Gamma(m, h, j)\mathbf{v}^{(i)}(m, p_i, l) \neq \mathbf{0}$, if $h \geq \max(m < p_i(\text{impossible}), q_i - 1)$ or $j < q_i$.
What if $[j < q_i$ and $m < p_i]$?

THEOREM 1(d). *According to Grubb (Grubb, 1989)*

$$\Gamma(m, h, j) = \begin{bmatrix} \Gamma_{j+1} & \Gamma_j & \dots & \Gamma_{j+1-m} \\ \Gamma_{j+2} & & & \\ \vdots & & & \\ \Gamma_{j+1+h} & & & \end{bmatrix},$$

$$\mathbf{v}^{(i)}(m, m, 1) = (\mathbf{v}^{(i)^T})^T \in \mathbf{R}^{k(m+1)}.$$

$\Gamma(m, h, j)\mathbf{v}^{(i)}(m, m, 1) = \Gamma(m, h, j)\mathbf{v}^{(i)} \neq \mathbf{0}$, *if* $\mathbf{h} \geq \max\{(m < p_i$, *implying that* $k(m+1) < k(p_i+1)), (q_i - 1)\}$ *or* $j < q_i$.

For future reference, we note that eigen analysis of matrix $\mathbf{A}(m, h, j)$ defined below yields the SCM's. We start by performing the canonical correlation analysis between $\mathbf{Y}_{h,t-j-1}$ and $\mathbf{Y}^T_{m,t}$:

$$\beta(m, h, j) = \left(\mathbf{E}[\mathbf{Y}_{h,t-j-1}\mathbf{Y}^T_{h,t-j-1}]\right)^{-1}\Gamma(m, h, j),$$

$$\mathbf{A}(m, h, j) = \left(\mathbf{E}[\mathbf{Y}_{m,t}\mathbf{Y}^T_{m,t}]\right)^{-1}\Gamma^T(m, h, j)\beta(m, h, j),$$

where

$$\left(\mathbf{E}[\mathbf{Y}_{h,t-j-1}\mathbf{Y}^T_{h,t-j-1}]\right)^{-1} = \begin{bmatrix} \Gamma_0 & \Gamma_{-1} & \dots & \Gamma_{-h} \\ \Gamma_1 & \Gamma_0 & \dots & \Gamma_{-h+1} \\ \vdots & \vdots & \ddots & \vdots \\ \Gamma_h & \Gamma_{h-1} & \dots & \Gamma_0 \end{bmatrix},$$

$$\left(\mathbf{E}[\mathbf{Y}_{m,t}\mathbf{Y}_{m,t}^T]\right)^{-1} = \begin{bmatrix} \Gamma_0 & \dots & \Gamma_{-m} \\ \vdots & \ddots & \vdots \\ \Gamma_m & \dots & \Gamma_0 \end{bmatrix}.$$

Then $\mathbf{A}(m,h,j)\mathbf{W} = \Lambda\mathbf{W}$ gives the SCM:s, where Λ = diagonal matrix with eigenvalues of $\mathbf{A}()$ in decreasing order of magnitude and $\mathbf{w}$ = eigenvectors corresponding to Λ. The eigenvalues of $\mathbf{A}(m,h,j)$ are the squared canonical correlations of $\mathbf{Y}_{m,t}$ and $\mathbf{Y}_{h,t-j-1}$ and the number of zero eigenvalues and the associated right eigenvectors of $\mathbf{A}(m,h,j)$ correspond exactly with the number of zeros and associated right vectors of $\Gamma(m,h,j)$ (Tiao and Tsay, 1989, p.171).

2. *SCM Algorithm*

From the algorithmic viewpoint, §§4.5, 4.6 and 5.2.3 of (Tiao and Tsay, 1989) are essential. The strategy is to successively increment m, keeping $j+m$ constant and to study the eigenstructure of $\mathbf{A}(m,h,j)$ for each relevant (m,j) up to a limit, represented by the diagonal increment of zero eigenvalues. In particular, for a k-dimensional vector process, the proper model order is detected when the diagonal increment equals k. Exchangeable models, if any, may be detected by moving cross-diagonally in the root table, always keeping $m+j$ fixed at the value first satisfying the diagonal increment criterion.

We have interpreted the algorithmic procedure as follows:

STAGE 1. *Identifying an overall order* (p,q).

Step 0. Set $j = m = c = 0$, $n_0(m-1, j-1) = 0$.

Step 1. Set $h = m$, $\mathbf{A}(m,h,j)\mathbf{W}_1 = \Lambda_1\mathbf{W}_1$. Let $n_0(m,j)$ = number of zero eigenvalues.

If $n_0(m,j) - n_0(m-1,j-1) = k$, then
  let $(p_m, q_j) = (m,j)$,
  if $m = c$ go to Step 2
  if $m < c$ let $m = m+1$, $j = j-1$
  go to Step 1.
End if.

If $n_0(m,j) - n_0(m-1,\ j-1) < k$, then
  if $m < c$ let $m = m+1$, $j = j-1$,
  if $m = c$ let $c = c+1$, $m = 0$, $j = c$
  go to Step 1.
End if.

Step 2. Choose (p,q), e.g., $(p,q) = (m,j)$ which minimizes crit (m,j).

STAGE 2. *Identifying the scalar components.*

Step 0. Fixing parameter bounds. Set $j = m = 0$.

Step 1. Set $h = m+q-j$ and solve $\mathbf{A}(m,h,j)\mathbf{W}_1 = \Lambda_1\mathbf{W}_1 \in \mathbf{R}^{(h+1)k\times(m+1)k}$, where $\mathbf{W}_1$ = eigenvectors of $\mathbf{A}(m,h,j)$ corresponding to Λ_1, Λ_1 = eigenvalues of $\mathbf{A}(m,h,j)$, in decreasing order of magnitude.

Step 2. Form $\mathbf{G}(m,j)$ and $\mathbf{H}(m,j)$, where

$\mathbf{R}^{(m+1)k\times n} \ni \mathbf{G}(m,j)$ = all (n) eigenvectors of $\mathbf{A}(m,h,j)$ corresponding to zero eigenvalues,
$\mathbf{R}^{(m+1)k\times l} \ni \mathbf{H}(m,j)$ = all (l) eigenvectors of $\mathbf{A}(m,h,j)$ induced by SCM's of suborders of (m,j).
If $m = 0$, then $\mathbf{H}(0,j) = \mathbf{G}(0, \max(0, j-1))$. Otherwise,

$$\mathbf{H}(m,j) = \begin{bmatrix} \mathbf{0} \\ \mathbf{G}(m-1, \max(0, j-1)) \end{bmatrix},$$

where $\mathbf{0}$ is $k \times l$.

Step 3. Extract the newly detected SCM's. We suppress the arguments (m,j) of $\mathbf{H}(m,j)$ and $\mathbf{G}(m,j)$ and solve

$$\left[(\mathbf{G}^T\mathbf{G})^{-1}\mathbf{G}^T\mathbf{H}(\mathbf{H}^T\mathbf{H})^{-1}\mathbf{H}^T\mathbf{G}\right]\mathbf{W}_2 = \Lambda_2\mathbf{W}_2 \in \mathbf{R}^{n\times n},$$

where $\mathbf{W}_2 \in \mathbf{R}^{n\times n}$ are the right eigenvectors of the canonical correlation analysis between $\mathbf{G}$ and $\mathbf{H}$, ordered in correspondence to Λ_2,
$\Lambda_2 \in \mathbf{R}^{n\times n}$ are the eigenvalues in decreasing order of magnitude.

Step 4. Detect new SCM's.
Let $\mathbf{R}^{n\times b} \ni \mathbf{U}$ be the collection of newly detected SCM's of Step 3, i.e., corresponding to zero eigenvalues in Λ_2. Then, the eigenvectors associated with the newly detected SCM's of order (m,j) are given by:

$$\mathbf{GU} \in \mathbf{R}^{(m+1)k\times b}.$$

Step 5. Update the collection of zero eigenvectors $\mathbf{M}(c)$ of iteration c:

$$\mathbf{R}^{(m+1)k\times d} \ni \mathbf{M}(m+1,\ j+1) = \left[\begin{matrix} \mathbf{M}(m,j) \\ \mathbf{0} \end{matrix} \middle| \mathbf{GU}\right],$$

where $\mathbf{0} \in \mathbf{R}^{k\times(d-b)}$, $\mathbf{M}(m,j) \in \mathbf{R}^{(m+1)k\times(d-b)}$.

Step 6. If $d \geq k$, $m = p$ and $j = 0$, then go to Step 7. Otherwise:
If $j < q$, then let $j = j+1$, else let $j = 0$, $m = m+1$.
Go to Step 1.

Step 7. Let $\mathbf{M}$ = $k \times k$ top portion of $\mathbf{M}(m,j)$. Then $\mathbf{T} = \mathbf{M}^T$, and $\mathbf{y}_t = \mathbf{T}\mathbf{z}_t \in \mathbf{R}^{k\times 1}$ is the process of SCM's. of minimal orders. $\mathbf{M}(m,j)$ contains the collection of zero eigenvectors at location (m,j). If the number of zeros equals k for $m+j = c$, then the $k \times k$ top portion of $\mathbf{M}(m,j)$ provides the transformation matrix $\mathbf{T}$. The same holds for any other (m,j) location on the cth cross-diagonal.

REFERENCES

Bartlett, M.S. (1946). On the theoretical specification of sampling properties of autocorrelated time series. *J. Roy. Statist. Soc.* **B 8**, 27–41.

Beguin, J.M., Gourieroux, C., and Montfort, A. (1981). The applicability of the corner method: A reply. *J. Operat. Res. Soc.* **32**, 1042–1045.

Fox, G., Johnson, M., Lyzenga, G., Otto, S., Salmon, J., and Walker, D. (1988). *Forecasting Economic Time Series.* Second Ed. Prentice Hall, Englewood Cliffs.

Granger, C.W.J., and Newbold, P. (1986). *Forecasting Economic Time Series.* Second Ed. Academic Press, London.

Grubb, H. (1989). In the discussion of Tiao and Tsay, 1989. *J. Roy. Statist. Soc.* **B 51**, 202–203.

Hannan, E. and Rissanen, J. (1982). Recursive estimation of mixed autoregressive moving average order. *Biometrica.* **69**, 81–84. Correction, **70** (1983), 303.

Höglund, R. and Östermark, R. (1990). Automatic ARIMA modelling by the Cartesian ARIMA search algorithm. *J. Forecast.* **10**, 425–476.

Judge, G.G., Griffiths, W.E., Hill, R.C., Lütkepohl, H., and Lee, T.-C. (1985). *The Theory and Practice of Econometrics.* Second ed. (Wiley Series in Probability and Mathematical Statistics). John Wiley, New York.

Koreisha, S. and Pukkila, T. (1990). A generalized least squares approach for estimation of autoregressive moving average models. *J. Time Series Anal.* **11**, 139–151.

Koreisha, S. and Yoshimoto, G. (1991). A comparison among identification procedures for autoregressive moving average models. *Intern. Statist. Rev.* **59**, 37–57.

Petrucelli, J.D. and Davies, N. (1983). Some restrictions on the use of the corner method hypothesis tests. *Comm. Statist.* **B 13**, 543–551.

Pukkila, T., Koreisha, S., and Kallinen, A. (1990). The identification of ARMA models. *Biometrica* **77**, 537–548.

Seber, G. (1977). *Linear Regression Analysis.* John Wiley, New York.

Spliid, H. (1983). A fast estimation method for the vector autoregressive moving average model with exogenous variables. *J. Amer. Statist. Ass.* **78**, 843–849.

Tiao, G.C. and Tsay, R.S. (1989). Model specification in multivariate time series (with discussion). *J. Roy. Statist. Soc.* **B 51**, 157–213.

Tsay, R.S. and Tiao, G.C. (1984). Consistent estimates of autoregressive parameters and extended sample autocorrelation function for stationary and nonstationary ARMA-models. *J. Amer. Statist. Ass.* **79**, 84–86.

Östermark, R. and Höglund, R. (1988). The Cartesian ARIMA Search Algorithm. Åbo Akademi University A:261.

Östermark, R. (1989). Predictability of Finnish and Swedish stock returns. *OMEGA Int. J. Mgmt Sci.* **17**, 223–236.

Frontiers in Pure and Applied Probab. 1, pp. 86 – 94
H. Niemi et al. (Eds.)
1993 TVP/VSP

A BRIEF ACCOUNT OF THE THEORY OF HOMOGENEOUS GAUSSIAN DIFFUSIONS IN FINITE DIMENSIONS

M. JACOBSEN
Institute of Mathematical Statistics, University of Copenhagen,
5 Universitetsparken, DK-2100 Copenhagen Ø, Denmark

ABSTRACT

Without proofs, a number of results from the theory of homogeneous Gaussian diffusions are presented. In particular, the structure of the transition probabilities is studied, the diffusions are constructed as solutions to stochastic differential equations and a characterization is presented of those of the diffusions for which the transitions have nonsingular covariances. Also there is a discussion of coordinatewise, possibly time-dependent, affine transformations that carry one homogeneous Gaussian diffusion into another, the stationary diffusions are determined, and there is a discussion of the problem of equivalence between the measures induced by two diffusions on the space of continuous paths (in finite time).

1. INTRODUCTION

This paper summarizes some results about timehomogeneous Gaussian diffusions in finite dimensions. The results are only stated and for proofs and further details, the reader is referred to (Jacobsen, 1991).

A basic reference to the work discussed here is the paper (Doob, 1944) on stationary Gaussian processes which are timehomogeneous Markov. For such processes in continuous time, Doob gave a classification into three deterministic classes (deterministic in the sense that the processes are completely determined by their value at time 0) and one further class, M, the members of which he represented in the form

$$X_t = \int_0^t e^{(t-s)B} D dW_s + e^{tB} X_0, \tag{1}$$

where W is a driving Brownian motion of the same dimension as X, B is a square matrix with characteristic roots that all have strictly negative real parts and D is symmetric and positive semidefinite.

Doob noticed that type M processes may have smooth components with sample paths that are at least one time differentiable. This led him to define Nth order

Markov processes in continuous time, i.e., real valued processes with $N-1$ pathwise derivatives such that the future evolution of the process given the past at time s, depends only on the value at time s of the process itself and the $N-1$ derivatives. These higher order Markov processes were studied intensively by a number of authors, e.g. (Hida, 1960; Mandrekar, 1974; Okabe, 1974). A much more recent reference continuing Doob's work is a paper (Gzyl, 1987) (see below for a corrected version of his result). Other than that, timehomogeneous Gaussian diffusions were not much discussed in the literature, and mostly they appeared in connection with examples. It is this lack of a systematic treatment that led to the author's paper (Jacobsen, 1991) and the present brief account.

2. SUMMARY OF THE RESULTS

The class of processes to be studied here arises by considering one-parameter semigroups of Gaussian transition probabilities on $\mathbf{R}^n$. More precisely, if $p_t(x,\cdot)$ for $t>0$, $x\in\mathbf{R}^n$ denotes $N(\alpha(t)+\beta(t)x,\ \Sigma(t))$, the Gaussian distribution on $\mathbf{R}^n$ with expectation vector $\alpha(t)+\beta(t)x$ and covariance matrix $\Sigma(t)$, then the Chapman–Kolmogorov equations

$$p_{t+s}(x,\cdot)=\int_{\mathbf{R}^n} p_t(x,dy)p_s(y,\cdot)$$

are easily seen to hold for all $s,t>0$, $x\in\mathbf{R}^n$ if and only if

$$\alpha(t+s)=\alpha(t)+\beta(t)\alpha(s), \tag{2}$$

$$\beta(t+s)=\beta(t)\beta(s), \tag{3}$$

$$\Sigma(t+s)=\Sigma(t)+\beta(t)\Sigma(s)\beta^T(t). \tag{4}$$

If it is assumed that the functions α, β and Σ are continuous on $(0,\infty)$ and satisfy the boundary conditions

$$\lim_{t\downarrow 0}\alpha(t)=0,\quad \lim_{t\downarrow 0}\beta(t)=I,\quad \lim_{t\downarrow 0}\Sigma(t)=0$$

(so that $p_t(x,\cdot)$ converges weakly to the distribution degenerate at x as $t\downarrow 0$), then the solutions to the functional equations (2), (3), and (4) take the forms

$$\beta(t)=e^{tB}, \tag{5}$$

$$\alpha(t)=\int_0^t e^{sB}A\,ds, \tag{6}$$

$$\Sigma(t)=\int_0^t e^{sB}Ce^{sB^T}\,ds \tag{7}$$

with B an arbitrary $n \times n$-matrix, A an arbitrary $n \times 1$-column vector and C an arbitrary $n \times n$ symmetric, positive semidefinite matrix (in which case $\Sigma(t)$ given by (7) is always a covariance matrix). Notice that the parameters B, A and C are the derivatives at 0 of the functions β, α and Σ.
We shall call a timehomogeneous Gaussian Markov process with transition probabilities $N(\alpha(t)+\beta(t)x,\ \Sigma(t))$, where α, β and Σ are given by (5)–(7), an (A, B, C)-*diffusion.* (It will be seen below that such a process is indeed a diffusion).
It may be shown that any homogeneous Gaussian diffusion on $\mathbf{R}^n$ (in particular all paths are continuous and the finite-dimensional distributions are Gaussian), such that all transitions $p_t(x,\cdot)$ are Gaussian of full rank n, is necessarily an (A, B, C)-diffusion. However, there exists homogeneous Gaussian diffusions that are not (A, B, C)-diffusions (see Example 1.23 in (Jacobsen, 1991)).

Example 2.1. In dimension 1, the parameters are scalars $a, b \in \mathbf{R}$ and $c \geq 0$ and one finds

$$\beta(t) = e^{tb},$$

$$\alpha(t) = \frac{a}{b}(e^{tb} - 1),$$

$$\Sigma(t) = \frac{c}{2b}(e^{2tb} - 1).$$

For $c > 0$, $b = 0$ this gives a one-dimensional scaled Brownian motion with constant drift a, while for $c > 0$, $b \neq 0$ one obtains one-dimensional Ornstein–Uhlenbeck processes in their recurrent ($b < 0$) or transient ($b > 0$) form.
(A, B, C)-diffusions are easily constructed as solutions to stochastic differential equations. In particular the construction shows that (A, B, C)-diffusions are indeed diffusions.
To set up the construction, let $(\Omega, \mathcal{F}, \mathcal{F}_t, \mathbf{P})$ be a filtered probability space satisfying the usual conditions and supporting a BM(n')–process (a n'-dimensional standard Brownian motion) W with, in particular, $W_{t+s} - W_s$ independent of $\mathcal{F}_s$ for all $s, t \geq 0$.
Let the parameters (A, B, C) be given, assume, to avoid trivial cases, that $C \neq 0$ and let D be a $n \times n'$-matrix such that $DD^T = C$ (so that $n' \geq$ rank (C) and typically one would use $n' =$ rank (C)).

THEOREM 2.2. *Given a $\mathcal{F}_0$-measurable, $\mathbf{R}^n$-valued Gaussian random variable X_0, the* SDE

$$dX_t = (A + BX_t)\, dt + D dW_t \tag{8}$$

has a unique strong solution with initial value X_0, which is an (A, B, C)-diffusion with respect to the filtration $(\mathcal{F}_t)$.

One way of showing that the solution to (8) is an (A, B, C)-diffusion is to use Ito's formula and the functional equations (2)–(4) to identify the characteristic functions of the transition probabilities.
The SDE (8) is so simple that it may be solved explicitly (cf. (1) above),

$$X_t = e^{tB}X_0 + \int_0^t e^{(t-s)B}(A + BDW_s)\,ds + DW_t.$$

This expression makes it natural to study the process

$$V = (e^{-tB}X_t)_{t\geq 0},$$

which is easily shown to have independent but in general nonstationary Gaussian increments. The increments are stationary, making V an $(A, 0, C)$-diffusion, if and only if

$$BA = 0, \quad BC + CB^T = 0.$$

In order to identify an (A, B, C)-diffusion, it is useful to have the following simple generalization of Lévy's characterization of BM(n) as a local martingale W with quadratic variation $[W, W]_t = tI$.

THEOREM 2.3. *Let X be an adapted, continuous process on $(\Omega, \mathcal{F}, \mathcal{F}_t, \mathbf{P})$ and suppose that X_0 is Gaussian. Then X is an (A, B, C)-diffusion if and only if the process U, where*

$$U_t = X_t - X_0 - tA - \int_0^t BX_s\,ds$$

is a continuous local martingale with $[U, U]_t = tC$. If X is an (A, B, C)-diffusion, U is automatically a martingale.

The next result relates to the structure of the transition probabilities. First of all it may be shown that for an (A, B, C)- diffusion either all the transition covariances $\Sigma(t)$ for $t > 0$ are nonsingular or they are all singular. In the former case we call the (A, B, C)-diffusion *nonsingular* and then have the following characterization.

THEOREM 2.4. *An (A, B, C)-diffusion is nonsingular if and only if*

$$\text{rank}\,(D, BD, \ldots, B^{n-1}D) = n, \tag{9}$$

where D is any matrix such that $DD^T = C$.

The quantity in (9) is the dimension of the subspace of $\mathbf{R}^n$ spanned by the columns of the matrices B^kD, $0 \leq k \leq n-1$.
The condition (9) appears in several contexts in the literature. It is the condition for the infinitesimal generator of the transition semigroup to be hypoelliptic, see (Ichihara and Kunita, 1974) or (Chaleyat–Maurel and Elie, 1981). It is also the condition for the pair of matrices (B, D) to be controllable, see (Zakai and Snyders, 1970), §4.2 in (Davis, 1977) or §5.6 A in (Karatzas and Shreve, 1988). Theorem 2.4 is contained in Proposition 6.2 in (Ichihara and Kunita, 1974) and Proposition 4.2.5 in (Davis, 1977).
It is an important consequence of Theorem 2.4 that an (A, B, C)- diffusion may be nonsingular even though C is singular. However, it is only the finite-dimensional

distributions of such a process that are nonsingular Gaussian. When viewed in continuous time the process becomes singular in the following sense.

PROPOSITION 2.5. *Let X be a nonsingular (A, B, C)-diffusion with C singular. Then there exists a nonsingular $n \times n$-matrix F such that with*

$$\begin{pmatrix} Y_1 \\ Y_2 \end{pmatrix} = FX,$$

where Y_1 contains the $n' = \text{rank}(C)$ first and Y_2 the $n - n'$ last components of FX, the process Y_2 is adapted to the filtration $(\mathcal{G}_t)$ generated by the process Y_1 and the random variable $Y_{2,0}$. Also, Y_1 solves an Ito-equation of the form

$$dY_{1,t} = Z_t\, dt + dW_t,$$

where W is a BM(n')-process and W and Z are $(\mathcal{G}_t)$-adapted.

Nonsingular (A, B, C)-diffusions with a singular C also enjoy characteristic properties related to the smoothness of the sample paths: we shall say that an (A, B, C)-diffusion X has a *smooth component* if there exists a column vector $a \neq 0$ such that the paths of the process $a^T X$ are continuously differentiable. We shall say that a smooth component a is of *degree* $d \geq 1$ provided the dth derivative of $t \to a^T(X_t - X_0)$ exists and has a nonvanishing quadratic variation.

THEOREM 2.6. *Let X be a nonsingular (A, B, C)-diffusion. Then*
(a) *X has a smooth component if and only if C is singular;*
(b) *for $a \neq 0$, $a^T X$ is smooth of degree d if and only if*

$$a^T D = a^T BD = \ldots = a^T B^{d-1} D = 0, \quad a^T B^d D \neq 0.$$

To illustrate why smooth components arise, suppose that $a \neq 0$ with $a^T D = 0$. Then from (8)

$$d(a^T X)_t = a^T(A + BX_t)\, dt,$$

so the paths of $a^T X$ are differentiable.

Example 2.7. Consider the (A, B, C)-diffusion with $A = 0$, $b_{ij} = \delta_{i,j+1}$ and $c_{ij} = 1$ for $i = j = 1$ and $c_{ij} = 0$ otherwise. We can then take $D^T = (1\ 0\ \ldots\ 0)$ and (9) is quickly verified so the diffusion is nonsingular. Also, from (8) it follows that with W a BM(1)-process,

$$dX_{1,t} = dW_t, \quad dX_{i,t} = X_{i-1,t}\, dt \qquad (2 \leq i \leq n).$$

Thus, with initial condition 0, $X_1 = W$ while the other components are successive integrals of W with X_i smooth of degree i.

Let again X denote a nonsingular (A, B, C)-diffusion and let φ: $[0, \infty) \to \mathbf{R}^n$ and ψ: $[0, \infty) \to \mathbf{R}^{n \times n}$ be continuous vector and matrix-valued functions. Define a new continuous process $\widetilde{X}$ by

$$\widetilde{X}_t = \varphi(t) + \psi(t) X_t. \tag{10}$$

If all matrices $\psi(t)$ are nonsingular, $\widetilde{X}$ is a Gaussian diffusion with, in general, non-homogeneous transitions. Obviously, if φ and ψ are constant with ψ nonsingular, $\widetilde{X}$ is some $(\widetilde{A}, \widetilde{B}, \widetilde{C})$-diffusion.

PROPOSITION 2.8. *If X is an (A, B, C)-diffusion, E is a n-column vector and F is a $n \times n$ nonsingular matrix, then $\widetilde{X}_t = E + FX_t$ defines an $(\widetilde{A}, \widetilde{B}, \widetilde{C})$-diffusion, where*

$$\widetilde{A} = FA - FBF^{-1}E, \quad \widetilde{B} = FBF^{-1}, \quad \widetilde{C} = FCF^T.$$

If instead E is $m \times 1$, F is $m \times n$, where $m \neq n$ is allowed, then $\widetilde{X}_t = E + FX_t$ will be Markov if rank $(F) = n$ but need not be if rank $(F) < n$ (for instance if $m < n$). By Theorem 2.3 it follows easily however that $\widetilde{X}$ is an $(\widetilde{A}, \widetilde{B}, \widetilde{C})$-diffusion for some $(\widetilde{A}, \widetilde{B}, \widetilde{C})$ if and only if

$$B : \ \ker(F) \to \ker(F),$$

where $\ker(F) = \{x\colon\ Fx = 0\}$.
Returning now to (10), even if φ and ψ are time dependent it is sometimes possible to make $\widetilde{X}$ timehomogeneous Markov.
The first result shows how such a transformation may remove the parameter A.

PROPOSITION 2.9. *If X is an (A, B, C)-diffusion and*

$$\varphi(t) = -\int_0^t e^{sB} A \, ds, \quad \psi(t) = I,$$

then $\widetilde{X}$ is a $(0, B, C)$-diffusion.

Combining Propositions 2.8 and 2.9 one finds that by a suitable coordinatewise affine transformation, a nonsingular (A, B, C)-diffusion may be carried into a $(0, \widetilde{B}, \widetilde{C})$-diffusion, where $\widetilde{C}$ is diagonal of the form

$$\widetilde{C} = \mathrm{diag}\,(1 \ldots 1\, 0 \ldots 0)$$

with the number of 1's equal to rank (C).
Using a timedependent ψ, a homogeneous $\widetilde{X}$ may be obtained in for instance the following manner.

PROPOSITION 2.10. *Let $b \in \mathbf{R}$ and let Q be a skewsymmetric $n \times n$-matrix. If X is a $(0, bI, I)$-diffusion, then*

$$\widetilde{X}_t = e^{tQ} X_t$$

is a $(0, bI + Q, I)$-diffusion.

With Q skewsymmetric the $\psi(t) = e^{tQ}$ form a semigroup of orthogonal transformations. We call $\widetilde{X}$ a *rotating Brownian motion* if $b = 0$ and a *rotating Ornstein–Uhlenbeck* process if $b \neq 0$.

There is an algebraic result (see (Schmidt, 1986)) which states that if a $n \times n$-matrix B is normal ($BB^T = B^T B$), then there exists an orthogonal O such that

$$OBO^T = \begin{pmatrix} B_1 & \dots & 0 \\ \vdots & \ddots & \vdots \\ 0 & \dots & B_n \end{pmatrix}, \tag{11}$$

where each B_i is a scalar or a skewsymmetric 2×2-matrix. This observation was used by Gzyl (Gzyl, 1987) to give a product representation of certain stationary Gaussian diffusions. There appears to be a mistake in his paper, since he ends up with deterministic processes of the types originally studied by Doob. In our setup, Gzyl's result follows immediately from the preceding results and remarks and takes the form:

PROPOSITION 2.11. *Suppose X is a (O, B, I)-diffusion, where B is normal. Then there exists an orthogonal O such that $\widetilde{X} = OX$ is a $(O, \widetilde{B}, I)$-diffusion, where $\widetilde{B} = OBO^T$ is given by* (11).

If for instance $X_0 \equiv x_0$ is degenerate, the result shows that after a suitable orthogonal transformation, X splits into stochastically independent components which are either one-dimensional Brownian motions or Ornstein–Uhlenbecks' or two-dimensional rotating Brownian motions or Ornstein–Uhlenbecks'.
In his paper, Doob (Doob, 1944) studied only strictly stationary processes as is reflected by the condition he imposed on B, vizually that all the characteristic roots of B must have strictly negative real parts, see (1) and the remarks following (1). For (A, B, C)-diffusions, the conditions for stationarity take the following form, see also (Zakai and Snyders, 1970) or §6 in (Ichihara and Kunita, 1974), where we call the diffusion stationary if the transition semigroup (p_t) admits of an invariant probability π,

$$\pi = \int \pi(dx) p_t(x, \cdot)$$

for all t.

THEOREM 2.12. *A nonsingular (A, B, C)-diffusion is stationary if and only if all characteristic roots of B have strictly negative real parts. In that case the invariant probability π is uniquely determined as the Gaussian law with mean vector μ and nonsingular covariance matrix Γ, where*

$$A + B\mu = 0, \quad C + B\Gamma + \Gamma B^T = 0,$$

and furthermore, for all $x \in \mathbf{R}^n$, $p_t(x, \cdot) \to \pi$ weakly as $t \to \infty$.

We shall conclude with a brief discussion of the problem of when the measures induced by two (A, B, C)-diffusions on the space of continuous paths are equivalent, when the processes are observed on a finite time-interval.
Let Ω^0 denote the space of continuous paths w: $[0, \infty) \to \mathbf{R}^n$, write X_t^0 for the projection $X_t^0(w) = w(t)$ and introduce the σ-algebra $\mathcal{F}^0 = \sigma(X_t^0)_{t \geq 0}$ and the filtration $(\mathcal{F}_t^0)$, where $\mathcal{F}_t^0 = \sigma(X_s^0)_{s \leq t}$.

Now, fix an initial state $x^0 \in \mathbf{R}^n$ and let $P, \widetilde{P}$ denote the distribution of, respectively, an (A, B, C)-diffusion and an $(\widetilde{A}, \widetilde{B}, \widetilde{C})$-diffusion, both starting from x^0. Thus P and $\widetilde{P}$ are probabilities on $(\Omega^0, \mathcal{F}^0)$ and we denote the restrictions to $\mathcal{F}_t^0$ by P^t and $\widetilde{P}^t$.
Since P^t and $\widetilde{P}^t$ are Gaussian, either $\widetilde{P}^t \sim P^t$ or $\widetilde{P}^t \perp P^t$. The final theorem gives conditions, necessary and sufficient for equivalence.

THEOREM 2.13. (a) *If C is nonsingular, then for any $t > 0$, $\widetilde{P}^t \sim P^t$ if and only if $\widetilde{C} = C$ and in that case*

$$\frac{d\widetilde{P}^t}{dP^t} = \exp\left(\int_0^t \left(\widetilde{Z}_s^T - Z_s^T\right) C^{-1}\, dX_s^0 - \frac{1}{2}\int_0^t \left(\widetilde{Z}_s^T C^{-1} \widetilde{Z}_s - Z_s^T C^{-1} Z_s\right) ds\right),$$

where

$$Z_s = A + BX_s^0, \quad \widetilde{Z}_s = \widetilde{A} + \widetilde{B} X_s^0.$$

(b) *If C is singular and (A, B, C)-diffusions are nonsingular, then $\widetilde{P}^t \sim P^t$, for any $t > 0$, if and only if $\widetilde{C} = C$ and*

$$a^T \widetilde{A} = a^T A, \quad a^T \widetilde{B} = a^T B$$

for all $a \in \mathbf{R}^n$ with $a^T D = 0$.

If X is an (A, B, C)-diffusion, then $[X - X_0, X - X_0]_t = tC$, so the condition $\widetilde{C} = C$ is always necessary for $\widetilde{P}^t \sim P^t$. That it is also sufficient when C is nonsingular is well known, see Theorem 7.7 in (Liptser and Shiryaev, 1977) or §10 in (Kabanov, Liptser, and Shiryaev, 1980), where also the expression for the Radon–Nikodym derivative may be found. If however, C is singular, as Proposition 2.5 shows, under P there are constraints on the behaviour of the sample paths, and for $\widetilde{P}^t \sim P^t$ to hold, these constraints must be valid under $\widetilde{P}$ as well. Notice that the condition in (b) implies in particular that the diffusion has the same smooth components under P and $\widetilde{P}$, cf. Theorem 2.6. For a special case where $\widetilde{P}^t \sim P^t$ and C is singular, see Corollary 2.1.2 in (Le Breton and Musiela, 1985).

REFERENCES

Le Breton, A. and Musiela, M. (1985). Some parameter estimation problems for hypoelliptic homogeneous Gaussian diffusions. In: *Sequential Methods in Statistics*. Ed. by A. Zielinski (Banach Center Publications 16), Polish Scientific Publishers, Warsaw, pp. 337–356.

Chaleyat-Maurel, M. and Elie, L. (1981). Diffusions Gaussiennes. Géodésiques et diffusions en temps petit. *Séminaire de Probabilités Université de Paris* VII. *Astérisque* 84–85, pp. 255–279.

Davis, M.H.A. (1977). *Linear Estimation and Stochastic Control.* Chapman and Hall, London.

Doob, J.L. (1944). The elementary Gaussian processes. *Ann. Math. Statist.* **15**, 229–282.

Gzyl, H. (1987). Characterization of vector valued, Gaussian, stationary Markov processes. *Statist. Probab. Letters.* **6**, 17–19.

Hida, T. (1960). Canonical representation of Gaussian processes and their applications. *Mem. Coll. Sci. Kyoto* (A) **33**, 109–155.

Ichihara, K. and Kunita, H. (1974). A classification of the second order degenerate elliptic operators and its probabilistic characterization. *Z. Wahrscheinlichkeitstheor. verw. Geb.* **30**, 235–254.

Ikeda, N. and Watanabe, S. (1989). *Stochastic Differential Equations and Diffusion Processes.* Second ed. North-Holland/Elsevier, Amsterdam.

Jacobsen, M. (1991). *Homogeneous Gaussian Diffusions in Finite Dimensions.* Preprint 3. University of Copenhagen, Institute of Mathematical Statistics, Copenhagen.

Kabanov, Yu.M., Liptser, R.S., and Shiryaev, A.N. (1980). Absolute continuity and singularity of locally absolutely continuous probability distributions. II. *Math. USSR Sbornik.* **36**, 31–58.

Karatzas, I. and Shreve, S.E. (1988). *Brownian Motion and Stochastic Calculus.* (Graduate Texts in Mathematics 113), Springer-Verlag, New York.

Liptser, R.S. and Shiryaev, A.N. (1977). *Statistics of Random Processes.* I, *General Theory.* (Applications of Mathematics 5), Springer-Verlag, New York.

Mandrekar, V. (1974). On the multiple Markov property of Lévy–Hida for Gaussian processes. *Nagoya Math. J.* **54**, 69–78.

Okabe, Y. (1974). On the structure of splitting fields of stationary Gaussian processes with finite multiple Markovian property. *Nagoya Math. J.* **54**, 191–213.

Schmidt, E.J.P.G. (1986). An alternative approach to canonical forms of matrices. *Amer. Math. Monthly.* **93**, 176–184.

Zakai, M. and Snyders, J. (1970). Stationary probability measures for linear differential equations driven by white noise. *J. Different. Equat.* **8**, 27–33.

Frontiers in Pure and Applied Probab. 1, pp. 95 – 102
H. Niemi et al. (Eds.)
1993 TVP/VSP

A NOTE ON FIRST PASSAGES IN BRANCHING BROWNIAN MOTIONS

I. KAJ and P. SALMINEN

Department of Mathematics, Uppsala University, Thunbergsvägen 3, S-75238 Uppsala, Sweden
Department of Mathematics, Åbo Academy University, Fänriksgatan 3B, SF-20500 Åbo, Finland

ABSTRACT

Consider a realization of a one-dimensional branching Brownian motion generated by a single particle at the origin. Fix $x > 0$ and count any particle visiting x as the first among the members of its line of descent from the initial particle. The resulting first-passage process indexed by x is a Markov branching process which can be characterized in terms of the original branching mechanism. In this note we study the first-passage process with emphasis on supercritical properties.

1. INTRODUCTION

We consider a Markov branching Brownian motion on the real line with an initial particle located at the origin at time zero. Each particle, independently, is assigned an exponentially distributed lifelength with common rate $\alpha > 0$ and performs a Brownian motion starting at the position, where its parent died. At time of death a particle is removed and replaced with a random number of new particles according to the offspring distribution $\{p_k\colon\ k \geq 0\}$.

To this type of branching tree we associate an integer-valued random process $Z = \{Z_x\colon\ x \geq 0\}$, $Z_0 = 1$, indexed by the space variable x restricted to the positive halfline. This process counts for each $x > 0$ the number of particles in the tree with the property that they have *no ancestor who ever visited the space location x but during their own life a first visit to x indeed occured.*

In (Kaj and Salminen, 1993) the process Z is defined rigorously as a functional of a branching Brownian motion $X = \{X_t\colon\ t > 0\}$, $X_0 = 0$, and it is shown that under the assumption of subcritical or critical branching Z is a continuous time Galton–Watson process. Its branching rate is $\sqrt{2\alpha}$ and the offspring distribution $\{r_k\colon\ k \geq 0\}$ is determined in terms of the given probabilities p_k. The object here is to extend this result to cover also supercritical branching and to demonstrate that features of X relevant for the supercritical case nicely carry over to Z. This includes extinction sets, Harris transformation and martingale convergence.

2. THE FIRST-PASSAGE PROCESS

We shall only repeat the construction of X and Z briefly and refer the reader to (Kaj and Salminen, 1993), which is intended to be selfcontained but to a large extent is based on the works (Neveu, 1987) and (Chauvin, 1991).

A *particle* $u = (j_1, \ldots, j_n)$ is represented by a sequence of reproduction numbers $j_i \geq 1$ which signify that u is the j_nth child of $(j_1, \ldots, j_{n-1})$, who is the j_{n-1}th child of ... the j_1th child of the initial particle, denoted $\{0\}$. A *tree* ω is a set of particles which contains $\{0\}$ and all their ancestors and progeny. Particles are partially ordered by $u < v$ if u is an ancestor of v. To each particle u associate a lifelength ζ^u and a spatial motion γ^u. Let Ω^0 denote the set of all *marked trees* ω^0 defined as

$$\omega^0 = \{\omega\,,\, \{(\zeta^u, \gamma^u)\}\colon\; u \in \omega\}\,,$$

where ω runs over all trees. Now we let $\mathcal{F}^0$ denote the smallest σ-algebra on Ω^0 which makes all marked trees measurable and $\mathbf{P}$ be the probability measure on $(\Omega^0, \mathcal{F}^0)$ such that all ζ^u are exponential with parameter α, all γ^u standard Brownian motions starting at the location where the parent of u died, the branching is governed by the offspring generating function $F(u) := \sum_k p_k u^k$ and all particles evolve independently. We assume that the offspring distribution is proper $(F(1) = 1)$ and impose the *nonexplosion hypothesis*: For every $\varepsilon > 0$

$$\int_{1-\varepsilon}^{1} (u - F(u))^{-1}\, du = \infty\,. \tag{1}$$

The branching Brownian motion X is the canonical process on the triple $(\Omega^0, \mathcal{F}^0, \mathbf{P})$. In addition we let $N = \{N_t\colon\; t > 0\}$ denote the process which counts the total number of particles present at time t irrespective of their position in space.

Further, in order to define Z, let σ^u_y denote the time of the first visit of a particle u to the point $y \geq 0$ and set $\sigma^u_y = \infty$ if u never visits y. Then

$$\tau^u_y := \begin{cases} \sigma^u_y, & \text{if } \sigma^u_y < \infty \quad \text{and there is no} \quad v < u \quad \text{with} \quad \sigma^v_y < \infty, \\ +\infty, & \text{otherwise,} \end{cases}$$

is the time of the first visit to y for those particles u which do not have a predecessor v that already did visit y. Now for a marked tree ω^0 let

$$L_x(\omega^0) = \{u \in \omega\colon\; \tau^u_x(\omega^0) < \infty\}$$

be the set of particles with this first passage property in common and define the (rightcontinuous) *first passage process* associated with X as

$$Z = \{Z_x = |L_{x+}|\colon\; x \geq 0\}\,,$$

where $|\{\cdot\}|$ denotes the number of elements in $\{\cdot\}$.

3. SUPERCRITICAL BRANCHING TREES

In this section we collect some well-known properties, emphasizing supercritical aspects, of a continuous time Galton–Watson process $Y = \{Y_t : t \geq 0\}$, with branching rate η and offspring generating function H, say (see, e.g., (Athreya and Ney, 1972, Chapter III)). The basic example in our context is the total size process N with $N_0 = 1$, branching rate α and offspring generating function F. Our aim is to show that the first-passage process constitutes a further example and then to compare the properties of the two.

We include the defective case, $H(1) \le 1$, we suppose that $H(u)/H(1)$ satisfies (1) and interpret $p_\infty := 1 - H(1) \ge 0$ as the probability of a jump to an additional state $\{+\infty\}$ (hence properties referring to $H(1) < 1$ does not apply to the given offspring generating function F). Define

$$q = \inf\{u \ge 0\colon\ H(u) = u\}\,.$$

Of course, if Y is (sub)critical, then $q = 1$. For a given $t_0 > 0$ consider the initial-value problems

$$\varphi' = \eta(H(\varphi) - \varphi)\,, \qquad \varphi(-t_0) = 1\,, \tag{2}$$

where the parameter t_0 is positive. If $p_\infty > 0$ there is a unique decreasing solution φ on $[-t_0, \infty[$ with $\varphi'(-t_0) = -\eta(1 - H(1)) < 0$ and values in $]q, 1]$, $\varphi(\infty) = q$. If H is proper ($H(1) = 1$) the modified system

$$\varphi' = \eta(H(\varphi) - \varphi)\,, \qquad \varphi(-\infty) = 1\,, \tag{2'}$$

has a solution $\varphi \in]q, 1[$, $\varphi(\infty) = q$, defined on the whole real line. Moreover, in both the defective and the proper case the related system

$$\varphi' = \eta(H(\varphi) - \varphi)\,, \qquad \varphi(-t_0) = 0\,, \tag{3}$$

has an increasing unique solution taking values in $[0, q[$, $\varphi(\infty) = q$, with $\varphi'(-t_0) = \eta p_0$ ($p_0 = 0$ if and only if $q = 0$ and then $\varphi \equiv 0$).

Put $\overline{\varphi}(t, u) = \mathbf{E}[u^{Y_t}]$, $0 \le u \le 1$. The solutions of (2), (2$'$) and (3) determine $\overline{\varphi}(t, u)$ by the relation

$$\overline{\varphi}(t, \varphi(s)) = \varphi(t + s)\,, \quad s \ge -t_0\,, \quad t > 0\,.$$

(For this characterization of Y see also (Neveu, 1987)). For $q < u \le 1$ and $t_0 < \infty$ the solutions of (2) apply and then $\overline{\varphi}(t, 1) = \mathbf{P}(Y_t < \infty)$ and $\varphi(0) = \mathbf{P}(Y_{t_0} < \infty)$. On the other hand, (2$'$) yields $\overline{\varphi}(t, 1) = 1$, $t \ge 0$. Hence this is the relevant φ for a proper supercritical branching process, i.e., $\mathbf{P}(Y_t < \infty) = 1$ for all t and $H(1) = 1$. Taking $\varphi = q$ we obtain $\overline{\varphi}(t, q) = q$ for all $t \ge 0$. Lastly, if $0 \le u < q$ the solutions of (3) prescribe $\overline{\varphi}(t, u)$. Here $\overline{\varphi}(t, 0) = \mathbf{P}(Y_t = 0)$ and the choice of t_0 is reflected in $\varphi(0) = \mathbf{P}(Y_{t_0} = 0)$.

Let

$$A = \{Y_t \to \infty \text{ as } t \to \infty\}\,, \quad B = \{Y_t = 0 \text{ for some } t\}\,.$$

Then $\mathbf{P}(A \cup B) = 1$ and the probability of the *extinction set* B is $\mathbf{P}(B) = q$. If $p_\infty > 0$, then $A = \{Y_t = +\infty \text{ for some } t\}$ *a.s.* The supercritical branching process can be decomposed in two parts corresponding to the sets A and B, see (Athreya and Ney, 1972, Chapter I.12). We recall this idea briefly:

(i) For $t \geq 0$ define a new process $Y_t^{(1)}$ as follows. If $\omega^0 \in A$ let $Y_t^{(1)}(\omega^0)$ be the number of particles among the $Y_t(\omega^0)$ existing at time t which has an infinite number of descendents, if $\omega^0 \in B$ let $Y_t^{(1)}(\omega^0) = 0$. Then the law of $\{Y_t^{(1)}, t \geq 0\}$ conditioned on A is the law of a continuous time Galton–Watson process with the offspring generating function

$$\widetilde{H}(u) := \frac{H((1-q)u+q)-q}{1-q}, \quad 0 \leq u \leq 1.$$

This process is called the *Harris transformation* of Y. Note that $Y^{(1)}$ is obtained from Y by "erasing all the branches which become extinct". This point of view is adopted in (Neveu, 1986). In (Salminen, 1992) the construction is done for (rather) general Markov branching processes and so called terminal lines. It is quite obvious that the above described erasure procedure can also be performed for the branching Brownian motion X. The resulting process, called the Harris-transformation of X, is again a branching Brownian motion but now the generating function of the offspring distribution is given by $\widetilde{H}$.
(ii) The law of the process $Y_t^{(2)} = Y_t - Y_t^{(1)}$ can be obtained from the law of Y by conditioning on extinction. Here, the offspring generating function of $Y^{(2)}$ is

$$\widehat{H}(u) := \frac{H(qu)}{q}, \quad 0 \leq u \leq 1.$$

A similar transformation in the case of X leads again to a branching Brownian motion with $\widehat{H}$ as the generating function of the offspring distribution. Note that $\widetilde{q} = 0$ and $\widehat{q} = 1$, using self-explanatory notation.
We shall denote by Y_t^* the branching process with the proper offspring generating function $H(u)/H(1)$. It is obtained from Y_t by conditioning on the event that jumps to $\{+\infty\}$ do not occur. For ease of notation we may keep the star and identify Y and Y^* in the case $p_\infty = 0$. For example, we have

$$\mathbf{E}[Y_t^*] = \exp\left\{\eta\left(\frac{H'(1)}{H(1)} - 1\right)t\right\}. \tag{4}$$

Also note that the extinction probability increases from q to the smallest q^* with $H(q^*) = q^* H(1)$ and that $\widetilde{H}$ and $\widehat{H}$ change accordingly if we shift from Y to Y^*. If Y^* is supercritical ($H'(1) > H(1)$), then $Y_t^* / \mathbf{E}[Y_t^*]$, $t \geq 0$, is a non-negative martingale such that the limit $W = \lim_{t\to\infty} Y_t^* / \mathbf{E}[Y_t^*]$ exists *a.s.* and either $W = 0$ *a.s.* or $\mathbf{E}[W] = 1$, the latter holding if and only if

$$\sum_k p_k \, k \log k < \infty. \tag{5}$$

4. PROPERTIES OF THE FIRST PASSAGE PROCESS

We are now prepared to formulate a somewhat more general result than Theorem 1 in (Kaj and Salminen, 1993). The part of the proof which is covered there is only sketched below.

THEOREM. *The first passage process Z_x associated with a branching Brownian motion X with creation rate α and offspring generating function F with extinction probability $0 \le q \le 1$ is a continuous time Galton–Watson process, such that the creation rate of Z equals $\sqrt{2\alpha}$, the extinction probability of Z is q and the extinction sets of Z and N are a.s. equal and the offspring generating function $G(v) = \sum_k r_k v^k$ for Z is given by*

$$G(v) = v + \operatorname{sgn}\{q - v\}\sqrt{2\int_v^q (F(u) - u)\,du}\,, \quad 0 \le v \le 1. \tag{6}$$

In particular, in the supercritical case G is always defective with

$$r_\infty := 1 - G(1) = \sqrt{2\int_q^1 (u - F(u))\,du} > 0\,.$$

Proof. The Markov and branching properties and hence the conclusion that Z is a continuous time Galton–Watson process follow from the strong Markov property at the stopping line $\tau_x = \{\tau_x^u\}$. The proof of the fact that the branching rate equals $\sqrt{2\alpha}$ is based on the observation that the first jump of Z occurs after a time which has the same distribution as $\sup_{t \le \tau} B_t$, where B is a standard Brownian motion and τ is an exponential random variable with mean $1/\alpha$. In the rest of the proof we put $\eta_\alpha := \sqrt{2\alpha}$.
Consider now the extinction sets of Z and N. Since the branching Brownian motion, if it does not become extinct, visits all the points in $\mathbf{R}$ with probability one it follows that

$$\{N_t = 0 \text{ for some } t\} \subseteq \{Z_x = 0 \text{ for some } x\} \quad \text{a.s.}.$$

The reverse inclusion is obviously true and, hence, the extinction sets are a.s. equal.
Put $\overline{\psi}(x,u) = \mathbf{E}[u^{Z_x}]$, $x \ge 0$, $0 \le u \le 1$. From the above it follows for $0 < u < 1$ that $\lim_{x\to\infty} \overline{\psi}(x,u) = q$. Without presuming the condition $F'(1) \le 1$ we can still follow the derivations in (Kaj and Salminen, 1993) to the point of showing that $\overline{\psi}(x,u)$, $x \ge 0$, satisfies

$$\psi''(x) = -\eta_\alpha^2 (F(\psi(x)) - \psi(x))\,. \tag{7}$$

Using $\lim_{x\to\infty} \psi'(x) = 0$, we obtain

$$\psi'(x)^2 = -2\int_x^\infty \psi'(y)\psi''(y)\,dy = 2\eta_a^2 \int_x^\infty \psi'(y)(F(\psi(y)) - \psi(y))\,dy\,.$$

By a change-of-variable,

$$\psi'(x)^2 = 2\eta_a^2 \int_{\psi(x)}^q (F(u) - u)\,du\,.$$

(Here we use $\psi(\infty) = q$ and allow integration from the right to the left of an interval.) Thus, since ψ is monotone,

$$\psi'(x) = \eta_\alpha(G(\psi(x)) - \psi(x)), \tag{8}$$

where G is as given in (6). It is easily seen that $v \mapsto G(v)$ is a moment generating function, which is defect in the case $q < 1$. Hence equation (8) is of the type (2) or (3), whereas the type (2′) does not apply. In fact, it is known that there are no decreasing (travelling wave) solutions of equation (7) on $\mathbf{R}$ with $\psi(-\infty) = 1$ and $\psi(\infty) = q < 1$, see (Chauvin, 1991, Theorem 3.1).
According to the previous section we then obtain solutions of (8) which uniquely determine $\overline{\psi}(x, u)$, $x \geq 0$, by the relation

$$\overline{\psi}(x, \psi(y)) = \psi(x + y), \quad x \geq 0,$$

where y is allowed to vary over an interval $[-x_0, \infty)$, $x_0 > 0$.
We proceed stating some further properties of Z and of the conditioned process Z^* with offspring generating function $G(u)/G(1)$. The proof of the first corollary is straightforward from (6).

COROLLARY 1. (i) *The process Z conditioned on extinction has the offspring generating function*

$$\widehat{G}(v) = v + \sqrt{2\int_v^1 (\widehat{F}(u) - u)\, du}\,, \quad 0 \leq v \leq 1,$$

i.e., it is also obtained from X conditioned on extinction.

(ii) *The Harris transformation of Z has the offspring generating function*

$$\widetilde{G}(v) = v - \sqrt{2\int_0^v (\widetilde{F}(u) - u)\, du}\,, \quad 0 \leq v \leq 1,$$

i.e., it is also obtained from the Harris transformation of X.

COROLLARY 2. *If X (or, equivalently, N) is supercritical ($q < 1$), then Z^* is supercritical with finite mean*

$$\mathbf{E}[Z_x^*] = \exp\left\{\sqrt{2\alpha}\left(\frac{r_\infty}{1 - r_\infty}\right) x\right\}.$$

The martingale limit $W = \lim_{x\to\infty} Z_x^* / \mathbf{E}[Z_x^*]$ exists a.s. and is nontrivial.

Proof. Since $G'(1) = 1 > G(1)$, Z is supercritical with finite mean obtained from (4).
To prove the last statement we need to establish (5) for the probabilities $\{r_k/G(1) : k \geq 1\}$. However, this instance of (5) is equivalent to (Athreya and Ney, 1972, Chapter I.10)

$$\int_0^1 v^{-2}(G(1 - v) - G(1) + G'(1)v)\, dv < \infty. \tag{9}$$

By a Taylor expansion (recall that we have assumed $F(1) = 1$ throughout)

$$G(1-v) - G(1) + G'(1)v \sim \int_0^v (1 - F(1-u) + u)\,du\,, \quad \text{as } v \to 0\,.$$

Hence (9) is equivalent to

$$\int_0^1 \frac{1}{u}(1 - F(1-u))\,du < \infty\,.$$

Let U be a random variable with the distribution $\{p_k\}$. Changing variable we see that (9) is equivalent to

$$\int_0^c \frac{1}{u}(1 - F(e^{-u}))\,du = \mathbf{E}\left[\int_0^{cU} \frac{1}{u}(1 - e^{-u})\,du\right] < \infty\,,$$

for some positive constant c. Clearly this is the same as $\mathbf{E}[\log_+ U] < \infty$ ($\log_+ x := \log(x \vee 1)$), which in its turn is the same as

$$\sum_k p_k \log_+ k < \infty\,. \tag{10}$$

We finish the proof by showing that (10) follows from (1). Actually, the argument below shows that a nonexplosive branching process has more than the logarithmic moment required here. We use a reformulation of the nonexplosion hypothesis due to (Doney, 1984). He proves that (1) is equivalent to

$$\sum_{n=1}^{\infty}(nM(n))^{-1} = \infty, \quad M(n) := \sum_{k=0}^{n}\sum_{i=k+1}^{\infty} p_i\,.$$

Let b_n be a sequence such that $1/b_n$ is convex, decreasing and $\sum_n b_n^{-1} < \infty$. Since the divergent sequence $(nM(n))^{-1}$ is convex and decreasing to zero there is a n_0 such that $b_n > nM(n)$, all $n > n_0$. Hence

$$\sum_{k=n_0}^{\infty} \mathbf{P}(U > k) \sum_{n=k}^{\infty} nb_n^{-2} \le \sum_{n=n_0}^{\infty} nb_n^{-2}M(n) < \sum_{n=n_0}^{\infty} b_n^{-1} < \infty\,.$$

In particular, take $b_n = n^{3/2}$ to see that (10) holds.

REFERENCES

Athreya, K.B. and Ney, P.E. (1972). *Branching Processes.* Springer-Verlag, Berlin, Heidelberg, New York.

Chauvin, B. (1991). Product martingales and stopping lines for branching Brownian motion. *Ann. Probab.* **19**, 1195–1205.

Doney, R.A. (1984). A note on some results of Schuh. *J. Appl. Probab.* **21**, 192–196.

Kaj, I. and Salminen, P. (1993). On a first passage problem for branching Brownian motions. *Ann. Appl. Probab.* **3**, No 1.

Neveu, J. (1986). Erasing a branching tree. *Adv. Appl. Probab. Spec. Suppl.*, 101–108.

Neveu, J. (1987). Multiplicative martingales for spatial branching processes. In: *Seminar in Stochastic Processes.* Ed. by Çinlar, Chung and Getoor. Birkhäuser, Boston, Basel, pp.223–242.

Salminen, P. (1992). Cutting Markovian trees. *Ann. Acad. Sci. Fennicae, Series A* **17**, 123–137.

Frontiers in Pure and Applied Probab. 1, pp. 103 - 111
H. Niemi et al. (Eds.)
1993 TVP/VSP

RANDOM GRAPHS WITH MARKED CYCLES

V. F. KOLCHIN
Steklov Mathematical Institute, Vavilov Street 42, 117966 Moscow, GSP-1, Russia

ABSTRACT

We consider a random graph with marked cycles. Every cycle of the graph is marked with probability which depends on the cycle length. We study the asymptotic behaviour of the probability that the graph has no unmarked cycle. As a special case we find the asymptotic behaviour of the probability that the graph has exactly k cycles and the lengths of these cycles belong to a fixed set, and the probability that the lengths of all cycles of the graph belong to a fixed set.

1. INTRODUCTION

We consider a random graph $G_{n,T}$ with n labelled vertices and T edges. These edges are obtained by T independent trials. In each trial an edge between any two vertices i and j appears with probability $2n^{-2}$ and a loop at any vertex i appears with probability n^{-2}, $i, j = 1, \ldots, n$. We consider the number of cycles $\nu_{n,T}$ in the graph $G_{n,T}$. The behaviour of the number of cycles in different classes of random graphs is well investigated (see, for example, (Erdős and Rényi, 1960; Takács, 1988; Kolchin, 1990, 1986a; Kolchin and Khokhlov, 1990; Balakin, Kolchin, and Khokhlov, 1991)). If $n, T \to \infty$ in such a way that $2T/n \to \lambda$, $0 < \lambda < 1$, then with probability tending to one the graph $G_{n,T}$ has no connected components with more than one cycle and the distribution of the number of cycles $\nu_{n,T}$ converges to the Poisson distribution with parameter $\Lambda = \frac{1}{2}\log(1-\lambda)$. Denote $B_{n,T}$ the set of all graphs with n labelled vertices and T edges which have no connected component with more than one cycle, cycles of lengths one and two are allowed.
We mark the cycles of the graph $G_{n,T}$ according to the following rule. If a realization of $G_{n,T}$ belongs to the set $B_{n,T}$, then each cycle of length r is marked with probability p_r independently of the other cycles. If the graph has a connected component with more than one cycle, then we do not mark any cycle.
Denote $p_{n,T}(k)$ the probability that the number of cycles $\nu_{n,T}$ in $G_{n,T}$ is equal to k and there is no unmarked cycle. The probability that there is no unmarked cycle in $G_{n,T}$ is equal to

$$p_{n,T} = \sum_{k=0}^{\infty} p_{n,T}(k).$$

Let b_m be the number of mappings of the set $\{1,\ldots,m\}$ into itself whose graphs are connected, and let $b_m^{(r)}$ be the number of mappings of the set $\{1,\ldots,m\}$ into itself whose graphs are connected and each of them has a cycle of length r.
The explicit expressions for b_m and $b_m^{(r)}$ are well known. By the formula for the number of rooted trees we obtain that

$$b_m^{(r)} = \frac{m!}{(m-r)!} m^{m-r-1} \tag{1}$$

and consequently

$$b_m = \sum_{r=1}^{m} b_m^{(r)} = (m-1)! \sum_{k=0}^{m-1} \frac{m^k}{k!}.$$

Let

$$b(x) = \sum_{m=1}^{\infty} \frac{b_m x^m}{m!}, \qquad B_m = \sum_{r=1}^{m} b_m^{(r)} p_r, \qquad B(x) = \sum_{m=1}^{\infty} \frac{B_m x^m}{m!}.$$

In the present paper we prove the following assertions.

THEOREM. *If $n, T \to \infty$ in such a way that $2T/n \to \lambda$, $0 < \lambda < 1$, then for any fixed $k = 0, 1, \ldots$*

$$p_{n,T}(k) = \frac{(B(\alpha))^k \sqrt{1-\lambda}}{2^k k!} (1 + o(1))$$

and

$$p_{n,T} = e^{B(\alpha)/2} \sqrt{1-\lambda}\, (1 + o(1)),$$

where $\alpha = \lambda e^{-\lambda}$.

The following result is an immediate consequence of the theorem.
Let R be a subset of natural numbers, and let $p_{n,T}^{(R)}(k)$ be the probability that $G_{n,T} \in B_{n,T}$ and the number $\nu_{n,T}$ of cycles in $G_{n,T}$ is equal to k and all the cycle lengths belong to the set R, and let

$$p_{n,T}^{(R)} = \sum_{k=0}^{\infty} p_{n,T}^{(R)}(k).$$

Set $p_r = 1$ for $r \in R$, and $p_r = 0$ for $r \notin R$,

$$B_m^{(R)} = \sum_{r=1}^{\infty} b_m^{(r)} p_r = \sum_{r \in R} b_m^{(r)}, \qquad B^{(R)}(x) = \sum_{m=1}^{\infty} \frac{B_m^{(R)} x_m}{m!}.$$

Remind that $\alpha = \lambda e^{-\lambda}$. It is not difficult to see that

$$B^{(R)}(\alpha) = \sum_{r \in R} \frac{\lambda^r}{r}.$$

Indeed, according to (1)

$$B^{(R)}(\alpha) = \sum_{m=1}^{\infty}\sum_{r\in R}\frac{b_m^{(r)}\alpha^m}{m!} = \sum_{r\in R}\sum_{m=1}^{\infty}\frac{b_m^{(r)}\alpha^m}{m!}$$
$$= \sum_{r\in R}\alpha^r\sum_{m=r}^{\infty}\frac{m^{m-r-1}\alpha^{m-r}}{(m-r)!} = \sum_{r\in R}\alpha^r\sum_{n=0}^{\infty}\frac{(n+r)^{n-1}\alpha^n}{n!}.$$

Using the well-known equality (see, for example, (Pólya and Szegő, 1964, Chapter 5, Problem 210; Riordan, 1968))

$$\sum_{n=0}^{\infty}\frac{(n+r)^{n-1}\alpha^n}{n!} = \frac{e^{r\lambda}}{r}, \qquad (2)$$

we obtain that

$$B^{(R)}(\alpha) = \sum_{r\in R}\frac{\lambda^r}{r}.$$

COROLLARY. *If $n, T \to \infty$ in such a way that $2T/n \to \lambda$, $0 < \lambda < 1$, then for any fixed $k = 0, 1, \ldots$*

$$p_{n,T}^{(R)}(r) = \frac{\Lambda_R^k}{k!}e^{-\Lambda}(1+o(1)), \qquad p_{n,T}^{(R)} = e^{-\Lambda+\Lambda_R}(1+o(1)),$$

where

$$\Lambda = -\frac{1}{2}\log(1-\lambda), \qquad \Lambda_R = \frac{B^{(R)}(\alpha)}{2} = \sum_{r\in R}\frac{\lambda^r}{2r}.$$

Remark that the results of the corollary could be obtained with the help of results on the joint distribution of the random variables $\alpha_{r_1}, \ldots, \alpha_{r_s}$, $1 \le r_1 < \ldots < r_s$, where α_r is the number of cycles of length r in $G_{n,T}$. In (Kolchin and Khokhlov, 1990) it is shown that if $n, T \to \infty$ in such a way that $2T/n \to \lambda$, $0 < \lambda < 1$, then the joint distribution of the random variables $\alpha_{r_1}, \ldots, \alpha_{r_s}$ for any fixed $1 \le r_1 < \ldots < r_s$ converges to the distribution of s independent random variables which have the Poisson distributions with parameters $\lambda_{r_1}, \ldots, \lambda_{r_s}$ correspondingly, where $\lambda_r = \lambda^r/(2r)$. Using this result, (2) and some additional estimates, it is not difficult to obtain the results of the corollary.

2. AUXILIARY RESULTS

Let $F_{n,T}$ be the number of all forests of unrooted trees with n labelled vertices and T edges. It is clear that the number of trees in such a forest is $N = n - T$. In a series of papers V. E. Britikov investigated properties of random forests of unrooted trees. We use the following result from (Britikov, 1988).

LEMMA 1. *If $n, T \to \infty$ in such a way that $2T/n \to \lambda$, $0 < \lambda < 1$, then*

$$F_{n,T} = \frac{n^{2T}\sqrt{1-\lambda}}{2^T T!}(1+o(1)).$$

This result makes it possible to obtain relatively simple proofs of theorems on the Poisson convergence of the distribution of the number of cycles in random graphs of various types. This possibility is realized in (Kolchin, 1986a) for the random graph $G_{n,T}$ considered in the present paper. In (Kolchin, 1990) the following assertion is proved.

LEMMA 2. *If* $2T/n \leq 1$, *then there exists a constant* c *such that*

$$F_{n,T} \leq \frac{cn^{2T+1}}{2^T T!}.$$

Remind that $B_{n,T}$ is the set of all graphs with n labelled vertices and T edges which have no connected components with more than one cycle. In (Kolchin, 1986a) we prove the following result.

LEMMA 3. *If* $n, T \to \infty$ *in such a way that* $2T/n \to \lambda$, $0 \leq \lambda < 1$, *then*

$$\mathbf{P}\{G_{n,T} \in B_{n,T}\} \to 1.$$

Further we need the following assertion.

LEMMA 4. *If* $n, T \to \infty$ *in such a way that there exist constants* a_0 *and* a_1 *such that* $0 < a_0 \leq 2T/n \leq a_1 < 1$, *then for* m, $m_0 \leq m \leq T$, *where* m_0 *is sufficiently large,*

$$a_{n,T}(m) = e^{2Tm/n}\left(1-\frac{1}{T}\right)\cdots\left(1-\frac{m-1}{T}\right)\left(1-\frac{m}{n}\right)^{2(T-m)}\left(1-\frac{1}{n}\right)\cdots\left(1-\frac{m-1}{n}\right) \leq 1.$$

Proof. We represent the logarithm of $a_{n,T}(m)$ in the following form:

$$\begin{aligned}\log a_{n,T}(m) &= \frac{2Tm}{n} + \sum_{i=1}^{m-1} \log\left(1-\frac{i}{T}\right) + 2(T-m)\log\left(1-\frac{m}{n}\right) + \sum_{i=1}^{m-1} \log\left(1-\frac{i}{n}\right) \\ &= -\sum_{k=1}^{\infty} \frac{1}{kT^k}\sum_{i=1}^{m-1} i^k - \sum_{k=2}^{\infty} \frac{2T}{k}\left(\frac{m}{n}\right)^k + \sum_{k=1}^{\infty} \frac{2m}{k}\left(\frac{m}{n}\right)^k - \sum_{k=1}^{\infty} \frac{1}{kn^k}\sum_{i=1}^{m-1} i^k.\end{aligned}$$

Taking into account the inequality

$$\sum_{i=1}^{m-1} i^k \geq \frac{(m-1)^{k+1}}{k+1}$$

we obtain the estimate

$$\log a_{n,T}(m)$$
$$\le \sum_{k=1}^{\infty}\frac{2m}{k}\left(\frac{m}{n}\right)^k-\sum_{k=1}^{\infty}\frac{2T}{k+1}\left(\frac{m}{n}\right)^{k+1}-\sum_{k=1}^{\infty}\frac{(m-1)^{k+1}}{k(k+1)T^k}-\sum_{k=1}^{\infty}\frac{(m-1)^{k+1}}{k(k+1)n^k}$$
$$=\sum_{k=1}^{\infty}\left(\frac{m^{k+1}}{k(k+1)n^k}\left(2(k+1)-\frac{2Tk}{n}-\left(1-\frac{1}{m}\right)^{k+1}\frac{n^k}{T^k}-\left(1-\frac{1}{m}\right)^{k+1}\right)\right).$$

Set $a=2T/n$. It is not difficult to see that for sufficiently large m for all k

$$a_k=2(k+1)-ak-\left(1-\frac{1}{m}\right)^{k+1}\frac{2^k}{a^k}-\left(1-\frac{1}{m}\right)^{k+1}\le 0.$$

Indeed, since $0<a_0\le a\le a_1<1$, for sufficiently large m

$$\left(1-\frac{1}{m}\right)^{k+1}\frac{2^k}{a^k}\le 2^k,\qquad k\ge 1,$$

and therefore $a_k\le 2(k+1)-2^k$. From here we easily see that $a_k\le 0$ for sufficiently large m and all $k\ge 3$. In addition

$$a_1=4-a-\left(1-\frac{1}{m}\right)^2\frac{2}{a}-\left(1-\frac{1}{m}\right)^2\le 3-a-\frac{2}{a}+\frac{4}{am}+\frac{2}{m},$$

$$a_2=6-2a-\left(1-\frac{1}{m}\right)^3\frac{4}{a^2}-\left(1-\frac{1}{m}\right)^3\le 5-2a-\frac{4}{a^2}+\frac{4}{a^2m}+\frac{4}{m},$$

and consequently, $a_1<0$ and $a_2<0$ for sufficiently large m, since for $0<a_0\le a\le a_1<1$

$$3-a-2/a<0,\qquad 5-2a-4/a^2<0.$$

3. PROOF OF THE MAIN RESULT

It is possible to give an explicit formula for $p_{n,T}(k)$.

LEMMA 5. *The following equalities are true:*

$$p_{n,T}(0)=F_{n,T}T!\left(\frac{2}{n^2}\right)^T,$$

and for any natural k, $k\le\min(T,n)$,

$$p_{n,T}(k)=\frac{T!}{2^kk!}\left(\frac{2}{n^2}\right)^T\sum_{m=k}^{n}\binom{n}{m}F_{n-m,T-m}\sum_{m_1+\ldots+m_k=m}\frac{m!\,B_{m_1}\ldots B_{m_k}}{m_1!\ldots m_k!},$$

where

$$B_m = \sum_{r=1}^{m} b_m^{(r)} p_r.$$

Proof. For $k = 0$ the assertion is obvious. Denote $c_n^{(r)}$ the number of connected graphs with n labelled vertices and exactly one cycle of length r. It is clear that

$$c_n^{(1)} = b_n^{(1)}, \quad c_n^{(2)} = b_n^{(2)}, \quad c_n^{(3)} = b_n^{(3)}/2, \quad r \geq 3. \tag{3}$$

Denote $C_{n,T}$ the event that there is no unmarked cycle in the graph $G_{n,T}$. We can represent the event $\{\nu_{n,T} = k,\ G_{n,T} \in B_{n,T},\ C_{n,T}\}$ as a sum of disjoint events each of which means that in T independent trials we obtain a fixed sequence of edges forming a graph with k cycles which consists of trees and k components with exactly one marked cycle. Then we find that

$$\begin{aligned} p_{n,T}(k) &= \mathbf{P}\{\nu_{n,T} = k,\ G_{n,T} \in B_{n,T},\ C_{n,T}\} \\ &= \sum_{m=k}^{n} \binom{n}{m} \sum_{m_1+\ldots+m_k=m} \frac{m!}{m_1!\ldots m_k!\,k!} \\ &\times \sum_{r_1=1}^{m_1} c_{m_1}^{(r_1)} p_{r_1} \cdots \sum_{r_k=1}^{m_k} c_{m_k}^{(r_k)} p_{r_k} F_{n-m,T-m} T! \left(\frac{2}{n^2}\right)^T \frac{1}{2^{s_1+s_2}}, \end{aligned}$$

where $s_1 = s_1(r_1, \ldots, r_k)$ is the number of units among $r_1, \ldots, r_k$ and $s_2 = s_2(r_1, \ldots, r_k)$ is the number of two's among $r_1, \ldots, r_k$. The factor 2^{-s_1} appears since the probability $2n^{-2}$ must be replaced with n^{-2} exactly s_1 times. The factor 2^{-s_2} appears since any transposition of the numbers of trials with identical outcomes gives the same realization.
Using relation (3) we obtain the assertion of the lemma.

Proof of Theorem. We divide the sum from Lemma 5 into the parts S_1 and S_2 with the indices m, $k \leq m < M$, and the indices $m \geq M$, correspondingly. Take $M = T^{1/4}$. It is clear that for any x from the domain of convergence of the series $\sum_{m=1}^{\infty} B_m x^m/m!$ we have

$$\sum_{m\geq M} \sum_{m_1+\ldots+m_k=m} \frac{B_{m_1}x^{m_1}\ldots B_{m_k}x^{m_k}}{m_1!\ldots m_k!} \leq (B(x))^{k-1} \sum_{m\geq M/k} \frac{B_m x^m}{m!}. \tag{4}$$

For the generating functions

$$b(x) = \sum_{m=1}^{\infty} \frac{b_m x^m}{m!}, \qquad B(x) = \sum_{m=1}^{\infty} \frac{B_m x^m}{m!}$$

we have the obvious inequality $B(x) \leq b(x)$ since

$$B_m = \sum_{r=1}^{m} b_m^{(r)} p_r \leq \sum_{r=1}^{m} b_m^{(r)} = b_m. \tag{5}$$

In addition,

$$b_m = (m-1)! \sum_{k=0}^{m-1} \frac{m^k}{k!} \le (m-1)!\, e^m.$$

From this inequality it follows that

$$\sum_{m \ge M/k} \frac{b_m x^m}{m!} \le \sum_{m \ge M/k} (ex)^m. \tag{6}$$

We shall use the notations

$$\theta(x) = \sum_{n=1}^{\infty} \frac{n^{n-1} x^n}{n!}, \qquad a(x) = \sum_{n=0}^{\infty} \frac{n^n x^n}{n!}.$$

It is not difficult to show (see, for example, (Kolchin, 1986b)) that

$$b(x) = \log a(x), \qquad a(x) = (1-\theta(x))^{-1}.$$

Set $a = 2T/n$ and $x = ae^{-a}$ for $a < 1$. Then

$$\theta(x) = a, \qquad b(x) = -\log(1-a).$$

Under the conditions of the theorem $a = 2T/n \to \lambda$, $0 < \lambda < 1$, and for $x = ae^{-a}$ there exists $q < 1$ such that $ex = ae^{1-a} \le q < 1$ beginning with some n. Therefore

$$\sum_{m \ge M/k} \frac{b_m x^m}{m!} \le \frac{1}{1-q} q^{M/k}. \tag{7}$$

Using Lemmas 2, 4 and estimates (4), (5), (6) and (7) we find that

$$\begin{aligned}
S_2 &= \frac{T!}{2^k k!} \left(\frac{2}{n^2}\right)^T \sum_{m \ge M} \sum_{m_1+\ldots+m_k=m} \binom{n}{m} F_{n-m,T-m} \frac{m!\, B_{m_1} \ldots B_{m_k}}{m_1! \ldots m_k!} \\
&\le cT! \left(\frac{2}{n^2}\right)^T \sum_{m \ge M} \sum_{m_1+\ldots+m_k=m} \frac{n^{m+1}(n-m)^{2(T-m)} b_{m_1} \ldots b_{m_k}}{2^{T-m}(T-m)!\, m_1! \ldots m_k!} \\
&\le c_1 \sum_{m \ge M} \sum_{m_1+\ldots+m_k=m} n(ae^{-a})^m \frac{b_{m_1} \ldots b_{m_k}}{m_1! \ldots m_k!} \\
&\le c_1 n (B(x))^{k-1} \sum_{m \ge M/k} \frac{b_m x^M}{m!} \le \frac{c_1 n}{1-q} q^{T^{1/4}/k} \le c_2 T q^{T^{1/4}/k},
\end{aligned}$$

where c_1 and c_2 are some constants. Thus, $S_2 \to \infty$ under the conditions of the theorem.

According to Lemma 1, if $n, T \to \infty$, $2T/n \to \lambda$, $0 < \lambda < 1$, then

$$T!\, F_{n-m,T-m} = \frac{T!\,(n-m)^{2(T-m)}\sqrt{1-\lambda}}{2^{T-m}(T-m)!}(1+o(1)) = \frac{n^{2T} x^m \sqrt{1-\lambda}}{2^T n^m}(1+o(1))$$

uniformly with respect to $m \leq M = T^{1/4}$. Therefore for any fixed $k = 1, 2, \ldots$

$$S_1 = \frac{T!}{2^k k!}\left(\frac{2}{n^2}\right)^T \sum_{m=k}^{M-1} \sum_{m_1+\ldots+m_k=m} m!\binom{n}{m} F_{n-m,T-m} \frac{B_{m_1}\ldots B_{m_k}}{m_1!\ldots m_k!}$$
$$= \frac{\sqrt{1-\lambda}}{2^k k!} \sum_{m=k}^{M-1} \sum_{m_1+\ldots+m_k=m} \frac{B_{m_1}x^{m_1}\ldots B_{m_k}x^{m_k}}{m_1!\ldots m_k!}(1+o(1)).$$

Taking into account the estimate of S_2 we find that

$$S_1 = \frac{\sqrt{1-\lambda}}{2^k k!} \sum_{m=k}^{\infty} \sum_{m_1+\ldots+m_k=m} \frac{B_{m_1}x^{m_1}\ldots B_{m_k}x^{m_k}}{m_1!\ldots m_k!}(1+o(1)) + o(1)$$
$$= \frac{\sqrt{1-\lambda}}{2^k k!}(B(x))^k(1+o(1)).$$

Uniting the estimates of S_1 and S_2 we obtain that under the conditions of the theorem

$$p_{n,T}(k) = \mathbf{P}\{\nu_{n,T} = k,\ G_{n,T} \in B_{n,T},\ C_{n,T}\} = \frac{(B(x))^k\sqrt{1-\lambda}}{2^k k!}(1+o(1)). \quad (8)$$

From here we obtain the assertion of the theorem for $k \geq 1$, since

$$x = a^{-a} = (2T/n)e^{-2T/n} \to \lambda e^{-\lambda} = \alpha$$

and $B(x) \to B(\alpha)$. With the help of Lemma 1 and the representation from Lemma 5 we find that

$$p_{n,T}(0) = \sqrt{1-\lambda}\,(1+o(1)). \quad (9)$$

The result concerning the probability $p_{n,T}$ is an immediate consequence of (8) and (9). Denote $p_{n,T}^{(\mathbf{N})}(k)$ the probability that the graph $G_{n,T}$ has k components with exactly one cycle and has no unmarked cycle provided that all the probabilities p_r are equal to one, i.e., the admissible set R coincides with $\mathbf{N} = \{1, 2, \ldots\}$. In that case

$$B(\alpha) = b(\alpha) = 2\Lambda = -\log(1-\lambda), \quad (10)$$

and in accordance with (8) and (9)

$$p_{n,T}^{(\mathbf{N})}(k) \to \frac{\Lambda^k e^{-\Lambda}}{k!}, \qquad k = 0, 1, \ldots.$$

In order to find the asymptotic behaviour of $p_{n,T}$ it is sufficient to take the limit under the sum sign in

$$p_{n,T} = \sum_{k=1}^{\infty} p_{n,T}(k). \quad (11)$$

Let us check whether this operation is valid. It is sufficient to see that for an arbitrary $\varepsilon > 0$ there exists K such that

$$\sum_{k=K+1}^{\infty} p_{n,T}(k) \le \varepsilon. \tag{12}$$

If we choose K such that

$$\sum_{k=K+1}^{\infty} \frac{\Lambda^k e^{-\Lambda}}{k!} \le \frac{\varepsilon}{2}$$

and for the fixed K we choose the parameter n_0 such that for $n \ge n_0$

$$\left| \sum_{k=0}^{K} p_{n,T}^{(\mathrm{N})}(k) - \sum_{k=0}^{K} \frac{\Lambda^k e^{-\Lambda}}{k!} \right| \le \frac{\varepsilon}{2},$$

then for $n \ge n_0$

$$\left| \sum_{k=K+1}^{\infty} p_{n,T}^{(\mathrm{N})}(k) - \sum_{k=K+1}^{\infty} \frac{\Lambda^k e^{-\Lambda}}{k!} \right| = \left| \sum_{k=0}^{K} p_{n,T}^{(\mathrm{N})}(k) - \sum_{k=0}^{K} \frac{\Lambda^k e^{-\Lambda}}{k!} \right| \le \frac{\varepsilon}{2},$$

and consequently,

$$\sum_{k=K+1}^{\infty} p_{n,T}^{(\mathrm{N})}(k) \le \varepsilon.$$

Since $p_{n,T}(k) \le p_{n,T}^{(\mathrm{N})}(k)$, we obtain estimate (12) and prove the possibility to interchange the limit and sum signs in (11).

REFERENCES

Balakin, G.V., Kolchin, V.F., and Khokhlov, V.I. (1991). Hypercycles in a random hypergraph. *Discrete Math. Appl.* **3**, 563–570.

Britikov, V.E. (1988). Asymptotics of the number of forests of nonrooted trees. *Math. Notes.* **43**, 672–684.

Erdős, P. and Rényi, A. (1960). On the evolution of random graphs. *Publ. Math. Inst. Hungar. Acad. Sci. Ser. A.* **5**, 17–61.

Kolchin, V.F. (1986a). On the behaviour of a random graph near a critical point. *Theory Probab. Appl.* **31**, 439–451.

Kolchin, V.F. (1986b). *Random Mappings.* Optimization Software, New York.

Kolchin, V.F. (1990). On the number of cycles in a random graph. In: *Probability Problems of Discrete Mathematics*, MIEE, Moscow, pp. 3–8.

Kolchin, V.F., and Khokhlov, V.I. (1990). On the number of cycles in nonequiprobable random graphs. *Discrete Math. Appl.* **2**, 137–145.

Pólya, G., and Szegő, G. (1964). *Aufgaben und Lehrsätze aus der Analysis.* Springer–Verlag, Berlin.

Riordan, J. (1968). *Combinatorial Identities.* John Wiley, New York.

Takács, L. (1988). On the limit distribution of the number of cycles in a random graph. *J. Appl. Probab.* **25A**, 359–376.

Frontiers in Pure and Applied Probab. 1, pp. 112 – 134
H. Niemi et al. (Eds.)
1993 VSP/TVP

LOCAL LIMIT THEOREM ON CONVERGENCE OF MARKOV CHAINS TO DIFFUSION PROCESSES

V. D. KONAKOV
Central Economics and Mathematical Institute, Krasikov Street 32, 117421 Moscow, Russia

ABSTRACT

Let $X^{(\Delta)}$ be a family of Markov chains in $\mathbf{R}^\nu$ whose distributions converge weakly as $\Delta \to 0$ to the distribution of a diffusion process X. Wide sufficient conditions are given which ensure convergence in variation of the finite-dimensional distributions of $X^{(\Delta)}$ to those of X. The "parametrix" method is used in combination with multi-dimensional local theorems.

The goal of the present paper is to establish local theorems for natural families of Markov chains, i.e., to prove the uniform convergence of the densities of their finite-dimensional distributions to the corresponding densities of finite-dimensional distributions of some nondegenerate diffusion process. Theorems on the weak convergence of the distributions of a family of Markov chains to the distributions of diffusion processes were first considered by Khinchin (Khinchin, 1936) and in the general form by Skorokhod (Skorokhod, 1961) (see also the extensive bibliography in (Gikhman and Skorokhod, 1982)). The weak convergence of the distributions of a sequence of semimartingales to the distributions of diffusion type processes were obtained in (Liptser and Shiryaev, 1983) (the convergence of Markov chains was considered there as a special case).

The main result of the present paper (see Theorem) contains sufficient conditions for the uniform convergence of the transition densities of Markov chains to the transition densities of the limit diffusion process. The considered family of Markov chains is large enough and the main result has many applications (we mention only applications to conditional Markov chains, to induced order statistics, to stochastic approximation and to local invariance principles). To obtain our main result we use the "parametrix" technique well-known in the theory of partial differential equations, but apparently not used before in the given context. The main idea of this method is in the following: it is well-known that the transition density $p(s,t,x,y)$ of a Markov diffusion process being the fundamental solution of the Cauchy problem for the parabolic equation (2), may be represented in the series form (8), where the main term is the gaussian transition density $\tilde{p}^{(0)}(s,t,x,y)$ which is the fundamental solution of Cauchy problem (5) for the equation with coefficients "frozen" at the point y. For the L-step transition density $p(0, L\Delta, x, y)$ of a Markov chain, we obtain a discrete analogue of (8) (series (20)), where the main term is the density $p^{(0)}(0, L\Delta, x, y)$ of the sum of L independent random vectors with the mean vector and the covariance matrix "frozen" at the point y. The main difficulty is to prove that we can majorize series (20) and then apply the dominated convergence theorem. The main technique used is asymptotic expansions in multidimensional local limit theorems.

Suppose that a stochastic differential equation

$$dy_t = \sigma(t, y_t)dW_t + c(t, y_t)dt, \qquad y_0 = x, \quad t \in [0, T] \tag{1}$$

is given in $\mathbf{R}^\nu$ and there exist constants $C_i > 0$, $i = 1, 2$, $0 < \alpha \leq 1$, and $\varepsilon > 0$ such that for the vector field $c(t, x)$ and the matrix field $a(t, x)$ the following conditions hold:

1) $\|a_{ij}(t, x_1) - a_{ij}(t, x_2)\| < C_1|x_1 - x_2|^\alpha$, $|c_i(t, x_1) - c_i(t, x_2)| \leq C_1|x_1 - x_2|^\alpha$, $i, j = 1, 2, \ldots, \nu$, $\alpha > 0$, $C_1 < \infty$, $a(u, x) = \sigma(u, x)\sigma^*(u, x)$.
2) $\sup_{t,x} \|a(t, x)\| \leq C_2$, $\sup_{t,x} |c(t, x)| \leq C_2$, $C_2 < \infty$.
3) $\forall \theta \in \mathbf{R}^\nu$, $\langle \theta, a(t, x)\theta\rangle \geq \varepsilon|\theta|^2$, $\varepsilon > 0$, $(t, x) \in [0, T] \times \mathbf{R}^\nu$, $a(t, x) = a^*(t, x)$.
4) The matrix field $a(t, \cdot\,)$ and the vector field $c(t, \cdot\,)$ are continuous in t.

It is well-known that there exists, unique strictly, positive density function $p(u, t, x, z)$, $0 \leq u < t$, which is a solution of the Cauchy problem for the parabolic equation

$$\frac{\partial p}{\partial u} + L_u p = 0, \qquad 0 \leq u < t, \quad p(u, u, x, z) = \delta(x - z), \tag{2}$$

where $L_u = 1/2(a(u, x)\nabla, \nabla) + (c(u, x), \nabla)$,

$$\nabla = (\partial/\partial x_1, \ldots, \partial/\partial x_\nu), \qquad a(u, x) = \sigma(u, x)\sigma^*(u, x).$$

We now consider the lattice time $t = 0, \Delta, 2\Delta, \ldots, n\Delta = T$ with

5) $\Delta \to 0$.

Let $p_{t,x}(\cdot\,)$ be a family of densities in $\mathbf{R}^\nu$, depending on the parameter (t, x), $t \in [0, \infty)$, $x \in \mathbf{R}^\nu$, and

6) $\int p_{t,x}(u)u\,du = 0$, $\forall (t, x) \in [0, \infty) \times \mathbf{R}^\nu$, $\int p_{t,x}(u)u_i u_j\,du = a_{ij}(t, x)$, $i, j = 1, 2, \ldots, \nu$, $a(t, x) = \sigma(t, x)\sigma^*(t, x)$.
7) There exists a function $\Psi(x)$, $x \in \mathbf{R}^\nu$, with $\operatorname{ess\,sup}_x |\Psi(x)| < \infty$, $\int |x|^{4\nu+1} \times \Psi(x)dx < \infty$ and $p_{t,x}(u) \leq \Psi(u)$, $\forall (t, x)$.
8) $|p_{t,z}(u) - p_{t,y}(u)| \leq |z - y|^\alpha \Psi(u)$, $\alpha > 0$.

Suppose that the one-step transition densities have the form

$$p^{(\Delta)}(i\Delta, (i+1)\Delta, x, z) = \Delta^{-\nu/2} p_{i\Delta,x}\Big[\big(z - x - c(i\Delta, x)\Delta\big)/\Delta^{1/2}\Big]. \tag{3}$$

We now formulate the main result of the paper.

THEOREM. *Suppose that a sequence of Markov chains has one-step transition density (3), $\Delta \to 0$ and $p_{t,x}(\cdot\,)$ satisfy (6)–(8). Suppose also that a matrix field $a(t, x)$ and a vector field $c(t, x)$ satisfy 1)–4). Then*

$$\lim_{\Delta \to 0} p^{(\Delta)}(0, T, x, y) = p(0, T, x, y),$$

where $p(0,T,x,y)$ is a solution of the Cauchy problem for parabolic equation (2). The convergence is uniform on compact subsets of $\mathbf{R}^{2\nu}$.

To prove this theorem, we consider an approximating diffusion process

$$dx_t^0 = \sigma(t,y)dW_t + c(t,y)dt, \quad t \in [0,T], \tag{4}$$

whose transition probability has a positive density $\bar{p}^{(0)}(u,t,x,z)$ which is a solution of the Cauchy problem for the parabolic equation

$$\frac{\partial \bar{p}^{(0)}}{\partial u} + L_u^{(0)} \bar{p}^{(0)} = 0, \quad 0 \le u < t,$$

$$\bar{p}^{(0)}(t,t,x,z) = \delta(x-z), \qquad L_u^{(0)} = \frac{1}{2}\big(a(u,y)\nabla,\nabla\big) + \big(c(u,y),\nabla\big). \tag{5}$$

The densities $p(0,T,x,y)$ and $\bar{p}^{(0)}(0,T,x,y)$ are related by the "parametrix" integral equation

$$p(0,T,x,y) - \bar{p}^{(0)}(0,T,x,y) = \int_0^T du \int_{\mathbf{R}^\nu} p(0,u,x,z)(L_u - L_u^{(0)})\bar{p}^{(0)}(u,T,z,y)dz. \tag{6}$$

The validity of (6) follows from (2) and (5):

$$p(0,T,x,y) - \bar{p}^{(0)}(0,T,x,y) = \int_0^T du \frac{\partial}{\partial u} \int_{\mathbf{R}^\nu} p(0,u,x,z)\bar{p}^{(0)}(u,T,z,y)dz$$

$$= \int_0^T du \int_{\mathbf{R}^\nu} \big[L_u^* p(0,u,x,z)\bar{p}^{(0)}(u,T,z,y) - L_u^{(0)}\bar{p}^{(0)}(u,T,z,y)p(0,u,x,z)\big]dz$$

$$= \int_0^T du \int_{\mathbf{R}^\nu} p(0,u,x,z)(L_u - L_u^{(0)})\bar{p}^{(0)}(u,T,z,y)dz.$$

The fundamental solution $\bar{p}^{(0)}(u,t,x,y)$, $u < t$, of the Cauchy problem (5) has the following form

$$\bar{p}^{(0)}(u,t,x,y) = (2\pi)^{-\nu/2}(\det\Sigma_{u,t})^{-1/2} \exp\Big\{-\frac{1}{2}\big(y - m_{u,t}(x)\big), \Sigma_{u,t}^{-1}\big(y - m_{u,t}(x)\big)\Big\},$$

where

$$m_{u,t}(x) = x + \int_u^t c(v,y)dv, \qquad \Sigma_{u,t} = \int_u^t a(v,y)dv. \tag{7}$$

Let us introduce the kernel $\overline{H}^{(0)}(u,T,z,y) = (L_u - L_u^{(0)})\,\bar{p}^{(0)}(u,T,z,y)$ and the convolution operation "#"

$$f \# g(0,u,x,y) = \int_0^u dv \int_{\mathbf{R}^\nu} f(0,v,x,z)g(v,T,z,y)dz,$$

then by iterating we get the (formal) Levi sum for p (see (McKean and Singer, 1967))

$$p(0,T,x,y) = \bar{p}^{(0)}(0,T,x,y) + \sum_{m\geq 1} \bar{p}^{(0)} \# \underbrace{\overline{H}^{(0)} \# \ldots \# \overline{H}^{(0)}}_{m\text{-fold}}(0,T,x,y). \quad (8)$$

LEMMA 1. *Under conditions 1)–4) series* (8) *converges uniformly on compact subsets of* $\mathbf{R}^{2\nu}$ *and is a solution of Cauchy problem* (2). *For the kernel* $\overline{H}^{(0)}$, *the terms of* (8), *and* $p(0,T,x,y)$ *the following estimates are valid*

$$|\overline{H}^{(0)}(u,T,z,y)| \leq C_1(T-u)^{-(\nu+2-\alpha)/2} \exp\Big\{ -C_2|y-m_{u,T}(z)|^2(T-u)^{-1}\Big\};$$

$$\Big|\bar{p}^{(0)} \# \underbrace{\overline{H}^{(0)} \# \ldots \# \overline{H}^{(0)}}_{m\text{-fold}}(0,T,x,y)\Big| \leq C_3^m T^{-\nu/2+m\alpha/2}$$

$$\times \exp\Big\{ -C_4|y-m_{0,T}(x)|^2/T\Big\}\Big/\Gamma\Big(1+m\frac{\alpha}{2}\Big);$$

$$|p(0,T,x,y)| \leq \sum_{m\geq 0} C_3^m T^{m\alpha/2-\nu/2} \exp\Big\{ -C_4|y-m_{0,T}(x)|^2/T\Big\}\Big/\Gamma\Big(1+m\frac{\alpha}{2}\Big).$$

Proof. It follows from (7) and (2), (3) that

$$C_1(t-u)^{-1}\|\theta\|^2 \leq \langle \Sigma_{u,t}^{-1}\theta,\theta\rangle \leq C_2(t-u)^{-1}\|\theta\|^2, \quad (9)$$

where $t \in K$, K is a compact subset of $\mathbf{R}$, C_1 and C_2 depend only on K. We have now

$$\overline{H}^{(0)}(u,T,z,y) = (L_u - L_u^{(0)})\bar{p}^{(0)}(u,T,z,y) = 2^{-1}\sum_{i,j=1}^{\nu}(a_{ij}(u,z)$$

$$-a_{ij}(u,y))\frac{\partial^2 \bar{p}^{(0)}(u,T,z,y)}{\partial z_i \partial z_j} + \sum_{i=1}^{\nu}(c_i(u,z)-c_i(u,y))\frac{\partial \bar{p}^{(0)}(u,T,z,y)}{\partial z_i}. \quad (10)$$

We have from 1) and 2)

$$\big\|a_{ij}(u,z)-a_{ij}(u,y)\big\| \leq \Big\|a_{ij}(u,y)-a_{ij}\big(u,m_{u,T}(z)\big)\Big\|$$

$$+\Big\|a_{ij}(u,m_{u,T}(z))-a_{ij}(u,z)\Big\| \leq C\Big[|y-m_{u,T}(z)|^\alpha + |m_{u,T}(z)-z|^\alpha\Big], \quad (11)$$

and, analogously,

$$|c_i(u,z)-c_i(u,y)| \leq C\Big[|y-m_{u,T}(z)|^\alpha + |m_{u,T}(z)-z|^\alpha\Big]. \quad (12)$$

It follows from 2) and 7) that

$$|m_{u,T}-z| \leq C|T-u|, \quad (13)$$

where C depends only on T. By the definition of $\bar{p}^{(0)}$,

$$\frac{\partial \bar{p}^{(0)}(u,T,z,y)}{\partial z_i} = (2\pi)^{-\nu/2}(\det\Sigma_{u,T})^{-1/2}\exp\Big\{-2^{-1}(y$$

$$-m_{u,T}(z))^*\Sigma_{u,T}^{-1}(y-m_{u,T}(z))\Big\}\sum_{k=1}^{\nu}\sigma_{u,T}^{ik}(y_k-m_{u,T}^k(z)), \tag{14}$$

and

$$\frac{\partial^2 \bar{p}^{(0)}(u,T,z,y)}{\partial z_i\partial z_j} = (2\pi)^{-\nu/2}(\det\Sigma_{u,T})^{-1/2}\exp\Big\{-2^{-1}(y$$

$$-m_{u,T}(z))^*\Sigma_{u,T}^{-1}(y-m_{u,T}(z))\Big\}\Big[\sum_{l=1}^{\nu}\sigma_{u,T}^{il}(y_l-m_{u,T}^l(z))\sum_{k=1}^{\nu}\sigma_{u,T}^{ik}(y_k$$

$$-m_{u,T}^k(z))-\sigma_{u,T}^{ij}\Big], \tag{15}$$

where $\Sigma_{u,t}^{-1}=\|\sigma_{u,t}^{ij}\|$, $i,j=1,\dots,\nu$, $m_{u,t}(z)=\big(m_{u,t}^1(z),\dots,m_{u,t}^\nu(z)\big)$.
It follows from (9)–(15) and from (2), (7) that

$$|\overline{H}^{(0)}(u,T,z,y)| \le C\Big(|y-m_{u,T}(z)|^\alpha+|T-u|^\alpha\Big)$$

$$\times\Big[|y-m_{u,T}(z)|(T-u)^{-1}+|y-m_{u,T}(z)|^2(T-u)^{-2}+(T-u)^{-1}\Big]$$

$$\times(T-u)^{-\nu/2}\exp\Big\{-C_1|y-m_{u,T}(z)|^2(T-u)^{-1}\Big\}\le C(T-u)^{-(\nu+2-\alpha)/2}$$

$$\times\exp\{-C_1|v|^2\}\big[|v|^{1+\alpha}|T-u|^{1/2}+|v|^{2+\alpha}+|v|^\alpha+|v|(T-u)^{(\alpha+1)/2}$$

$$+v^2(T-u)^{\alpha/2}+(T-u)^{\alpha/2}\big]\le C_2(T-u)^{-(\nu+2-\alpha)/2}$$

$$\times\exp\Big\{-C_3|y-m_{u,T}(z)|^2/(T-u)\Big\}, \tag{16}$$

where $v=\big(y-m_{u,T}(z)\big)/(T-u)^{1/2}$. A direct computation easily gives the estimate

$$\bar{p}^{(0)}(0,u,x,z)\le Cu^{-\nu/2}\exp\Big\{-C|z-m_{0,u}(x)|^2u^{-1}\Big\}. \tag{17}$$

It follows from this estimate and (16), (17), (9) that

$$|\bar{p}^{(0)}\#\overline{H}^{(0)}(0,T,x,y)|\le\int_0^T du\int_{\mathbf{R}^\nu}|\bar{p}^{(0)}(0,u,x,z)\overline{H}^{(0)}(u,T,z,y)|dz$$

$$\le C_1\int_0^T du\int_{\mathbf{R}^\nu}(T-u)^{\alpha/2-1}(\det\Sigma_{u,T})^{-1/2}$$

$$\times\exp\Big\{-C\Big\langle\Sigma_{u,T}^{-1}(y-m_{u,T}(z)),(y-m_{u,T}(z))\Big\rangle\Big\}$$

$$\times(\det\Sigma_{0,u})^{-1/2}\exp\Big\{-C\Big\langle\Sigma_{0,u}^{-1}(z-m_{0,u}(x)),(z-m_{0,u}(x))\Big\rangle\Big\}dz$$

$$=(2C)^{\nu/2}C_1\int_0^T(T-u)^{\alpha/2-1}T^{-\nu/2}\exp\Big\{-C\Big\langle\Sigma_{0,T}^{-1}(y$$

$$-m_{0,T}(x)),(y-m_{0,T}(x))\Big\rangle\Big\}du=C_4B(1,\alpha/2)T^{(\alpha-\nu)/2}$$

$$\times\exp\Big\{-C\Big\langle\Sigma_{0,T}^{-1}(y-m_{0,T}(x)),(y-m_{0,T}(x))\Big\rangle\Big\}. \tag{18}$$

To obtain (18) we used the Chapman–Kolmogorov equation for Gaussian family with parameters defined in (7) and made the change of variables $v=1-u/T$. We have now from (18) and (16)

$$|\,\bar p^{(0)}\#\overline{H}^{(0)}\#\overline{H}^{(0)}(0,T,x,y)|\le\int_0^T u^{\alpha/2}T^{\alpha/2-1}du\,C_4^2B(1,\alpha/2)$$

$$\times T^{-\nu/2}\exp\Big\{-C\Big\langle\Sigma_{0,T}^{-1}(y-m_{0,T}(x)),(y-m_{0,T}(x))\Big\rangle\Big\}=C_4^2B(1,\alpha/2)$$

$$\times T^{\alpha-\nu/2}\int_0^1 v^{\alpha/2}(1-v)^{\alpha/2-1}dv\exp\Big\{-C\Big\langle\Sigma_{0,T}^{-1}(y-m_{0,T}(x)),$$

$$(y-m_{0,T}(x))\Big\rangle\Big\}=C_4^2\Gamma^2(\alpha/2)T^{\alpha-\nu/2}\big(\Gamma(1+\alpha)\big)^{-1}$$

$$\times\exp\Big\{-C\Big\langle\Sigma_{0,T}^{-1}(y-m_{0,T}(x)),(y-m_{0,T}(x))\Big\rangle\Big\}.$$

From this it is not hard to deduce by induction that

$$\Big|\bar p^{(0)}\#\underbrace{\overline{H}^{(0)}\#\ldots\#\overline{H}^{(0)}}_{m\text{-fold}}(0,T,x,y)\Big|\le C_5^mT^{-\nu/2+m\alpha/2}$$

$$\times\exp\Big\{C\big|y-m_{0,T}(x)\big|^2T^{-1}\Big\}\Big/\Gamma\Big(1+m\frac{\alpha}{2}\Big);$$

and hence

$$|p(0,T,x,y)|\le\sum_{m\ge0}C_5^mT^{m\alpha/2-\nu/2}$$

$$\times\exp\Big\{-C\big|y-m_{0,T}(x)\big|^2T^{-1}\Big\}\Big/\Gamma\Big(1+m\frac{\alpha}{2}\Big). \tag{19}$$

This completes the proof.

We now consider two families of Markov chains $x_t^{(n)}$ and $y_t^{(n)}$ in $\mathbf{R}^\nu$ with lattice time $0, \Delta, \dots, n\Delta$ and one-step transition densities $p^{(\Delta)}(\cdot,\cdot,x,y)$ and $p^{(0)}(\cdot,\cdot,x,y)$. Let us introduce the operators

$$A_{s,t}f(x) = \left[\int p^{(\Delta)}(s,t,x,u)f(u)du - f(x)\right]/(t-s),$$

$$A_{s,t}^{(0)}f(x) = \left[\int p^{(0)}(s,t,x,u)f(u)du - f(x)\right]/(t-s),$$

$$A_{s,t}^{*}f(u) = \left[\int p^{(\Delta)}(s,t,x,u)f(x)dx - f(u)\right]/(t-s),$$

$$(A_{s,t}^{(0)})^{*}f(u) = \left[\int p^{(0)}(s,t,x,u)f(x)dx - f(u)\right]/(t-s).$$

We get identities that are discrete analogues of Lévy series (8):

$$p^{(\Delta)}(0,n\Delta,x,y) - p^{(0)}(0,n\Delta,x,y)$$

$$= \sum_{j=0}^{n-1} \Delta \int_{\mathbf{R}^\nu} \left[p^{(0)}(j\Delta,n\Delta,z,y)A_{j\Delta,(j+1)\Delta}^{*}p^{(\Delta)}(0,j\Delta,x,z)\right.$$

$$\left.-p^{(\Delta)}(0,(j+1)\Delta,x,z)A_{j\Delta,(j+1)\Delta}^{(0)}p^{(0)}((j+1)\Delta,n\Delta,z,y)\right]dz$$

$$\times \sum_{j=0}^{n-1} \Delta \int_{\mathbf{R}^\nu} \left[A_{j\Delta,(j+1)\Delta}p^{(0)}(j\Delta,n\Delta,z,y)p^{(\Delta)}(0,j\Delta,x,z)\right.$$

$$\left.-A_{j\Delta,(j+1)\Delta}^{(0)}p^{(0)}((j+1)\Delta,n\Delta,z,y)p^{(\Delta)}(0,(j+1)\Delta,x,z)\right]dz$$

$$= \Delta \int_{\mathbf{R}^\nu} A_{0,\Delta}p^{(0)}(0,n\Delta,z,y)p(0,0,x,z)dz$$

$$-\Delta \int_{\mathbf{R}^\nu} A_{(n-1)\Delta,n\Delta}^{(0)}p^{(0)}(n\Delta,n\Delta,z,y)p^{(\Delta)}(0,n\Delta,x,z)dz$$

$$+\sum_{j=1}^{n-1} \int_{\mathbf{R}^\nu} [\Delta A_{j\Delta,(j+1)\Delta} - \Delta A_{(j-1)\Delta,j\Delta}^{(0)}]p^{(0)}(j\Delta,n\Delta,z,y)p^{(\Delta)}(0,j\Delta,x,z)dz$$

$$= \int_{\mathbf{R}^\nu} p^{(\Delta)}(0,\Delta,x,u)p^{(0)}(0,n\Delta,u,y)du - p^{(0)}(0,n\Delta,x,y)$$

$$- \int_{\mathbf{R}^\nu} p^{(0)}((n-1)\Delta,n\Delta,u,y)p^{(\Delta)}(0,n\Delta,x,u)du + p^{(\Delta)}(0,n\Delta,x,y)$$

$$+\sum_{j=1}^{n} \left[\int_{\mathbf{R}^\nu} (\Delta A_{j\Delta,(j+1)\Delta} - \Delta A_{(j-1)\Delta,j\Delta}^{(0)})p^{(0)}(j\Delta,n\Delta,z,y)p^{(\Delta)}(0,j\Delta,x,z)dz\right]$$

$$-p^{(\Delta)}(0,(n+1)\Delta,x,y)+\int\limits_{\mathbf{R}^\nu} p^{(0)}((n-1)\Delta,n\Delta,u,y)p^{(\Delta)}(0,n\Delta,x,u)du$$

and hence

$$p^{(\Delta)}(0,(n+1)\Delta,x,y)-p^{(0)}(0,(n+1)\Delta,x,y)$$
$$=\overline{H}^{(0)}\oplus p^{(\Delta)}(0,n\Delta,x,y)+R(0,(n+1)\Delta,x,y),$$

where

$$f\oplus g(0,n\Delta,x,y)=\sum_{j=1}^{n}\Delta\int\limits_{\mathbf{R}^\nu} g(0,j\Delta,x,z)f(j\Delta,n\Delta,z,y)dz,$$

$$R(0,(n+1)\Delta,x,y)=\int p^{(\Delta)}(0,\Delta,x,u)p^{(0)}(0,n\Delta,u,y)du-p^{(0)}(0,(n+1)\Delta,x,y),$$

$$\overline{H}^{(0)}(j\Delta,n\Delta,z,y)=[A_{j\Delta,(j+1)\Delta}-A^{(0)}_{(j-1)\Delta,j\Delta}]p^{(0)}(j\Delta,n\Delta,z,y),\quad j=1,2,\ldots.$$

By iterating the last relation, we get the discrete analogue of (8)

$$p^{(\Delta)}(0,L\Delta,x,y)=(p^{(0)}+R)(0,L\Delta,x,y)$$

$$+\sum_{m=1}^{L}\underbrace{\overline{H}^{(0)}\oplus\ldots\oplus\overline{H}^{(0)}}_{m\text{-fold}}\oplus(p^{(0)}+R)(0,(L-m)\Delta,x,y). \tag{20}$$

We now consider a trajectory of the Markov chain $y_t^{(n)}$, $t=0,\Delta,2\Delta,\ldots$, and let Y_i be the position of a Markov particle at the moment $i\Delta$, $i=1,2,\ldots$, $Y_0=x$. Then

$$Y_1=x+c(\Delta,y)\Delta+\xi_{\Delta,y},$$
$$Y_2=Y_1+c(2\Delta,y)\Delta+\xi_{2\Delta,y}=x+c(\Delta,y)\Delta+\xi_{\Delta,y}+c(2\Delta,y)\Delta+\xi_{2\Delta,y},$$
$$Y_3=Y_2+c(3\Delta,y)\Delta+\xi_{3\Delta,y}=x+c(\Delta,y)\Delta+\xi_{\Delta,y}$$
$$+c(2\Delta,y)\Delta+\xi_{2\Delta,y}+c(3\Delta,y)\Delta+\xi_{3\Delta,y}.$$

From this it is not hard to deduce by induction that

$$Y_n=x+\sum_{i=1}^{n}c(i\Delta,y)\Delta+\sum_{i=1}^{n}\xi_{i\Delta,y}.$$

The random vectors $\xi_{k\Delta,y}$ are conditionally independent (conditioned by $Y_0=x$) and have the conditional density

$$\Delta^{-\nu/2}p_{k\Delta,y}(u/\Delta^{1/2}),\qquad k=1,2,\ldots,n.$$

Denoting $D_z^\nu=\partial^\nu/\partial z_1\partial z_2\ldots\partial z_\nu$, we have, for example, for $k=1$

$$D_z^\nu\mathbf{P}\{\xi_{\Delta,y}<z|Y_0=x\}=D_z^\nu\mathbf{P}_x\{Y_1<z+x+c(\Delta,y)\Delta\}$$

$$= p^{(0)}(0, \Delta, x, z + x + c(\Delta, y)\Delta) = \Delta^{-\nu/2} p_{\Delta,y}(z/\Delta^{1/2}).$$

The independence of $\xi_{i\Delta,y}$ and $\xi_{j\Delta,y}$ follows from the Chapman–Kolmogorov equation. For example,

$$D_z^\nu \mathbf{P}\{\xi_{\Delta,y} + \xi_{2\Delta,y} < z | Y_0 = x\} = p^{(0)}(0, 2\Delta, x, z + x + c(\Delta, y)\Delta + c(2\Delta, y)\Delta)$$

$$= \int_{\mathbf{R}^\nu} \Delta^{-\nu/2} p_{\Delta,y}\Big[(u - x - c(\Delta, y)\Delta)\Delta^{-1/2}\Big]\Delta^{-\nu/2}$$

$$\times p_{2\Delta,y}\Big[(z + x + c(\Delta, y)\Delta - u)/\Delta^{1/2}\Big] du$$

$$= \int_{\mathbf{R}^\nu} \Delta^{-\nu/2} p_{\Delta,y}[\Delta^{-1/2}(z - \mu)]\Delta^{-\nu/2} p_{2\Delta,y}[\mu\Delta^{-1/2}] d\mu.$$

The independence of the others $\xi_{\cdot,y}$'s may be proved quite similarly.
Denoting by $m_{j,n}(x)$ the sum $x + \sum_{i=j}^n c(i\Delta, y)\Delta$, we have

$$Y_n = m_{1,n}(x) + \sum_{i=1}^n \xi_{i\Delta,y}.$$

Using the conditions imposed on the vector field $c(t, x)$, one can easily obtain

$$\lim_{\Delta \to 0} \big| m_{j,n}(x) - m_{j\Delta,n\Delta}(x) \big| = 0. \tag{21}$$

The covariance matrix of Y_n (conditioned by $Y_{j-1} = x$) is equal to

$$V_{j,n} = \sum_{i=j}^n a(i\Delta, y)\Delta, \qquad j = 1, 2, \ldots,$$

and the conditions imposed on the matrix field $a(t, x)$ imply

$$\lim_{\Delta \to 0} \| V_{j,n} - \Sigma_{j\Delta,n\Delta} \| = 0. \tag{22}$$

Our next step is to show that the terms of series (20) are integral sums of the corresponding terms of (8). We shall prove that series (20) (with $L = n$) is dominated by a series not depending on Δ. Thus, passing to the limit (as $\Delta \to 0$) termwise in (20), we obtain the conclusion of our theorem.
Using (21) and (22), we obtain that $p^{(\Delta)}(\cdot, \cdot, x, y)$ in (20) converges to the Gaussian density $\bar{p}^{(0)}(\cdot, \cdot, x, y)$ in (8), moreover, from Taylor's expansion and from condition 6) on $p_{t,x}$, we have for $f \in \mathbf{C}^2(\mathbf{R}^\nu)$ and $t - s = \Delta$ (t and s are lattice points)

$$A_{s,t} f = \Delta^{-1} \int p^{(\Delta)}(s, t, x, u)(f(u) - f(x)) du$$

$$= \int p_{s,x}(v)\Big[f(x + c(s, x)\Delta + v\Delta^{1/2}) - f(x)\Big] dv \Delta^{-1} = L_s f + O(\Delta^{1/2}),$$

where

$$L_s = \frac{1}{2}\langle a(s,x)\nabla, \nabla\rangle + \langle c(s,x), \nabla\rangle, \qquad \nabla = \Big(\frac{\partial}{\partial x_1}, \dots, \frac{\partial}{\partial x_\nu}\Big).$$

Similarly, $A^{(0)}_{s,t} f = L^{(0)}_s f + O(\Delta^{1/2})$ and, hence,

$$H^{(\Delta)}(j\Delta, (n-m)\Delta, z, y) = \big(L_{j\Delta} - L^{(0)}_{(j-1)\Delta} + O(\Delta^{1/2})\big)p^{(0)}(j\Delta, (n-m)\Delta, z, y)$$

as $\Delta \to 0$. Let us now consider $H^{(\Delta)} \oplus (p^{(0)} + R)(0, (n-1)\Delta, x, y)$. We find that

$$\begin{aligned} &H^{(\Delta)} \oplus (p^{(0)} + R)(0, (n-1)\Delta, x, y) \\ &= \sum_{j=1}^{n-1} \Delta \int_{\mathbf{R}^\nu} (p^{(0)} + R)(0, j\Delta, x, z) \times H^{(\Delta)}(j\Delta, n\Delta, z, y)dz \\ &\to \int_0^T du \int_{\mathbf{R}^\nu} p^{(0)}(0, u, x, z)H^{(0)}(u, T, z, y)dz \\ &= \bar{p}^{(0)} \# \overline{H}^{(0)}(0, T, x, y) \quad \text{as} \quad \Delta \to 0. \end{aligned}$$

(It follows from the definition of $R(\cdot)$ that $R(0, j\Delta, x, z) \to 0$ as $\Delta \to 0$ if $j\Delta > \delta > 0$.) The convergence of the other terms of series (20) to the corresponding terms of (8) can be established similarly. For example,

$$H^{(\Delta)} \oplus H^{(\Delta)} \oplus (p^{(0)} + R)(0, (n-2)\Delta, x, y) = \sum_{i=1}^{n-2} \Delta \int_{\mathbf{R}^\nu} \sum_{k=1}^{i} \Delta \int_{\mathbf{R}^\nu} (p^{(0)} + R)(0, k\Delta, x, \rho)$$

$$\times H^{(\Delta)}(k\Delta, n\Delta, r, z)dr H^{(\Delta)}(i\Delta, n\Delta, z, y)dz \to \bar{p}^{(0)} \# \overline{H}^{(0)} \# \overline{H}^{(0)}(0, T, x, y).$$

Let us now construct a dominant series for (20).

By the definition of $\overline{H}^{(0)}(i\Delta, n\Delta, z, y)$ it satisfies the relation

$$\overline{H}^{(0)}(i\Delta, n\Delta, z, y) = \Delta^{-1}\Big[q^{(1)}_{j,n}(y|z) - q^{(2)}_{j,n}(y|z)\Big],$$

where the conditional density of $(X_n | X_{j-1} = z)$

$$q^{(1)}_{j,n}(y|z) = \int p^{(\Delta)}(j\Delta, (j+1)\Delta, z, u)p^{(0)}(j\Delta, n\Delta, u, y)du$$

corresponds to the trajectory $X_{j-1} = z$, $X_j = z + c(j\Delta, z)\Delta + \xi_{j\Delta, z}$, $X_{j+1} = X_j + c((j+1)\Delta, y)\Delta + \xi_{(j+1)\Delta}$, $X_n = X_{n-1} + c(n\Delta, y)\Delta + \xi_{n\Delta, y}$, and $q^{(2)}_{j,n}$ is the conditional density of $(Y_n | Y_{j-1})$

$$q^{(2)}_{j,n}(y|z) = \int p^{(0)}((j-1)\Delta, j\Delta, z, u)p^{(0)}(j\Delta, n\Delta, u, y)du.$$

We prove now the validity of the following asymptotic expansions for the densities $q_{j,n}^{(i)}(y|z)$, $i=1,2$:

$$q_{j,n}^{(i)}(y|z) = \det V_{j,n}^{-1/2} \sum_{r=0}^{4\nu-2} (n-j+1)^{-r/2} P_r(-\varphi:\ \{\bar\chi_{\beta,y}\}) \Big(V_{j,n}^{-1/2}(y-m_{j,n}(z))\Big)$$

$$+O\left((n-j+1)^{-(4\nu-1)/2} \det V_{j,n}^{-1/2}\left[1+\left|V_{j,n}^{-1/2}(y-m_{j,n}(z))\right|^{4\nu+1}\right]^{-1}\right), \qquad (23)$$

$$q_{j,n}^{(1)}(y|z) = \det (V_{j,n}+\delta V_{j,n})^{-1/2} \sum_{r=0}^{4\nu-2} (n-j+1)^{-r/2}$$

$$\times P_r(-\varphi:\ \{\bar\chi_{\beta,y}+\delta\bar\chi_\beta\}) \Big((V_{j,n}+\delta V_{j,n})^{-1/2}(y-m_{j,n}(z)-\delta m_{j,n})\Big)$$

$$+O\Bigg((n-j+1)^{-(4\nu-1)/2} \det (V_{j,n}+\delta V_{j,n})^{-1/2}$$

$$\times\left[1+\left|(V_{j,n}+\delta V_{j,n})^{-1/2}(y-m_{j,n}(z)-\delta m_{j,n})\right|^{4\nu+1}\right]^{-1}\Bigg), \qquad (24)$$

where

$$P_r(-\varphi:\ \{\bar\chi_{\beta,y}\})(u) = \widetilde P_r(-D:\ \{\bar\chi_{\beta,y}\})\varphi(u;0,E),$$

$$-D = (-D_1,\dots,-D_\nu), \qquad D_j = \partial/\partial u_j,$$

$$\widetilde P_r(-D:\ \{\bar\chi_{\beta,y}\}) = \sum_{m=1}^{r} (m!)^{-1}\Big\{ \sum_{j_1,\dots,j_m}{}^{*} \Big[\sum{}^{**} (\beta_1!\dots\beta_m!)^{-1}$$

$$\times \bar\chi_{\beta_1,y}\bar\chi_{\beta_m,y}(-D)^{\beta_1+\dots+\beta_m}\Big]\Big\},$$

$\varphi(u;m,\Sigma)$ is the density of the normal distribution with the vector m of means and with the covariance matrix Σ, $\sum^*$ is the summation over all positive integers $j_1,\dots,j_m$ such that $\sum_{i=1}^m j_i = r$, $\sum^{**}$ is the summation over m-tuples of non-negative integer vectors $(\beta_1,\dots,\beta_m)$ such that $|\beta_i| = j_i+2$ $(1\le i\le m)$, and $\widetilde P_0(-D:\ \{\bar\chi_{\beta,y}\}) = 1$. For fixed r, m, $(j_1,\dots,j_m)$, $(\beta_1,\dots,\beta_m)$, we have

$$(-D)^{\beta_1+\dots+\beta_m}\varphi(u;0,E) = G_{\beta_1+\dots+\beta_m}(u)\varphi(u;0,E),$$

where $G_{\beta_1+\dots+\beta_m}(u)$ is the Hermite polynomial in $u=(u_1,\dots,u_\nu)$ of total degree $|\beta_1+\dots+\beta_m| = \sum_{i=1}^m j_i + 2m = r+2m$. Finally,

$$\delta V_{j,n} = \big[a(j\Delta,z)-a(j\Delta,y)\big]\Delta, \qquad \delta m_{j,n} = \big[c(j\Delta,z)-c(j\Delta,y)\big]\Delta, \qquad (25)$$

$\bar\chi_{\beta,y}$ and $\bar\chi_{\beta,y}+\delta\bar\chi_\beta$ are the arithmetic means of the cumulants of order $\beta = (\beta_1,\dots,\beta_\nu)$, $3\le|\beta|\le 4\nu+1$, of the random vectors $V_{j,n}^{-1/2}\eta_{j+1,y},\dots,V_{j,n}^{-1/2}\eta_{n+1,y}$ and $(V_{j,n}+\delta V_{j,n})^{-1/2}\eta_{j+1,z}$, $(V_{j,n}+\delta V_{j,n})^{-1/2}\eta_{j+2,y},\dots,(V_{j,n}+\delta V_{j,n})^{-1/2}\eta_{n+1,y}$

respectively, where $\eta_{p,x} = ((n-j+1)\Delta)^{1/2}\xi_{(p-1)\Delta,x}/\Delta^{1/2}$, $p = j+1,\dots,n+1$. To establish the validity of (23) and (24), we should to verify that Theorem 19.3 from (Bhattacharya and Rao, 1976) is applicable. We consider the series of random vectors

$$(n-j+1)^{-1/2}V_{j,n}^{-1/2}(\eta_{j+1,y} + \ldots + \eta_{n+1,y})$$

and

$$(n-j+1)^{-1/2}(V_{j,n} + \delta V_{j,n})^{-1/2}(\eta_{j+1,z} + \ldots + \eta_{n+1,y}).$$

It follows from (2) and (3) that there exist $\underline{c}$ and $\bar{c}$ such that for any $\theta \in \mathbf{R}^\nu$ the following inequalities

$$\underline{c}(n-j+1)\Delta \le \langle V_{j,n}\theta, \theta\rangle/|\theta|^2 \le \bar{c}(n-j+1)\Delta,$$

$$\underline{c}(n-j+1)\Delta \le \langle (V_{j,n} + \delta V_{j,n})\theta, \theta\rangle/|\theta|^2 \le \bar{c}(n-j+1)\Delta, \tag{26}$$

hold, where the constants $\underline{c}$ and $\bar{c}$ depend only on T. Then (26) implies that all eigenvectors of the matrix $B_p^{1/2} = V_{j,n}^{-1/2}((n-j+1)\Delta)^{1/2}$ belong to some segment $[c_1, c_2]$; taking into account 7), we obtain

$$\sum_{p=j+1}^{n+1} \mathbf{E}\|V_{j,n}^{-1/2}\eta_{p,y}\|^{4\nu+1}/(n-j+1) \le C \sum_{p=j+1}^{n+1} \int \|v\|^{4\nu+1} p_{(p-1)\Delta,y}(v)dv/(n-j+1)$$

$$\le C \int \|v\|^{4\nu+1}\Psi(v)dv < \infty,$$

i.e., condition (19.27) from (Bhattacharya and Rao, 1976) holds. Similarly,

$$\Big[\mathbf{E}\big\|(V_{j,n} + \delta V_{j,n})^{-1/2}\eta_{j+1,z}\big\|^{4\nu+1}$$

$$+ \sum_{p=j+2}^{n+1} \mathbf{E}\big\|(V_{j,n} + \delta V_{j,n})^{-1/2}\eta_{p,y}\big\|^{4\nu+1}\Big]/(n-j+1) < \infty.$$

Let us show that condition (19.29) from (Bhattacharya and Rao, 1976) is satisfied (with $p = 2$). Let us consider the function $g_{l,n}(t)$,

$$g_{l,n}(t) = \prod_{p=l+1}^{l+2} \Big|\mathbf{E}\Big(\exp\{i\langle t, V_{j,n}^{-1/2}\eta_{j+p,y}\rangle\}\Big)\Big|, \qquad 0 \le l \le n-j-1, \quad n-j+1 \ge 3.$$

Clearly,

$$g_{l,n}(t) = \prod_{i=l+1}^{l+2} |\hat{p}_{j+i-1,y}(B_{j+i}^{1/2}t)|,$$

where $\hat{p}_{j+k,y}(\cdot)$ is the characteristic function corresponding to the density $p_{(j+k)\Delta,y}(\cdot)$. The Parseval equality and 7) imply

$$\int |\hat{p}_{j+i-1,y}(v)|^2 dv = \int p^2_{(j+i-1)\Delta,y}(x)dx \le \big(\sup_x \Psi(x)\big)\int \Psi(u)du < \infty,$$

$$\left[\int g_{l,n}(t)dt\right]^2 \leq \prod_{i=l+1}^{l+2} \det B_{j+i}^{-1/2} \int \left|\hat{p}_{j+i-1,y}(v)\right|^2 dv < \infty.$$

Thus, condition (19.29) is satisfied. Denote $u = B_{j+l+1}^{1/2} t$ and for any positive b consider the value

$$\delta(b) = \sup\left\{g_{l,n}(t)\colon |t| > b,\ 0 \leq l \leq n-j-1,\ n-j+1 \geq 3\right\}.$$

Inequalities $g_{l,n}(t) \leq \left|\hat{p}_{j+l,y}(u)\right|$ and $|t| > b$ imply $|u| > b_0(b) > 0$ and hence $\delta(b) \leq \sup\left\{\left|\hat{p}_{j+l,y}(u)\right|\colon |u| > b_0,\ 0 \leq l \leq n-j-1,\ n-j+1 \geq 3\right\}$. If there were a point $u^* \in \mathbf{R}^\nu$, $|u^*| > b_0$, such that $\left|\hat{p}_{j+l,y}(u^*)\right| = 1$, then the projection of a random vector distributed with the density $p_{(j+l)\Delta,y}(x)$ onto the direction defined by u^* should have a lattice distribution (see (Ibragimov and Linnik, 1965, p. 22)), which is impossible, and hence, for any $b > 0$, we have $\delta(b) < 1$ and condition (19.30) from (Bhattacharya and Rao, 1976) holds also. Thus, for sufficiently large $n-j+1$, we can write in standard notation the asymptotic expansions (23) and (24). Let us prove now some preliminary results.

Let $\Lambda = \|\lambda_{ij}\|$ be the covariance matrix of a nondegenerate normal distribution, and let $A = A(\lambda)$ be a solution of the equation $A^2 = \Lambda$ with $A = A^*$, and $a_{ij}(\Lambda)$ be the corresponding matrix elements. Suppose, further, that $n = (n_1, \dots, n_\nu)$ and let $G_n(u)$ be the Hermite polynomial of total degree $|n| = \sum_{i=1}^{\nu} n_i$.

LEMMA 2. *The following identity*

$$\frac{\partial\left[\varphi(x;0,\Lambda)G_n(\Lambda^{-1/2}x)\right]}{\partial\lambda_{ij}} = \frac{1}{2}\frac{\partial^2\left[\varphi(x;0,\Lambda)G_n(\Lambda^{-1/2}x)\right]}{\partial x_i \partial x_j}$$

$$-\sum_{k=1}^{\nu} n_k \sum_{l=1}^{\nu} \frac{\partial a_{kl}}{\partial\lambda_{ij}} \frac{\partial\left[\varphi(x;0,\Lambda)G_{n-e_k}(\Lambda^{-1/2}x)\right]}{\partial x_l}, \tag{27}$$

hold, where $e_k \in \mathbf{R}^\nu$ is the vector with kth coordinate 1 *and remaining coordinates* 0.

The proof can be found in (Konakov and Molchanov, 1984).

LEMMA 3. *Let $\mathfrak{L}(\nu)$ be the set of symmetric matrices and let $\mathfrak{D}_{\lambda^+,\lambda^-} \subset \mathfrak{L}(\nu)$, $0 < \lambda^- < \lambda^+ < \infty$, be the open subset characterized by $\Lambda \in \mathfrak{D}_{\lambda^+,\lambda^-} \Leftrightarrow \lambda^- E \leq \Lambda \leq \lambda^+ E$. Then, for any $k, l \leq \nu$, $i, j \leq \nu$, and $\Lambda \subset \mathfrak{D}_{\lambda^+,\lambda^-}$, we have*

$$\left|\frac{\partial a_{ij}(\Lambda)}{\partial\lambda_{kl}}\right| \leq (4\lambda^-)^{-1/2}. \tag{28}$$

The proof can be found in (Konakov and Molchanov, 1984).

LEMMA 4. *The following estimates hold:*

$$|\delta m_{j,n}| \leq C_1|z-y|^\alpha \Delta, \tag{29}$$

$$\|\delta V_{j,n}\| \le C_2|z-y|^\alpha \Delta, \tag{30}$$

$$|\delta \bar{\chi}_\beta| \le C_3|z-y|^\alpha/(n-j+1), \qquad |\beta| \le 4\nu+1. \tag{31}$$

Proof. Inequalities (29) and (30) follow immediately from (25) and from condition 1) on the coefficients $a_{ij}(\cdot,\cdot)$. Let $\varphi_\xi(t)$ be the characteristic function of a random vector ξ. We have by definition

$$i^{|\beta|}\bar{\chi}_{\beta,y} = (n-j+1)^{-1} \sum_{p=j+1}^{n+1} D^\beta \log \varphi_{B_p^{-1/2} V_{j,n}^{-1/2} \eta_{p,y}} (B_p^{1/2} t)\Big|_{t=0}, \tag{32}$$

$$\begin{aligned} i^{|\beta|}(\bar{\chi}_{\beta,y} + \delta\bar{\chi}_{\beta,y}) = (n-j+1)^{-1}\Bigg\{ & \sum_{p=j+1}^{n+1} D^\beta \log \varphi_{(B_p+\delta B_p)^{-1/2}(V_{j,n}+\delta V_{j,n})^{-1/2}\eta_{p,y}} \\ & \times ((B_p+\delta B_p)^{1/2} t)\Big|_{t=0} \\ & + D^\beta \log \varphi_{(B_{j+1}+\delta B_{j+1})^{-1/2}(V_{j,n}+\delta V_{j,n})^{-1/2}\eta_{j+1,z}} ((B_{j+1}+\delta B_{j+1})^{1/2} t)\Big|_{t=0} \\ & - D^\beta \log \varphi_{(B_{j+1}+\delta B_{j+1})^{-1/2}(V_{j,n}+\delta V_{j,n})^{-1/2}\eta_{j+1,y}} ((B_{j+1}+\delta B_{j+1})^{1/2} t)\Big|_{t=0}\Bigg\}, \end{aligned} \tag{33}$$

where $(B_p+\delta B_p)^{1/2} = (V_{j,n}+\delta V_{j,n})^{-1/2}((n-j+1)\Delta)^{1/2}$, $\beta = (\beta_1, \beta_2, \dots, \beta_\nu)$, $D^\beta = \partial^{\beta_1+\dots+\beta_\nu}/\partial t_1^{\beta_1} \dots \partial t_\nu^{\beta_\nu}$. Let $\mu_{\beta,\cdot}$ be the moment of order β of the random vector $(V_{j,n}+\delta V_{j,n})^{-1/2}\eta_{j+1,\cdot}$. It follows from 7), 8), and (26) that, for $|\beta| \le 4\nu+1$,

$$\begin{aligned} |\mu_{\beta,z} - \mu_{\beta,y}| &\le \int \left|(B_{j+1}+\delta B_{j+1})^{1/2} u\right|^\beta |p_{j\Delta,z}(u) - p_{j\Delta,y}(u)| du \\ &\le \int \|(B_{j+1}+\delta B_{j+1})^{1/2}\|^{|\beta|} |u|^{|\beta|} |z-y|^\alpha \Psi(u) du \le C|z-y|^\alpha \\ &\quad \times \int |u|^{4\nu+1} \Psi(u) du \le C_1 |z-y|^\alpha. \end{aligned}$$

By the boundness of $\mu_{\beta,x}$, for $|\beta| \le 4\nu+1$, using the relation of the moments to the cumulants of a random vector (cf. (Bhattacharya and Rao, 1976)) we see that the difference between the two last terms in braces in (33) does not exceed $C_2|z-y|^\alpha$. It follows from Lemma 3 that all elements of the matrix $(B_p+\delta B_p)^{1/2} - B_p^{1/2}$ do not exceed $C_3\|\delta B_p\|$, $\delta B_p = (A+\delta A)^{-1} - A^{-1}$, where

$$A = V_{j,n}/((n-j+1))\Delta, \qquad \delta A = \delta V_{j,n}/((n-j+1))\Delta.$$

If $C_4|z-y|^\alpha/(n-j+1) < 1$ for some $C_4 < \infty$, then (30) imply $\|A^{-1}\delta A\| < 1$ and expanding in the series we get $\|\delta B_p\| < C_5|z-y|^\alpha/(n-j+1)$. It follows from (26) and from the definition of inverse matrix that $\|\delta B_p\| \le C_6$, for some $C_6 < \infty$, and again if $C_4|z-y|^\alpha/(n-j+1)\geq 1$, then $\|\delta B_p\|\leq C_6 \leq C_6\ C_4|z-y|^\alpha/(n-j+1)$. Substracting (32) from (33), we get (31). This completes the proof.

Proof of Theorem. The difference between the terms in (23) and (24) corresponding to the same collection of r, m, $(j_1,\dots,j_m)$ and $(\beta_1,\dots,\beta_m)$ has the form

$$(m!\beta_1!\dots\beta_m!)^{-1}(n-j+1)^{-r/2}(\bar\chi_{\beta_1,y}+\delta\bar\chi_{\beta_1})\dots(\bar\chi_{\beta_m,y}+\delta\bar\chi_{\beta_m})\det(V_{j,n}+\delta V_{j,n})^{-1/2}$$
$$\times\varphi\Big((V_{j,n}+\delta V_{j,n})^{-1/2}(y-m_{j,n}(z)-\delta m_{j,n});0,E\Big)G_{\beta_1+\dots+\beta_m}\Big((V_{j,n}+\delta V_{j,n})^{-1/2}$$
$$\times(y-m_{j,n}(z)-\delta m_{j,n})\Big)-(m!\beta_1!\dots\beta_m!)^{-1}(n-j+1)^{-r/2}\bar\chi_{\beta_1,y}\dots\bar\chi_{\beta_m,y}$$
$$\times\det V_{j,n}^{-1/2}\varphi\Big(V_{j,n}^{-1/2}(y-m_{j,n}(z));0,E\Big)G_{\beta_1+\dots+\beta_m}\Big(V_{j,n}^{-1/2}(y-m_{j,n}(z))\Big). \quad (34)$$

First we note that the following estimate

$$|y-z|^\alpha\leq C\Big[\Big|(y-m_{j,n}(z))/((n-j+1)\Delta)^{1/2}\Big|^\alpha+1\Big]((n-j+1)\Delta)^{\alpha/2} \quad (35)$$

holds.

The validity of (35) follows from the relation

$$m_{j,n}(z)=z+\sum_{i=j+1}^{n+1}c((i-1)\Delta,y)\Delta, \quad (36)$$

which implies $|m_{j,n}(z)-z|\leq C(n-j+1)\Delta$ and hence (35). It follows from the boundness of $\bar\chi_{\alpha_i,y}$, $|\alpha_i|\leq 4\nu+1$, and from (26), (29), (31), (35) that (34) does not exceed

$$C\Big\{\Delta[(n-j+1)\Delta]^{\alpha/2-1}\Big[\varphi(y-m_{j,n}(z);0,c\Delta E)\Big]*\dots*\Big[\varphi(y-m_{j,n}(z);0,c\Delta E)\Big]$$
$$+\Big|\det(V_{j,n}+\delta V_{j,n})^{-1/2}\varphi\Big((V_{j,n}+\delta V_{j,n})^{-1/2}(y-m_{j,n}(z)-\delta m_{j,n});0,E\Big)$$
$$\times G_{\beta_1+\dots+\beta_m}\Big((V_{j,n}+\delta V_{j,n})^{-1/2}(y-m_{j,n}(z)-\delta m_{j,n})\Big)-\det V_{j,n}^{-1/2}\varphi\Big(V_{j,n}^{-1/2}(y$$
$$-m_{j,n}(z));0,E\Big)G_{\beta_1+\dots+\beta_m}\Big(V_{j,n}^{-1/2}(y-m_{j,n}(z))\Big)\Big|\Big\}, \quad (37)$$

where the constants $C>0$ and $c>0$ depend only on the fixed r, m, $(j_1,\dots,j_m)$, $(\beta_1,\dots,\beta_m)$ and on the points s and t. Using Lemmas 2, 3 and estimates (29), (26), we see that the second term in (37) does not exceed

$$C\Delta[(n-j+1)\Delta]^{\alpha/2-1}\Big[\varphi(y-m_{j,n}(z);0,c\Delta E)\Big]$$

$$* \ldots * \Big[\varphi\big(y - m_{j,n}(z); 0, c\Delta E\big)\Big]. \tag{38}$$

Let us find now the difference between the remainder terms in (24) and (23):

$$(n-j+1)^{-(4\nu-1)/2}\left|\frac{\det(V_{j,n}+\delta V_{j,n})^{-1/2}}{1+\Big|(V_{j,n}+\delta V_{j,n})^{-1/2}\big(y-m_{j,n}(z)-\delta m_{j,n}\big)\Big|^{4\nu+1}}\right.$$

$$\left.-\frac{\det V_{j,n}^{-1/2}}{1+\Big|V_{j,n}^{-1/2}\big(y-m_{j,n}(z)\big)\Big|^{4\nu+1}}\right|. \tag{39}$$

If $\Lambda = V_{j,n} + \delta V_{j,n} = \|\lambda_{pq}\|$, $\Lambda^{-1/2} = \|\mu_{pq}\|$, $\Lambda^{-1} = \|\lambda^{pq}\|$, then

$$\frac{\partial[\det \Lambda^{-1/2}]}{\partial \lambda_{pq}} = \frac{\partial[\det \Lambda^{-1/2}]}{\partial \mu_{ik}}\frac{\partial \mu_{ik}}{\partial \lambda^{pq}}\frac{\partial \lambda^{pq}}{\partial \lambda_{pq}}.$$

Using the estimates $|\mu_{pq}| \le C\big((n-j+1)\Delta\big)^{-1/2}$, $p = 1, \ldots, \nu$, $q = 1, \ldots, \nu$, after elementary transformations we have

$$\frac{\partial[\det \Lambda^{-1/2}]}{\partial \mu_{ik}} \le C\big[(n-j+1)\Delta\big]^{-(\nu-1)/2}.$$

It follows from Lemma 3 that $|\partial\mu_{ik}/\partial\lambda^{rl}| \le C\big[(n-j+1)\Delta\big]^{1/2}$. Finally, from the relation of Λ to Λ^{-1} we easily get

$$\left|\frac{\partial \lambda^{pq}}{\partial \lambda_{pq}}\right| \le C\big[(n-j+1)\Delta\big]^{-2},$$

whence

$$\left|\frac{\partial[\det \Lambda^{-1/2}]}{\partial \lambda_{pq}}\right| \le C\big[(n-j+1)\Delta\big]^{-(\nu-1)/2}. \tag{40}$$

Then (29), (30), and (31) imply that

$$\big(|\delta m_{j,n}| + \|\delta V_{j,n}\|\big) \le C\Delta\big[(n-j+1)\Delta\big]^{\alpha/2}\left[1+\left|\frac{y-m_{j,n}(z)}{\big((n-j+1)\Delta\big)^{1/2}}\right|^{\alpha}\right]. \tag{41}$$

We use now the elementary inequalities

$$\frac{|x|^{m-1}}{\big(1+|x|^m\big)^2} \le \frac{C}{\big(1+|x|^{m+1}\big)}, \qquad \frac{1+|x|^\alpha}{\big(1+|x|^m\big)} \le \frac{C}{\big(1+|x|^{m-1}\big)}, \quad 0 < \alpha \le 1,$$

and the inequality $|\partial\mu_{ik}/\partial\lambda_{pq}| \le C\big[(n-j+1)\Delta\big]^{-3/2}$, which obviously follows from the above considerations. If we differentiate (39) with respect to the covariances and means, we get from (40) and (41) that (39) does not exceed

$$C(n-j+1)^{-(4\nu-1)/2}\Delta\big[(n-j+1)\Delta\big]^{\alpha/2}\left[1+\left|\frac{y-m_{j,n}(z)}{\big((n-j+1)\Delta\big)^{1/2}}\right|^{\alpha}\right]\big[(n-j+1)\Delta\big]^{-\nu/2}$$

$$\times\Bigg[((n-j+1)\Delta)^{-1}\Bigg(1+\Bigg|\frac{y-m_{j,n}(z)}{((n-j+1)\Delta)^{1/2}}\Bigg|^{4\nu+1}\Bigg)^{-1}$$

$$+((n-j+1)\Delta)^{-3/2}\frac{|y-m_{j,n}(z)|}{1+\big|(y-m_{j,n}(z))/((n-j+1)\Delta)^{1/2}\big|^{4\nu+2}}$$

$$+((n-j+1)\Delta)^{-1/2}\Bigg(1+\Bigg|\frac{y-m_{j,n}(z)}{((n-j+1)\Delta)^{1/2}}\Bigg|^{4\nu+2}\Bigg)^{-1}\Bigg]$$

$$\leq\frac{|y-m_{j,n}(z)|}{1+\big|(y-m_{j,n}(z))/((n-j+1)\Delta)^{1/2}\big|^{4\nu+2}}$$

$$+\frac{((n-j+1)\Delta)^{-1/2}}{1+\big|(y-m_{j,n}(z))/((n-j+1)\Delta)^{1/2}\big|^{4\nu+2}}$$

$$\leq C\Delta(n-j+1)^{-(4\nu-1)/2}\frac{((n-j+1)\Delta)^{-\nu/2-1+\alpha/2}}{1+\big|(y-m_{j,n}(z))/((n-j+1)\Delta)^{1/2}\big|^{4\nu}}. \qquad (42)$$

It follows from (38) and (42), that for some C and c depending only on T and ν

$$|H^{(\Delta)}(j\Delta,n\Delta,z,y)|\leq C\big[(n-j+1)\Delta\big]^{\alpha/2-1}\Bigg\{\Big[\varphi(y-m_{j,n}(z);0,c\Delta E)\Big]$$

$$*\ldots*\Big[\varphi(y-m_{j,n}(z);0,c\Delta E)\Big]+(n-j+1)^{-(4\nu-1)/2}\big[(n-j+1)\Delta\big]^{-\nu/2}$$

$$\times\Bigg[1+\Bigg|\frac{(y-m_{j,n}(z))}{((n-j+1)\Delta)^{1/2}}\Bigg|^{4\nu-1}\Bigg]^{-1}\Bigg\}. \qquad (43)$$

Consider the function $L_\sigma(u_1,\ldots,u_\nu)=\prod_{r=1}^{\nu}l_\sigma(u_r)$, $l_\sigma(u_r)=2\sigma^3/(\pi(\sigma^2+u_r^2)^2)$, where $u=(u_1,u_2,\ldots,u_\nu)=y-z$, and the variance σ^2 will be chosen later. The following relations follow from the definition of convolution operation "$*$"

$$\Big[\Delta^{-\nu/2}L_\sigma((u-v_{j+1})/\Delta^{1/2})\Big]*\ldots*\Big[\Delta^{-\nu/2}L_\sigma((u-v_{n+1})/\Delta^{1/2})\Big]$$

$$=\Big[\prod_{r=1}^{\nu}\Delta^{-\nu/2}l_\sigma((u_r-v_{j+1,r})/\Delta^{1/2})\Big]*\ldots*\Big[\prod_{r=1}^{\nu}\Delta^{-\nu/2}l_\sigma((u_r-v_{n+1,r})/\Delta^{1/2})\Big]$$

$$=\prod_{r=1}^{\nu}\Big[\Delta^{-\nu/2}l_\sigma((u_r-v_{j+1,r})/\Delta^{1/2})\Big]*\ldots*\Big[\Delta^{-\nu/2}l_\sigma((u_r-v_{n+1,r})/\Delta^{1/2})\Big], \qquad (44)$$

where $v_p=c((p-1)\Delta,y)\Delta$, $v_p=(v_{p,1},\ldots,v_{p,\nu})$, $p=j+1,\ldots,n+1$. Thus, if ξ_i^r, $i=j+1,\ldots,n+1$, are independent random variables having the density

$\Delta^{-\nu/2} l_\sigma\big((u_r - v_{i,r})/\Delta^{1/2}\big)$, then in (44) we multiply the densities of the random variables $\xi^r_{j+1} + \ldots + \xi^r_{n+1}$. But by (36)

$$\mathbf{P}\{\xi^r_{j+1} + \ldots + \xi^r_{n+1} < u_r\} = \mathbf{P}\left\{ \sum_{p=j+1}^{n+1} (\zeta^r_p \Delta^{1/2} + v_{p,r}) < u_r \right\}$$

$$= \mathbf{P}\left\{ \sum_{p=j+1}^{n+1} \frac{\zeta^r_p}{\sigma(n-j+1)^{1/2}} < \frac{(y_r - m_{j,n,r}(z))}{\sigma\big((n+j+1)\Delta\big)^{1/2}} \right\}, \tag{45}$$

where $\zeta^r_p = (\xi^r_p - v_{p,r})/\Delta^{1/2}$, $m_{j,n}(z) = \big(m_{j,n,1}(z), \ldots, m_{j,n,\nu}(z)\big)$. Clearly, for any $r = 1, 2, \ldots, \nu$, the random variables ζ^r_p, $p = j+1, \ldots, n+1$, are independent and have the density $l_\sigma(\cdot)$. By (44) and (45)

$$\left[\Delta^{-\nu/2} L_\sigma\big((u - v_{j+1})/\Delta^{1/2}\big)\right] * \ldots * \left[\Delta^{-\nu/2} L_\sigma\big((u - v_{n+1})/\Delta^{1/2}\big)\right]$$

$$= \prod_{r=1}^{\nu} \left[\sigma^2(n-j+1)\Delta\right]^{-\nu/2} p_{Z_{j,n}} \left[\frac{(y_r - m_{j,n,r}(z))}{\sigma\big((n-j+1)\Delta\big)^{1/2}} \right], \tag{46}$$

where

$$Z_{j,n} = \sum_{p=j+1}^{n+1} \frac{\zeta^r_p}{\sigma(n-j+1)^{1/2}}.$$

Let us consider two cases $\left|(y - m_{j,n}(z))/\sigma\left[(n-j+1)\Delta\right]^{1/2}\right| > \widetilde{C}$ and $\left|(y - m_{j,n}(z))/\sigma\left[(n-j+1)\Delta\right]^{1/2}\right| \le \widetilde{C}$, where $\widetilde{C}$ will be chosen later. For the second case, reasoning as in Theorem 19.1 from (Bhattacharya and Rao, 1976), we see that according to the classical local limit theorem

$$\lim_{n\to\infty} \sup_x \left| p_{Z_{j,n}}(x) - \varphi(x; 0, 1) \right| = 0, \tag{47}$$

$r = 1, 2, \ldots, \nu$. Using (46) and (47), we find that for $n - j + 1$ large enough

$$\left[\Delta^{-\nu/2} L_\sigma\big((u - v_{j+1})/\Delta^{1/2}\big)\right] * \ldots * \left[\Delta^{-\nu/2} L_\sigma\big((u - v_{n+1})/\Delta^{1/2}\big)\right]$$

$$= \left[2\pi\sigma^2(n-j+1)\Delta\right]^{-\nu/2} \left[\exp\left\{ -\frac{1}{2} \left| \frac{y - m_{j,n}(z)}{\sigma\left[(n-j+1)\Delta\right]^{1/2}} \right|^2 \right\} + o(1) \right]$$

$$\ge \left[2\pi\sigma^2(n-j+1)\Delta\right]^{-\nu/2} \exp(-\widetilde{C}^2). \tag{48}$$

The asymptotic behavior (as $n - j + 1 \to \infty$) of the convolutions of the density $p_{Z_{j,n}}(x)$, $|x| > \widetilde{C}$, is known (see (Ibragimov and Linnik, 1965, Chapter XIV))

$$p_{Z_{j,n}}(x) = (2\pi)^{-1/2} \exp(-x^2/2) + 2\pi^{-1}(n-j+1)^{-1/2} x^{-4}$$

$$+ B(x)(n-j+1)^{-1/2} x^{-5}, \tag{49}$$

where $B(x)$ is a bounded function. Suppose $\left|(y-m_{j,n}(z))/\sigma\left[(n-j+1)\Delta\right]^{1/2}\right| > \widetilde{C}$ and let ν_1 be the number of components of the vector $(y - m_{j,n}(z))/\sigma\left[(n-j+1)\Delta\right]^{1/2}$ which are less in absolute magnitude than $\widetilde{C}/\nu^{1/2}$, $\nu_2 = \nu - \nu_1$. Obviously, $0 \leq \nu_1 \leq \nu - 1$. Let I_{ν_1} be the set of indices of such components and I_{ν_2} be the complement of the set I_{ν_1}. For the components from I_{ν_2} asymptotic expansion (49) holds. From (49) it follows that there is a $\widetilde{C} < \infty$ such that for $|x| > \widetilde{C}/\nu^{1/2}$

$$p_{Z_{j,n}}(x) \geq (2\pi)^{-1/2}\exp(-x^2/2) + 2^{-1}(n-j+1)^{-1/2}x^{-4}. \tag{50}$$

For the components from I_{ν_1} according to the classical local limit theorem (see (Bhattacharya and Rao, 1976, Remark to Theorem 19.2, p. 194))

$$p_{Z_{j,n}}(x) = (2\pi)^{-1/2}\exp(-x^2/2) + \frac{o(1)}{1+|x|^2}$$

and hence, for a fixed $\varkappa > 1$ and a sufficiently large $n - j + 1$, we get

$$p_{Z_{j,n}}(x) \geq \varkappa^{-2}(2\pi)^{-1/2}\exp\{-\varkappa^2 x^2/2\}. \tag{51}$$

By (46), (50), and (51)

$$\left[\Delta^{-\nu/2}L_\sigma((u-v_{j+1})/\Delta^{1/2})\right] * \ldots * \left[\Delta^{-\nu/2}L_\sigma((u-v_{n+1})/\Delta^{1/2})\right]$$

$$\geq \varkappa^{-2\nu_1}\prod_{i\in I_{\nu_1}}\left[(2\pi\sigma^2\tau)^{-1/2}\exp\left\{\frac{-\varkappa^2\left(y_i - m_{j,n,i}(z)\right)^2}{2\sigma^2\tau}\right\}\right]$$

$$\times\prod_{k\in I_{\nu_2}}\left[(2\pi\sigma^2\tau)^{-1/2}\exp\left\{\frac{-\left(y_k - m_{j,n,k}(z)\right)^2}{2\sigma^2\tau}\right\}\right]$$

$$+\varkappa^{-2\nu_1-\nu}\varkappa_1(n-j+1)^{-\nu_2/2}\tau^{-\nu/2}\prod_{k\in I_{\nu_2}}\left[\frac{y_k - m_{j,n,k}(z)}{\tau^{1/2}}\right]^{-4}, \tag{52}$$

where $\varkappa_1(\widetilde{C},\sigma,\{\gamma_p\}) > 0$, $\tau = (n-j+1)\Delta$. For a sufficiently large $n - j + 1$

$$(n-j+1)^{-(4\nu_2-1)/2}\tau^{-\nu/2}\left[1 + \left|\frac{(y - m_{j,n}(z)}{\tau^{1/2}}\right|^{4\nu}\right]^{-1}$$

$$\leq \varkappa_1(n-j+1)^{-\nu_2/2}\tau^{-\nu/2}\prod_{k\in I_{\nu_2}}\left[\frac{y_k - m_{j,n,k}(z)}{\tau^{1/2}}\right]^{-4}$$

hence the second term in braces in (43) does not exceed

$$\varkappa^{2\nu_1+\nu}\left[\Delta^{-\nu/2}L_\sigma((u-v_{j+1})/\Delta^{1/2})\right] * \ldots * \left[\Delta^{-\nu/2}L_\sigma((u-v_{n+1})/\Delta^{1/2})\right]$$

$$-\left[\varphi(y - m_{j,n}(z); 0, \sigma^2\varkappa^{-2}\Delta E)\right] * \ldots * \left[\varphi(y - m_{j,n}(z); 0, \sigma^2\varkappa^{-2}\Delta E)\right]. \tag{53}$$

Substituting the estimate (53) in (43) and choosing $\sigma^2 = c\varkappa^2$ (c is the same as in (43)), we get in the first case

$$\left|H^{(\Delta)}(j\Delta, n\Delta, z, y)\right| \le C\left[(n-j+1)\Delta\right]^{\alpha/2-1}\left[\Delta^{-\nu/2}L_\sigma((u-v_{j+1})/\Delta^{1/2})\right]$$

$$*\ldots*\left[\Delta^{-\nu/2}L_\sigma((u-v_{n+1})/\Delta^{1/2})\right]. \tag{54}$$

Substituting (48) in (43), choosing sufficiently large $\varkappa > 1$ and again using the classical local limit theorem, we see that (54) also holds in the second case, i.e., for any z, y.
The following slightly more general estimate can be obtained similarly, namely, for a sufficiently large $n-m-j+1$,

$$\left|H^{(\Delta)}(j\Delta, (n-m)\Delta, z, y)\right| \le C\left[(n-m-j+1)\Delta\right]^{\alpha/2-1}\left[\Delta^{-\nu/2}L_\sigma((u-v_{j+1})/\Delta^{1/2})\right]$$

$$*\ldots*\left[\Delta^{-\nu/2}L_\sigma((u-v_{n-m+1})/\Delta^{1/2})\right] \tag{55}$$

where $u = y-z$, $v_i = c((i-1)\Delta, y)\Delta$. We obtain similarly to (54) the following estimate for $p^{(0)}(0, j\Delta, x, z)$

$$|p^{(0)}(0, j\Delta, x, z)| \le C\left[\Delta^{-\nu/2}L_\sigma\Big((z-x-c(\Delta, y)\Delta)/\Delta^{1/2}\Big)\right]$$

$$*\ldots*\left[\Delta^{-\nu/2}L_\sigma\Big((z-x-c(j\Delta, y)\Delta)/\Delta^{1/2}\Big)\right]. \tag{56}$$

Using the relation

$$H^{(\Delta)}(j\Delta, (j+m)\Delta, z, y) = \Delta^{-1}\Big[\int p^{(\Delta)}(j\Delta, (j+1)\Delta, z, u)$$

$$\times p^{(0)}(j\Delta, (j+m)\Delta, u, y)du - p^{(0)}((j-1)\Delta, (j+m)\Delta, z, y)\Big],$$

we can write the remainder term $R(0, j\Delta, x, z)$ in the form

$$R(0, j\Delta, x, z) = \Delta H^{(\Delta)}(s^- + \Delta, s^- + j\Delta, x, z)$$

$$+p^{(0)}(s^-, s^- + j\Delta, x, z) - p^{(0)}(0, j\Delta, x, z), \tag{57}$$

where $s^- = s - \Delta$. By (55) and (57)

$$|(p^{(0)} + R)(0, j\Delta, x, z)| \le C\Big\{\Delta^{-\nu/2}L_\sigma\Big((z-x-c(s, y)\Delta)/\Delta^{1/2}\Big)$$

$$*\ldots*\Delta^{-\nu/2}L_\sigma\Big(\Big(z-x-c((j-1)\Delta, y)\Delta\Big)/\Delta^{1/2}\Big)\Big\}. \tag{58}$$

Using (55) and (58), one can estimate the convolutions in (20):

$$\Big|H^{(\Delta)} \oplus (p^{(0)} + R)(0, (n-m)\Delta, x, y)\Big|$$

$$\le \sum_{j=1}^{n-m} \Delta \int_{\mathbf{R}^\nu} |(p^{(0)} + R)(0, j\Delta, x, z)| \Big|H^{(\Delta)}(j\Delta, (n-m)\Delta, z, y)\Big| dz$$

$$\le C((n-m)\Delta)\,[(n-m-j+1)\Delta]^{\alpha/2-1}$$

$$\times \int_{\mathbf{R}^\nu} \Big[\Delta^{-\nu/2} L_\sigma\Big((z - x - c(s,y)\Delta)/\Delta^{1/2}\Big)$$

$$* \ldots * \Delta^{-\nu/2} L_\sigma\Big(\Big(z - x - c((j-1)\Delta, y)\Delta\Big)/\Delta^{1/2}\Big)\Big]$$

$$\times \Big[\Delta^{-\nu/2} L_\sigma\Big((y - z - c(j\Delta, y)\Delta)/\Delta^{1/2}\Big) * \ldots * \Delta^{-\nu/2}$$

$$\times L_\sigma\Big(\Big(y - z - c((n-m)\Delta, y)\Delta\Big)/\Delta^{1/2}\Big)\Big] dz \le C \sum_{j=1}^{n-m} \Big[1 - \sum_{l=1}^{j} \Delta / \sum_{l=1}^{n-m+1} \Delta\Big]^{\alpha/2-1}$$

$$\times \Delta \Big[\sum_{l=1}^{n-m+1} \Delta\Big]^{-1} \Big[\sum_{l=1}^{n-m+1} \Delta\Big]^{\alpha/2} \Delta^{-\nu/2}$$

$$\times L_\sigma\Big((y - x - c(s,y)\Delta)/\Delta^{1/2}\Big)$$

$$* \ldots * \Delta^{-\nu/2} L_\sigma\Big(\Big(y - x - c((n-m)\Delta, y)\Delta\Big)/\Delta^{1/2}\Big)$$

$$\le C_1 B(1, \alpha/2) T^{\alpha/2} \Delta^{-\nu/2} L_\sigma\Big((y - x - c(s,y)\Delta)/\Delta^{1/2}\Big)$$

$$* \ldots * \Delta^{-\nu/2} L_\sigma\Big(\Big(y - x - c((n-m)\Delta, y)\Delta\Big)/\Delta^{1/2}\Big), \qquad (59)$$

where $B(p, q)$ is the Beta function. In particular, for $m = 1$,

$$\Big|H^{(\Delta)} \oplus (p^{(0)} + R)(0, (n-1)\Delta, x, y)\Big| \le C_2 B(1, \alpha/2) T^{\alpha/2} \Delta^{-\nu/2}$$

$$\times L_\sigma\Big((y - x - c(s,y)\Delta)/\Delta^{1/2}\Big) * \ldots * \Delta^{-\nu/2} L_\sigma\Big(\Big(y - x - c((n-1)\Delta, y)\Delta\Big)/\Delta^{1/2}\Big). \qquad (60)$$

Analogously to (55), we obtain

$$\Big|H^{(\Delta)}(j\Delta, (n-m)\Delta, z, y)\Big| \le C\Big[\sum_{p=j+2}^{n-m+2} \Delta\Big]^{\alpha/2-1} \Delta^{-\nu/2} L_\sigma((u - v_{j+2})/\Delta^{1/2})$$

$$* \ldots * \Delta^{-\nu/2} L_\sigma\big((u - v_{n-m+2})/\Delta^{1/2}\big). \tag{61}$$

By (59) and (61)

$$\Big|H^{(\Delta)} \oplus H^{(\Delta)} \oplus (p^{(0)} + R)(0, (n-m)\Delta, x, y)\Big|$$

$$\leq \sum_{j=1}^{n-m} \Delta \int_{\mathbf{R}^\nu} \Big|\big[H^{(\Delta)} \oplus (p^{(0)} + R)\big](0, j\Delta, x, z)\Big|$$

$$\times \Big|H^{(\Delta)}(j\Delta, (n-m)\Delta, z, y)\Big| dz \leq C \sum_{j=1}^{n-m} \Delta \sum_{r=1}^{j} \Delta \Big[\sum_{t=r+1}^{j+1} \Delta\Big]^{\alpha/2-1} \sum_{p=j+2}^{n-m+2} \Delta$$

$$\times \int_{\mathbf{R}^\nu} \Big[\Delta^{-\nu/2} L_\sigma\Big(\big(z - x - c(s,y)\Delta\big)/\Delta^{1/2}\Big) * \ldots * \Delta^{-\nu/2} L_\sigma\Big(\big(z - x - c(j\Delta, y)\Delta\big)/\Delta^{1/2}\Big)\Big]$$

$$\times \Big[\Delta^{-\nu/2} L_\sigma\Big(\Big(y - z - c\big((j+1)\Delta, y\big)\Delta\Big)/\Delta^{1/2}\Big)$$

$$* \ldots * \Delta^{-\nu/2} L_\sigma\Big(\Big(y - z - c\big((n-m+1)\Delta, y\big)\Delta\Big)/\Delta^{1/2}\Big)\Big] dz$$

$$\leq C^2 B(1, \alpha/2) B(1 + \alpha/2, \alpha/2) T^\alpha \Delta^{-\nu/2} L_\sigma\Big(\big(y - x - c(s,y)\Delta\big)/\Delta^{1/2}\Big)$$

$$* \ldots * \Delta^{-\nu/2} L_\sigma\Big(\Big(y - x - c\big((n-m+1)\Delta, y\big)\Delta\Big)/\Delta^{1/2}\Big)$$

$$\leq C_1^2 T^\alpha \big(\Gamma(1+\alpha)\big)^{-1} \Delta^{-\nu/2} L_\sigma\Big(\big(y - x - c(s,y)\Delta\big)/\Delta^{1/2}\Big)$$

$$* \ldots * \Delta^{-\nu/2} L_\sigma\Big(\Big(y - x - c\big((n-m+1)\Delta, y\big)\Delta\Big)/\Delta^{1/2}\Big). \tag{62}$$

In particular, for $m = 2$

$$\Big|H^{(\Delta)} \oplus H^{(\Delta)} \oplus (p^{(0)} + R)(0, (n-2)\Delta, x, y)\Big|$$

$$\leq C_1^2 T^\alpha \big(\Gamma(1+\alpha)\big)^{-1} \Delta^{-\nu/2} L_\sigma\Big(\big(y - x - c(s,y)\Delta\big)/\Delta^{1/2}\Big)$$

$$* \ldots * \Delta^{-\nu/2} L_\sigma\Big(\Big(y - x - c\big((n-1)\Delta, y\big)\Delta\Big)/\Delta^{1/2}\Big). \tag{63}$$

It is not difficult to deduce by induction that

$$\Big|\underbrace{H^{(\Delta)} \oplus \ldots \oplus H^{(\Delta)}}_{m\text{-fold}} \oplus (p^{(0)} + R)(0, (n-m)\Delta, x, y)\Big|$$

$$\leq C_1^m T^{m\alpha/2} \big(\Gamma(1 + m\alpha/2)\big)^{-1} \Delta^{-\nu/2} L_\sigma\Big(\big(y - x - c(0,y)\Delta\big)/\Delta^{1/2}\Big)$$

$$*\ldots*\Delta^{-\nu/2}L_\sigma\Big(\big(y-x-c((n-1)\Delta,y)\Delta\big)/\Delta^{1/2}\Big) \tag{64}$$

whence, substituting (58) and (64) into (20), we get

$$\big|p^{(\Delta)}(0,n\Delta,x,y)\big| = \Big|(p^{(0)}+R)(0,n\Delta,x,y)$$

$$+\sum_{m=1}^{n}\underbrace{H^{(\Delta)}\oplus\ldots\oplus H^{(\Delta)}}_{m\text{-fold}}\oplus(p^{(0)}+R)(0,(n-m)\Delta,x,y)\Big|$$

$$\leq\Big[\sum_{m=0}^{\infty}C_1^m T^{m\alpha/2}\big(\Gamma(1+m\alpha/2)\big)^{-1}\Big]\Delta^{-\nu/2}L_\sigma\Big(\big(y-x-c(0,y)\Delta\big)/\Delta^{1/2}\Big)$$

$$*\ldots*\Delta^{-\nu/2}L_\sigma\Big(\big(y-x-c(n-1)\Delta,y)\Delta\big)/\Delta^{1/2}\Big). \tag{65}$$

Again applying the classical local limit theorem to the right-hand side in (65), we obtain a dominant for the series in (20) not depending on the parameter Δ and hence the proof of the theorem is completed by passing to the limit (as $\Delta\to 0$) termwise in (20).

REFERENCES

Bhattacharya, R. and Ranga Rao, R. (1976). *Normal Approximation and Asymptotic Expansions.* John Wiley, New York, London.

Gikhman, I. and Skorokhod, A. (1982). *Stochastic Differential Equations and Its Applications.* Naukova Dumka, Kiev (in Russian).

Ibragimov, I. and Linnik, Y. (1965). *Independent and Stationary Sequences of Random Variables.* Nauka, Moscow (in Russian).

Khinchin, A. (1936). *Asymptotic Laws of the Probability Theory.* ONTI, Moscow (in Russian).

Konakov, V. and Molchanov, S. (1984). On the convergence of Markov chains to diffusion processes. *Theory Probab. Math. Statist.* **31**, 51–64.

Liptser, R. and Shiryaev, A. (1983). Weak convergence of the sequence of semimartingales to the diffusion type process. *Math. USSR Sbornik* **121**, 176–200 (in Russian).

McKean, H. and Singer, I.Jr. (1967). Curvature and the eigen values of the Laplacian. *J. Different. Geometry* **1**, 43–69.

Skorokhod, A.V. (1961). *Studies in the Theory of Stochastic Process.* Kiev University, Kiev (in Russian).

Frontiers in Pure and Applied Probab. 1, pp. 135 - 144
H. Niemi et al. (Eds.)
1993 TVP/VSP

A NOTE ON THE STATIONARY MODIFICATION OF THE LOGLIKELIHOOD FUNCTION IN ASYMPTOTIC CONTINUOUS TIME SYSTEM IDENTIFICATION

T. KOSKI

Department of Applied Mathematics, Luleå University of Technology, S-95187 Luleå, Sweden

ABSTRACT

The mean square and almost sure convergence of an averaged loglikelihood function, modified by being computed on a finite segment of data using the stationary filtering gain, to the exact average loglikelihood function is established asymptotically for observations of a linear, time invariant, continuous time stochastic differential system.

1. INTRODUCTION

Consider the innovations representation of a process $y(\theta) = \{y(t;\theta) \mid 0 \leq t \leq T\}$ obtained by noise perturbed linear measurements of a linear time invariant stochastic differential equation for $x(\theta) = \{x(t;\theta) \mid 0 \leq t \leq T\}$, i.e.,

$$d\hat{x}(t;\theta \mid 0) = A(\theta)\hat{x}(t;\theta \mid 0)dt + P_t(\theta)C^* dv(t;\theta \mid 0), \tag{1}$$

$$dy(t;\theta) = C\hat{x}(t;\theta \mid 0)dt + dv(t;\theta \mid 0), \tag{2}$$

where t denotes time, θ is an element of a (compact) finite dimensional parameter space Θ, and for every θ and t the random variables (r.v's) $x(t;\theta \mid 0)$ and $y(t;\theta)$ belong to $\mathbf{R}^n$ and $\mathbf{R}^m$, respectively. For every θ, $A(\theta)$, and C (which is independent of θ) are $n \times n$ and $m \times n$-matrices, respectively, and are independent of t. C^* denotes the transposed matrix. The random variables (r.v's) $\hat{x}(t;\theta \mid 0)$ are the conditional expectations on a probability space with respect to some probability defined by

$$\hat{x}(t;\theta \mid 0) = \mathbf{E}[x(t;\theta) \mid \{y(s;\theta) \mid 0 \leq s \leq t\}] \tag{3}$$

so that $v(\theta \mid 0) = \{v(t;\theta \mid 0) \mid 0 \leq t \leq T\}$ is the $\mathbf{R}^m$-valued innovations process. The $n \times n$-matrix $P_t(\theta)$ in (1) can be expressed as

$$P_t(\theta) = \mathbf{E}[(x(t;\theta) - \hat{x}(t;\theta \mid 0))(x(t;\theta) - \hat{x}(t;\theta \mid 0))^*]. \tag{4}$$

We have $y(0;\theta) = v(0;\theta \mid 0) = 0$ and $P_0(\theta) = \Gamma$, where Γ is some positive definite matrix. Detailed derivations of (1)–(4) are found, e.g., in (Kallianpur, 1980) or (Knobloch and Kwakernaak, 1986).

The identification problem is to estimate the parameter θ given one realization of the process $y = \{y(t) \mid 0 \leq t \leq T\}$. (Here it is natural to drop θ from the description of the observation process, since a path of y is observed under one true unknown element θ_0.) If one wishes to solve the identification problem, e.g., by maximum likelihood technique (MLE), the appropriate averaged loglikelihood function $L_T(\theta)/T$ corresponding to system (1)–(2) is

$$\frac{1}{T}L_T(\theta) = \frac{1}{T}\Big\{\int_0^T [C\hat{x}(t;\theta \mid 0)]^* dy_t - \frac{1}{2}\int_0^T \|C\hat{x}(t;\theta \mid 0)\|^2 dt\Big\}, \tag{5}$$

where $\|\cdot\|$ is the Euclidean vector norm in $\mathbf{R}^m$. (One of the original derivations of (5) is found in the paper (Duncan, 1970).)

The asymptotic consistency of MLE is, however, investigated (see, e.g., (Bagchi, 1980; Koski, 1986; Tugnait, 1985) and their references) by invoking the random function

$$\frac{1}{T}\bar{L}_T(\theta) = \frac{1}{T}\Big\{\int_0^T [C\bar{x}(t;\theta \mid -\infty)]^* dy(t) - \frac{1}{2}\int_0^T \|C\bar{x}(t;\theta \mid -\infty)\|^2 dt\Big\}, \tag{6}$$

where $\bar{x}(t;\theta \mid -\infty)$ (a state reconstruction) is computed for $t \geq 0$ by means of

$$d\bar{x}(t;\theta \mid -\infty) = A(\theta)\bar{x}(t;\theta \mid -\infty)dt - P(\theta)C^*C\bar{x}(t;\theta \mid -\infty)dt + P(\theta)C^*dy(t) \tag{7}$$

and the matrix $P(\theta)$ is the stationary filtering error covariance matrix defined by

$$P(\theta) = \mathbf{E}[(x(t;\theta) - \hat{x}(t;\theta \mid -\infty))(x(t;\theta) - \hat{x}(t;\theta \mid -\infty))^*] , \tag{8}$$

where $\hat{x}(t;\theta \mid -\infty) = \mathbf{E}[x(t;\theta) \mid \{y(s) \mid -\infty < s \leq t\}]$, i.e., $\hat{x}(t;\theta \mid -\infty)$ is the state estimator which has reached the steady state. Modification (6)–(8) is also used in computation based on experimental data (see (Iliff and Maine, 1979)), since $P(\theta)$ in (8) can be obtained for any θ by solving the associated algebraic Riccati equation. In fact

$$\|P_t(\theta) - P(\theta)\|_n \to 0 \ \text{ as } \ t \to \infty \tag{9}$$

($\|\cdot\|_n$ is a matrix norm) under certain stabilizability and detectability conditions (see (Knobloch and Kwakernaak, 1986, pp.219–228) or (Kailath and Ljung, 1976)). Now, (9) means strictly speaking that the (Gaussian) measure induced on $\mathbf{R}^n$ by the error process $x(t;\theta) - \hat{x}(t;\theta \mid 0)$ converges weakly to a (Gaussian) measure on $\mathbf{R}^n$ with mean zero and covariance matrix $P(\theta)$. This form of convergence need not be adequate for inferring the properties of $L_T(\theta)$ from the properties of $\bar{L}_T(\theta)$. Indeed, the work (Caines and Rissanen, 1979) on convergence of MLE for discrete time ARMA-processes has taken a similar question into account.

In this paper we focus on this question in the context of (1)–(2) by proving, under a stability assumption, that

$$\sup_{\Theta\setminus\theta_0} \frac{1}{T}\left(\bar{L}_T(\theta) - L_T(\theta_0)\right) \to 0$$

in mean square and almost surely as $T \to \infty$, where θ_0 denotes the true parameter value, i.e., the one description of the process $y(\theta) = \{y(t;\theta) \mid 0 \leq t \leq T\}$ in terms of (1)–(2), where $v(\theta \mid 0)$ is a Wiener process. The argument is based on the fundamental work and ideas of Y. Baram (Baram, 1980) incorporated in Lemma 1 below.

2. THE MEAN SQUARE APPROXIMATION PROPERTY

First, an exponential stability assumption on the state transition matrix $\Phi_t(\theta) = e^{A(\theta)t}$ is introduced. In other words, it is supposed that for any $t \geq 0$ (with the matrix norm $\|A\|_n = \sup_{\|x\|=1} \|Ax\|$)

(A.1)

$$\|\Phi_t(\theta)\|_n \leq Me^{-\alpha t}, \tag{10}$$

where $M > 0$ and $\alpha > 0$ are constants independent of θ.

Furthermore, we assume that the matrix $A(\theta) - P(\theta)C^TC$ will also generate a stable transition matrix, i.e.,

(A.2) The pair (A, C) is detectable. (For this condition see (Knobloch and Kwakernaak, 1986, Satz 11.2) or (Kailath and Ljung, 1976)).

It is also convenient to assume that the state process $x = \{x(t;\theta) \mid 0 \leq t \leq T\}$ is for every θ a continuous stationary Gaussian Markov process, i.e.,

(A.3)

$$\mathbf{E}x(0;\theta)(x(0;\theta))^* = Q(\theta), \tag{11}$$

where $Q(\theta)$ solves the Lyapunov equation

$$A(\theta)Q + QA^*(\theta) = -I_n$$

(I_n is the identity matrix in $\mathbf{R}^n$. Assuming I_n in (11) is done for convenience and does not form a restriction.)

(A.4) For any θ and θ_0 and every $T > 0$,

$$L_T(\theta) = L_T(\theta_0) \quad \text{a.s.}$$

implies $\theta = \theta_0$.

This identifiability assumption is actually unnecessarily restrictive but, as will turn out in the sequel (see (13)), this is not critical here.

PROPOSITION 1. *Let* (A.1)–(A.4) *hold. Then*

$$\mathbf{E}\left|\sup_{\Theta\setminus\theta_0} \frac{1}{T}\left(\bar{L}_T(\theta) - L_T(\theta_0)\right)\right|^2 \to 0 \tag{12}$$

as $T \to \infty$.

The proof of this assertion is facilitated by the following observations. For any θ,

$$\bar{L}_T(\theta) - L_T(\theta_0) = (\bar{L}_T(\theta) - \bar{L}_T(\theta_0)) + (\bar{L}_T(\theta_0) - L_T(\theta_0)). \tag{13}$$

The term inside the first parentheses, or

$$\sup_{\Theta\setminus\theta_0} \left| \frac{1}{T} (\bar{L}_T(\theta) - \bar{L}_T(\theta_0)) \right|$$

is the quantity dealt with by the techniques of (Bagchi, 1980; Tugnait, 1985) (see also (Koski, 1987)), which yield the desired convergence to zero by virtue of (A.4). The second term may be written, using (2), as

$$\bar{L}_T(\theta_0) - L_T(\theta_0) = \int_0^T [C\bar{x}(t;\theta_0 \mid -\infty)]^* C\hat{x}(t;\theta_0 \mid 0)dt$$

$$+ \int_0^T [C\bar{x}(t;\theta_0 \mid -\infty)]^* dv(t;\theta_0 \mid 0) - \frac{1}{2}\int_0^T \|C\bar{x}(t;\theta_0 \mid -\infty)\|^2 dt$$

$$- \int_0^T [C\hat{x}(t;\theta_0 \mid 0)]^* dv(t;\theta_0 \mid 0) - \frac{1}{2}\int_0^T \|C\bar{x}(t;\theta_0 \mid 0)\|^2 dt$$

and rearranged as

$$\bar{L}_T(\theta_0) - L_T(\theta_0) = \int_0^T [C(\bar{x}(t \mid -\infty) - \hat{x}(t \mid 0))]^* dv(t \mid 0)$$

$$-\frac{1}{2}\int_0^T \|C(\bar{x}(t \mid -\infty) - \hat{x}(t \mid 0))\|^2 dt. \tag{14}$$

In the expression (14) the dependence on θ_0 has been omitted for simplicity, as will be done in the sequel also.
The approximation property is established for $(\bar{L}_T(\theta_0) - L_T(\theta_0))/T$ by means of representation (14) in a series of lemmas, which will also entail the almost sure convergence, as shown in the final section.

3. SOME RESULTS ON CONVERGENCE OF THE STATE RECONSTRUCTION

The key results lie in the following lemma and in its proof.

LEMMA 1. *Under* (A.1)–(A.3) *it follows that*

$$\|P_t - P\|_n \le 4Me^{-\alpha t}\operatorname{tr} Q \tag{15}$$

for M, α, *and* Q *as in* (10) *and* (11), *respectively, and* tr *denotes the trace of a matrix.*

Proof. By elementary calculations it follows from (4) and (8) that

$$P_t - P = \mathbf{E}[\hat{x}(t \mid -\infty)(\hat{x}(t \mid -\infty))^* - \hat{x}(t \mid 0)(\hat{x}(t \mid 0))^*]$$

$$= \mathbf{E}[(\hat{x}(t \mid -\infty) - \hat{x}(t \mid 0))(\hat{x}(t \mid -\infty))^*] \\ + \mathbf{E}[\hat{x}(t \mid 0)(\hat{x}(t \mid -\infty) - \hat{x}(t \mid 0))^*] \ . \tag{16}$$

Then

$$\|P_t - P\|_n \leq \mathbf{E}\|\hat{x}(t \mid -\infty) - \hat{x}(t \mid 0)\|\|\hat{x}(t \mid -\infty)\| \\ + \mathbf{E}\|\hat{x}(t \mid 0)\|\|\hat{x}(t \mid -\infty) - \hat{x}(t \mid 0)\| \tag{17}$$

by an application of the properties of the matrix norm $\|\cdot\|_n$, the triangle inequality and Jensen's inequality. Schwartz inequality yields

$$\mathbf{E}\|\hat{x}(t \mid -\infty) - \hat{x}(t \mid 0)\|\|\hat{x}(t \mid -\infty)\|$$

$$\leq \sqrt{\mathbf{E}\|\hat{x}(t \mid -\infty) - \hat{x}(t \mid 0)\|^2 \mathbf{E}\|\hat{x}(t \mid -\infty)\|^2} \ . \tag{18}$$

By a basic property of the filtering estimate we have

$$\mathbf{E}\|\hat{x}(t \mid -\infty)\|^2 \leq \mathbf{E}\|x(t)\|^2 = \operatorname{tr} Q$$

as well as

$$\mathbf{E}\|\hat{x}(t \mid 0)\|^2 \leq \mathbf{E}\|x(t)\|^2 = \operatorname{tr} Q \ . \tag{19}$$

Hence, we only need to consider the expression $\mathbf{E}\|\hat{x}(t \mid -\infty) - \hat{x}(t \mid 0)\|^2$. To that end, the auxiliary concepts, from (Baram, 1980) are introduced:

$$\mathcal{F}_t = \sigma(y_s \mid -\infty < s \leq t), \quad \mathcal{B}_t = \sigma(y_s \mid 0 \leq s \leq t)$$

i.e., $\mathcal{F}_t$ is the sigma-algebra generated by the process $y = \{y(s) \mid -\infty < s \leq t\}$ up to time t and $\mathcal{B}_t$ is the sigma-algebra generated by this process with truncated history. Note that $\mathcal{F}_t$ may be considered in system identification terms as an ideal object, in the same way as we may consider $\hat{x}(t \mid -\infty)$. By construction we have

$$\mathcal{B}_t \subseteq \mathcal{F}_t \ , \tag{20}$$

$$\hat{x}(t \mid -\infty) = \mathbf{E}[x_t \mid \mathcal{F}_t] \ , \tag{21}$$

$$\hat{x}(t \mid 0) = \mathbf{E}[x_t \mid \mathcal{B}_t] \ . \tag{22}$$

Consider now the r.v.'s $z(t)$ defined by

$$z(t) := x(t) - \mathbf{E}[x(t) \mid \mathcal{F}_0] \ . \tag{23}$$

(A.3) implies by, e.g., (Arató, 1982, pp.118 ff.)

$$x(t) = \Phi_t x(0) + \int_0^t \Phi_{t-s} dw_s \tag{24}$$

for some state noise Wiener process w. Hence the r.v. $\int_0^t \Phi_{t-s} dw_s$ is independent of x_0, in particular independent of $\mathcal{F}_0$. Therefore,

$$\mathbf{E}[x(t) \mid \mathcal{F}_0] = \mathbf{E}[\Phi_t x_0 \mid \mathcal{F}_0] \tag{25}$$

(note that $\mathbf{E}[\int_0^t \Phi_{t-s}dw_s \mid \mathcal{F}_0] = \mathbf{E}\int_0^t \Phi_{t-s}dw_s = 0$).
Furthermore, we have the inequality

$$\|\hat{x}(t \mid -\infty) - \hat{x}(t \mid 0)\|^2$$

$$= \{\|\hat{x}(t \mid -\infty) - \mathbf{E}(z(t) \mid \mathcal{F}_t) + \mathbf{E}(z(t) \mid \mathcal{F}_t)) - \hat{x}(t \mid 0)\|\}^2$$
$$\leq 2\{\|\hat{x}(t \mid -\infty) - \mathbf{E}(z(t) \mid \mathcal{F}_t)\|^2 + \|\hat{x}(t \mid 0) - \mathbf{E}(z(t) \mid \mathcal{F}_t)\|^2\} . \tag{26}$$

Here

$$\mathbf{E}\|\hat{x}(t \mid -\infty) - \mathbf{E}(z(t) \mid \mathcal{F}_t)\|^2$$
$$= \mathbf{E}\|\mathbf{E}[x(t) - z(t) \mid \mathcal{F}_t]\|^2 = \mathbf{E}\|\mathbf{E}[\mathbf{E}[\Phi_t x(0) \mid \mathcal{F}_0] \mid \mathcal{F}_t]\|^2$$
$$\leq \mathbf{E}\|\mathbf{E}[\Phi_t x(0) \mid \mathcal{F}_0]\|^2 \leq \|\Phi_t\|_n^2 \mathbf{E}\|x(0)\|^2 = \|\Phi_t\|_n^2 \mathrm{tr} Q \tag{27}$$

by (25), the smoothing properties of conditional expectation and Jensen's inequality. Furthermore, (20) yields

$$\mathbf{E}\|\hat{x}(t \mid 0) - \mathbf{E}(z(t) \mid \mathcal{F}_t)\|^2 = \mathbf{E}\|\mathbf{E}[x(t) \mid \mathcal{B}_t] - \mathbf{E}[z(t) \mid \mathcal{F}_t]\|^2$$
$$= \mathbf{E}\|\mathbf{E}[\mathbf{E}[x(t) \mid \mathcal{B}_t] \mid \mathcal{F}_t] - \mathbf{E}(z(t) \mid \mathcal{F}_t)\|^2 \tag{28}$$

in view of the property of conditional expectation that $\mathbf{E}(x(t) \mid \mathcal{B}_t) = \mathbf{E}(\mathbf{E}(x(t) \mid \mathcal{B}_t) \mid \mathcal{F}_t)$.
By construction $\mathcal{F}_t$ is the smallest sigma-algebra containing the product of $\mathcal{F}_0$ and $\mathcal{B}_t$. Also, $z(t)$ is Gaussian, hence independent of $\mathcal{F}_0$ in view of (23) and (24). Hence the argument of (Bauer, 1978, pp.295–296, Satz 54.4) yields that $\mathbf{E}[z(t) \mid \mathcal{F}_t]$ is $\mathcal{B}_t$-measurable. Hence we can write the expression inside the squared norm in (28) as

$$\mathbf{E}[\mathbf{E}(x(t) \mid \mathcal{F}_t) \mid \mathcal{B}_t] - \mathbf{E}(z(t) \mid \mathcal{F}_t) = \mathbf{E}[\mathbf{E}(x(t) \mid \mathcal{F}_t) \mid \mathcal{B}_t]$$
$$-\mathbf{E}[\mathbf{E}[(z(t) \mid \mathcal{F}_t]] \mid \mathcal{B}_t] = \mathbf{E}[\mathbf{E}\{x(t) - z(t) \mid \mathcal{F}_t\} \mid \mathcal{B}_t]$$

which by another application of Jensen's inequality yields in (28)

$$\mathbf{E}\|\mathbf{E}[\mathbf{E}\{x(t) - z(t) \mid \mathcal{F}_t\} \mid \mathcal{B}_t]\|^2$$

$$\leq \mathbf{E}(\mathbf{E}\|\mathbf{E}(x(t) - z(t)) \mid \mathcal{F}_t\|^2 \mid \mathcal{B}_t) = \mathbf{E}(\|\mathbf{E}[\mathbf{E}[\Phi_t x_0 \mid \mathcal{F}_0] \mid \mathcal{F}_t]\|^2)$$

by (23) and (25), and, furthermore,

$$\mathbf{E}\|\mathbf{E}[\mathbf{E}[\Phi_t x_0 \mid \mathcal{F}_0] \mid \mathcal{F}_t]\|^2 \leq \mathbf{E}\mathbf{E}\{\|\mathbf{E}\Phi_t x_0 \mid \mathcal{F}_0\|^2 \mid \mathcal{F}_t\}$$

$$= \mathbf{E}\|\mathbf{E}[\Phi_t x_0 \mid \mathcal{F}_0]\|^2 \leq \mathbf{E}\{\|\Phi_t\|_n^2 \mathbf{E}[\|x(0)\|^2 \mid \mathcal{F}_0]\}$$
$$= \|\Phi_t\|_n^2 \mathbf{E}\|x(0)\|^2 = \|\Phi_t\|_n^2 \mathrm{tr}\, Q \tag{29}$$

In view of (17), (18), (19), (27), and (29) it follows that $\|P_t - P\|_n \leq 4\mathrm{tr}Q\|\Phi_t\|_n$. From (10) above the assertion is established.
The bounds for $\mathbf{E}\|\hat{x}(t \mid -\infty) - \hat{x}(t \mid 0)\|^2$ obtained during the course of this proof yield now

LEMMA 2. *The following relation holds*

$$\mathbf{E}\|\bar{x}(t \mid -\infty) - \hat{x}(t \mid 0)\|^2 \leq Ge^{-\alpha t} \tag{30}$$

for a positive constant G *and* α *as in* (10).

Proof. We have

$$\mathbf{E}\|\bar{x}(t \mid -\infty) - \hat{x}(t \mid 0)\|^2 \leq 2\{\mathbf{E}\|\bar{x}(t \mid -\infty) - \hat{x}(t \mid -\infty)\|^2$$
$$+\mathbf{E}\|\hat{x}(t \mid -\infty) - \hat{x}(t \mid 0)\|^2\} .$$

First we note that it holds evidently by (26)–(29) in the proof of Lemma 1 that

$$\mathbf{E}\|\hat{x}(t \mid -\infty) - \hat{x}(t \mid 0)\|^2 \leq 2\|\Phi_t\|_n \mathrm{tr}\, Q . \tag{31}$$

Next, it should be observed that the steady state filtering estimate can be written as

$$\hat{x}(t \mid -\infty) = \int_{\infty}^{t} \exp\{(A - PC^*TC)(t-s)\} PC^* dy(s)$$
$$= \bar{x}(t \mid -\infty) + \int_{\infty}^{0} \exp\{(A - PC^*TC)(t-s)\} PC^* dy(s) \tag{32}$$

by definition of $\bar{x}(t \mid -\infty)$ in (7) above, where $\exp\{(A - PC^*C)t\}$ is the transition matrix corresponding to $(A - PC^*C)$. The second term in the right-hand side of (32) is equal to $\mathbf{E}[x(t) \mid \mathcal{F}_0]$ and can be represented in terms of innovations process $\{v(t \mid -\infty) \mid -\infty < t \leq 0\}$ corresponding to the observation process $\{y(t) \mid -\infty < t \leq 0\}$ as

$$\mathbf{E}[x(t) \mid \mathcal{F}_0] = \int_{-\infty}^{0} e^{A(t-s)} PC^* dv(s \mid -\infty) \tag{33}$$

using a well-known idea of filtering theory (see, e.g., (Kallianpur, 1980, pp. 263–266)). By virtue of (32) and (33) we have

$$\mathbf{E}\|\bar{x}(t \mid -\infty) - \hat{x}(t \mid \infty)\|^2 \leq \int_{-\infty}^{0} \|\Phi_{t-s} PC^*\|_n^2 ds \tag{34}$$

using a standard property for computing the expectation of the square of the norm of a stochastic integral like (33).
But in the right-hand side of (34) we have

$$\int_{-\infty}^{0} \|\Phi_{t-s} PC^*\|_n^2 ds \leq M\|PC^*\|_n^2 \int_{-\infty}^{0} e^{-2\alpha(t-s)} ds$$
$$= M\|PC^*\|_n^2 e^{-2\alpha t} . \tag{35}$$

Together with (31) this proves the lemma.
This yields immediately the mean square property for the stochastic integral term in (14).

LEMMA 3. *Under the standing assumptions we have*

$$\mathbf{E}\left|\frac{1}{T}\int_0^T (C[\bar{x}(t \mid -\infty) - \hat{x}(t \mid 0)])^* dv(t \mid 0)\right|^2 \to 0$$

as $T \to \infty$.

Proof. By the aforementioned property of stochastic integrals

$$\mathbf{E}\left|\frac{1}{T}\int_0^T (C[\bar{x}(t \mid -\infty) - \hat{x}(t \mid 0)])^* dv(t \mid 0)\right|^2$$

$$\leq \frac{1}{T^2}\|C\|_n^2 \int_0^T \mathbf{E}\|\bar{x}(t \mid -\infty) - \hat{x}(t \mid 0)\|^2 dt \ . \tag{36}$$

But by Lemma 2 the integral in this inequality is bounded above by an exponential, i.e.,

$$\int_0^T \mathbf{E}\|\bar{x}(t \mid -\infty) - \hat{x}(t \mid 0)\|^2 dt$$

$$\leq G \int_0^T e^{-\alpha t} dt = (G/\alpha)[1 - e^{-\alpha T}] \ . \tag{37}$$

From (36) and (37) the assertion is obvious.

LEMMA 4. *The following relation holds:*

$$\frac{1}{T}\int_0^T \|\bar{x}(t \mid -\infty) - \hat{x}(t \mid 0)\|^2 dt \to 0 \quad \text{a.s.}$$

as $T \to \infty$.

Proof. Let us set, for any $r > 0$ and $n = 1, 2, \ldots$,

$$\tau(n + r) := \bar{x}(n + r \mid -\infty) - \hat{x}(n + r \mid 0) \ .$$

Then we have by Chebyshev's inequality that

$$\mathbf{P}\{\|\tau(n + r)\| > \varepsilon\} \leq \frac{1}{\varepsilon^2}\mathbf{E}\|\tau(n + r)\|^2 \leq \frac{G}{\varepsilon^2} e^{-\alpha(n+r)}$$

in view of Lemma 2, (30). Hence

$$\sum_{n=1}^{\infty} \mathbf{P}\{\|\tau(n + r)\| > \varepsilon\} \leq \frac{G}{\varepsilon^2} \sum_{n=1}^{\infty} e^{-\alpha(n+r)} < \infty \ .$$

Thus it follows by Borel–Cantelli lemma (see (Bauer, 1978, p.168)) that for any $r > 0$

$$\lim_{n \to \infty} \|\tau(n + r)\| = 0 \ .$$

Now both $\{\hat{x}(t \mid 0) \mid 0 \leq t\}$ and $\{x(t \mid -\infty) \mid 0 \leq t\}$, have by construction (cf. (7)), continuous $\mathbf{R}^n$-valued sample paths with probability one. Hence there exists for any $\varepsilon > 0$ an n_ε such that if $n > n_\varepsilon$

$$\|\tau(n+r)\| < \varepsilon/2 \quad \text{a.s.},$$

where $r > 0$ can be chosen so that by continuity

$$\|\tau(t) - \tau(n+r)\| < \varepsilon/2 \quad \text{as} \quad t \to \infty,$$

with $t > n_\varepsilon$. Hence $\tau(t) \to 0$ as $t \to \infty$, almost surely. From this the assertion follows by standard calculus.

LEMMA 5. *The following relation holds:*

$$\mathbf{E}\left|\frac{1}{T}\int_0^T \|\bar{x}(t \mid -\infty) - \hat{x}(t \mid 0)\|^2 dt\right|^2 \to 0$$

as $T \to \infty$.

Proof. Let $\varphi(T) = (1/T)\int_0^T \|\bar{x}(t \mid -\infty) - \hat{x}(t \mid 0)\|^2 dt$. By virtue of Lemma 4 one can find, for any $\delta_1 > 0$, $T_1(\delta_1)$ such that $T > T_1(\delta_1)$ implies $\varphi(T) \leq \delta_1$ on a set with probability measure one. Hence, since $\varphi(T) \geq 0$

$$\mathbf{E} \mid \varphi(T) \mid^2 \leq \delta_1 \mathbf{E}\varphi(T) = \frac{\delta_1}{T}\int_0^T \mathbf{E}\|\bar{x}(t \mid -\infty) - \hat{x}(t \mid 0)\|^2 dt \,.$$

But by Lemma 2, (30), $(1/T)\int_0^T \mathbf{E}\|\bar{x}(t \mid -\infty) - \hat{x}(t \mid 0)\|^2 dt$ is smaller than any $\delta_2 > 0$ for sufficiently large T, $T > T_2(\delta_2)$. Thus we can choose for any $\varepsilon > 0$, $\delta_1 > 0$ and $\delta_2 > 0$ such that $\delta_1\delta_2 < \varepsilon$ and $\mathbf{E} \mid \varphi(T) \mid^2 < \varepsilon$ as soon as $T > \max(T_1(\delta_1), T_2(\delta_2))$.
Now the proof of Proposition 1 follows from Lemmas 3 and 5 and from (13) and (14).

4. THE ALMOST SURE APPROXIMATION PROPERTY

PROPOSITION 2. *The following relation holds:*

$$\frac{1}{T}\left[\bar{L}_T(\theta_0) - L_T(\theta_0)\right] \to 0 \quad \text{a.s.}$$

as $T \to \infty$.

Proof of this assertion follows from (14) and the fact that Lemma 4 above implies also

$$\frac{1}{T}\int_0^T [C(\bar{x}(t \mid -\infty) - \hat{x}(t \mid 0))]^* dv(t \mid 0) \to 0 \quad \text{a.s.}$$

as $T \to \infty$ by virtue of Lemmas 1 and 2 in (Dufková, 1977).

Acknowledgments
This work was done during a visit at the Computer and Automation Institute of the Hungarian Academy of Sciences (MTA SZTAKI) in Budapest. The author thanks Dr. L. Gerenczér from the Hungarian Academy of Sciences for discussions and comments, as well as Dr. A. Bagchi from TH Twente, The Netherlands, for inspiring conversations about identification.

REFERENCES

Arató, M. (1982). *Linear Stochastic Systems with Constant Coefficients.* Springer-Verlag, New York.

Bagchi, A. (1980). Consistent estimates of parameters in continuous time systems. In: *Analysis and Optimization of Stochastic Systems.* Ed. by O.L.R. Jacobs et al., Academic Press, New York, pp. 437–450.

Baram, Y. (1980). Truncation error bounds and convergence of least squares estimates. *SIAM J. Control Optimizat.* **18**, 346–351.

Bauer, H. (1978). *Wahrscheinlichkeitstheorie und Grundzüge der Maßtheorie.* W. de Gruyter, Berlin.

Caines, P.E. and Rissanen, J. (1979). Consistency of maximum likelihood estimation for ARMA processes. *Ann. Statist.* **7**, 297–314.

Dufková, V. (1977). On controlled one-dimensional diffusion processes with unknown parameter. *Adv. Appl. Probab.* **9**, 105–124.

Duncan, T. (1970). Likelihood functions for stochastic signals in white noise. *Information and Control* **16**, 303–310.

Iliff, K.W. and Maine, R.E. (1979). Observations on maximum likelihood estimation of aerodynamic characteristics from flight data. *J.Guidance Control* **2**, 228–234.

Kailath, T. and Ljung, L. (1976). The asymptotic behavior of constant-coefficient Riccati Differential Equations. *IEEE Trans. Automat. Control* **AC-21**, 385–388.

Kallianpur, G. (1980). *Stochastic Filtering Theory.* Springer-Verlag, New York.

Knobloch, H.W. and Kwakernaak, H. (1986). *Lineare Kontrolltheorie.* Akademie Verlag, Berlin.

Koski, T. (1987). Martingale methods for loglikelihood identification of some stochastic evolution equations. In: *Proc. of the Forth IFAC Symp. Control of Distributed Parameter Systems,* Pergamon Press, Oxford, pp. 11–17.

Tugnait, J.K. (1985). Continuous-time system identification on compact parameter sets. *IEEE Trans. Inform. Theory* **IT-31**, 652–659.

Frontiers in Pure and Applied Probab. 1, pp. 145 - 164
H. Niemi et al. (Eds.)
1993 TVP/VSP

CONVERGENCE OF FILTERED EXPERIMENTS TO THE EXPERIMENT GENERATED BY A SEMIMARTINGALE

D. O. KRAMKOV

Steklov Mathematical Institute, Vavilov Street 42, 117966 Moscow GSP-1, Russia

ABSTRACT

Let $\mathcal{E}^n, n \geq 1, \mathcal{E}$ be filtered statistical experiments and $Z^n, n \geq 1, Z$ be their respective families of density processes. In this paper we treat the problem of the convergence of the sequence $Z^n, n \geq 1$, to Z, where the limit experiment $\mathcal{E}$ is generated by some semimartingale X. In spite of the case, where the limit model $\mathcal{E}$ is H-deterministic, the convergence criterion here has to be expressed in terms of not only Hellinger processes, but semimartingale characteristics also.

INTRODUCTION

Let $\mathcal{E}^n$, $n \geq 1$, $\mathcal{E}$ be filtered statistical experiments and let Z^n, $n \geq 1$, Z be their respective families of density processes. In this paper we treat the problem of the convergence of the sequence Z^n, $n \geq 1$, to Z, when the limit experiment $\mathcal{E}$ is generated by some semimartingale X.

In the case, where the limit model $\mathcal{E}$ is H-deterministic (i.e., all the Hellinger processes of $\mathcal{E}$ are deterministic) these results have been proved by Greenwood and Shiryaev (Greenwood and Shiryaev, 1985), Vostrikova (Vostrikova, 1988), Memin (Memin, 1985) and Jacod (Jacod, 1989a); see also (Jacod and Shiryaev, 1987). In these papers the convergence criteria are expressed in terms of the Hellinger processes, a fundamental role being played by the following fact: if an experiment is H-deterministic, then the family of its Hellinger processes determines the laws of its density processes.

In our setting the Hellinger processes of the experiment $\mathcal{E}$ are random (more exactly they are functions of X). As a consequence the family of the Hellinger processes does not specify the distribution of Z. The first example of that type was given by Coquet (Coquet, 1990) for the case of counting processes. In §6 we present another example corresponding to the case of diffusion processes. In §4 we prove that the whole collection of predictable objects, which defines the distribution of Z, consists of the family of semimartingale characteristics of X and of the family of Hellinger processes (of order 1/2) between pairs of probability measures. Hence the convergence criterion has to be expressed in terms of both semimartingale characteristics and Hellinger processes.

Our method is similar to that used in the problem of convergence of stochastic integral processes (see (Jacod and Shiryaev, 1987, Chapter IX. 5)). We suppose that together with $\mathcal{E}^n$ there is a semimartingale X^n and we consider the problem of the convergence of the joint laws of (Z^n, X^n) to the law of (Z, X).

We shall need several results from the general theory of processes, for which we refer to (Jacod and Shiryaev, 1987; Liptser and Shiryaev, 1986). Let us just recall some notation, supposing that here $\mathcal{P}$ is a probability measure on a filtered space $(\Omega, \mathcal{F}, \mathbf{F} = (\mathcal{F}_t)_{t\geq 0})$: the filtration $\mathbf{F} = (\mathcal{F}_t)_{t\geq 0}$ is an increasing $(s \leq t \Rightarrow \mathcal{F}_s \subseteq \mathcal{F}_t)$ and right-continuous $(\mathcal{F}_t = \bigcap_{s>t} \mathcal{F}_s)$ family of σ-fields. We denote by $\mathcal{P}$ the predictable σ-field on $\Omega \times \mathbf{R}_+$. If X is a semimartingale and if H is a predictable process, we write $H \cdot X$ for the stochastic integral process $\int_0^{\cdot} H_s dX_s$ (when well-defined) and X^C denotes the "continuous martingale part" of X and $\mathcal{E}(X)$ stands for the Doleans–Dade exponential of X. If Y is another semimartingale, the "quadratic co-variation" is written as $[X, Y]$. If $\nu = \nu(\omega; dt \times dx)$ is a random measure on $\mathbf{R}_+ \times \mathbf{R}^d$ and if W is a $\mathcal{P} \times \mathbf{R}^d$-measurable function on $\Omega \times \mathbf{R}_+ \times \mathbf{R}^d$, we write $W * \nu$ for the stochastic integral process $\int_0^{\cdot} \int_{\mathbf{R}^d} W(s,x)\nu(ds \times dx)$.

Next, we recall the notion of characteristics of the $\mathcal{P}$-semimartingale X. At first we consider its jump measure $\mu = \mu^X$ on $\mathbf{R}_+ \times \mathbf{R}^d$:

$$\mu(\omega; dt \times dx) = \sum_{s:\Delta X_s(\omega)\neq 0} \varepsilon_{(s,\Delta X_s)}(dt \times dx).$$

Then the third characteristic ν of X is the compensator of μ, which is the unique random measure on $\Omega \times \mathbf{R}_+ \times \mathbf{R}^d$ such that for all non-negative $\mathcal{P} \times \mathbf{R}^d$-measurable functions W, the process $W * \nu$ is predictable, and $\mathbf{E}_{\mathcal{P}}(W * \mu) = \mathbf{E}_{\mathcal{P}}(W * \nu)$. Secondly, we fix a truncation function on $\mathbf{R}^d$, i.e., a bounded function $h : \mathbf{R}^d \longrightarrow \mathbf{R}^d$ with compact support and $h(x) = x$ in a neighborhood of 0. Then the first characteristic B is the unique predictable process with finite variation and $B_0 = 0$, such that $X - \sum_{s\leq \cdot}[\Delta X_s - h(\Delta X_s)] - B$ is a d-dimensional local martingale. Finally, the second characteristic $C = (C^{i,j})_{i,j\leq d}$ is defined by $C^{i,j} = [X^{i,c}, X^{j,c}]$. We denote by $\mathbf{C}_2(\mathbf{R}^d)$ the collection of bounded continuous functions $g : \mathbf{R}^d \longrightarrow \mathbf{R}^d$, which are equal to 0 in some neighborhood of $x = 0$ and which have a limit when x tends to infinity.

If $F = (F_t)_{t\geq 0}$ and $G = (G_t)_{t\geq 0}$ are positive nondecreasing functions we write $F \prec G$ if the difference $G - F$ is a nondecreasing function.

1. MAIN RESULT

We suppose that for $n \geq 1$ there are a filtered statistical experiment $\mathcal{E}^n = (\Omega^n, \mathcal{F}^n, \mathbf{F}^n, (P^n_\alpha : \alpha \in \Theta))$ and a $\mathbf{F}^n$-adapted d-dimensional process $X^n = (X^{n,i})_{i=1,\dots,d}$. We assume that for $\alpha \in \Theta$ the process X^n is a P^n_α-semimartingale and we denote by $(B^{n,\alpha}, C^{n,\alpha}, \nu^{n,\alpha})$ its predictable characteristics (the truncation function h being fixed). For $\alpha, \beta \in \Theta$ we write $Z^{n,\beta}(\alpha) = (Z^{n,\beta}_t(\alpha))_{t\geq 0}$ for the density process of P^n_β w.r.t. (with respect to) P^n_α (i.e., $Z^{n,\beta}_t(\alpha)$ is the density of the absolutely continuous part of P^n_β w.r.t. P^n_α in restriction to $\mathcal{F}^n_t$).

As a limit model for the sequence $(\mathcal{E}^n, X^n)_{n\geq 1}$ we will consider the experiment $\mathcal{E} = (\Omega, \mathcal{F}, \mathbf{F}, (P_\alpha : \alpha \in \Theta))$ and the $\mathbf{F}$-adapted process X such that

$$\begin{cases} (\Omega, \mathcal{F}, \mathbf{F}) & \text{is the canonical space } (D(\mathbf{R}^d), \mathcal{D}(\mathbf{R}^d), \mathbf{D}(\mathbf{R}^d)), \\ \text{and } X & \text{is the canonical process: } X_t(\gamma) = \gamma_t, \end{cases} \tag{1}$$

where $D(\mathbf{R}^d)$ is the Skorokhod space of right continuous with left-hand limits functions $\gamma : \mathbf{R}_+ \longrightarrow \mathbf{R}^d, \mathcal{D}_t^0(\mathbf{R}^d)$ is the σ-field, generated by the mappings $\gamma \longrightarrow \gamma(s)$, $s \leq t$, $\mathcal{D}(\mathbf{R}^d) = \bigcup_{t\geq 0} \mathcal{D}_t^0(\mathbf{R}^d)$, $\mathcal{D}_t(\mathbf{R}^d) = \bigcap_{s>t} \mathcal{D}_s^0(\mathbf{R}^d)$ and $\mathbf{D}(\mathbf{R}^d) = (\mathcal{D}_t(\mathbf{R}^d))_{t\geq 0}$.
We will assume that for $\alpha \in \theta$ there are the following predictable objects on $(D(\mathbf{R}^d),\ \mathcal{D}(\mathbf{R}^d), \mathbf{D}(\mathbf{R}^d))$, playing the role of characteristics of the limiting (left quasi-continuous) semimartingale X (the truncation function h being the same):
$B^\alpha = (B^{\alpha,i})_{i=1,\ldots,d}(\gamma)$ is a continuous predictable process of bounded variation, $B_0^\alpha = 0$;
$C^\alpha = (C^{\alpha,ij})_{i,j=1,\ldots,d}(\gamma)$ is a continuous process with values in the set of all $d \times d$ symmetric non-negative matrices, such that for $s < t$ the difference $C_t^\alpha - C_s^\alpha$ is a non-negative matrix, $C_0^\alpha = 0$;
$\nu^\alpha = \nu^\alpha(\gamma; dt \times dx)$ is a predictable random measure on $\mathbf{R}_+ \times \mathbf{R}^d$, such that $\nu^\alpha(\mathbf{R}_+ \times \{0\}) = 0$, $\nu^\alpha(\{t\} \times \mathbf{R}^d) = 0$ and $1 \wedge |x|^2 * \nu_t^\alpha < +\infty$, $t \geq 0$.
For $\alpha, \beta \in \Theta$ we associate with the characteristics $(B^\alpha, C^\alpha, \nu^\alpha)$ and $(B^\beta, C^\beta, \nu^\beta)$ the partial Hellinger process $H^{\mathbf{T},\alpha\beta} = (H_t^{\mathbf{T},\alpha\beta})_{t\geq 0}$ (see (Jacod, 1987, 1990; Jacod and Shiryaev, 1987, Chapter IV. 3)). The construction of $H^{\mathbf{T},\alpha\beta}$ is given in §4 (formula (28)).
We set as well for $\alpha \in \Theta$

$$\widetilde{C}^\alpha = C^\alpha + ff^T * \nu^\alpha.$$

We will assume that the prescribed collection of predictable characteristics satisfies the following technical conditions:

I_1) *(strong majorization condition)* for all $\alpha, \beta \in \Theta$ there is a continuous non-decreasing function $F(\alpha, \beta)$ such that for each $\gamma \in D(\mathbf{R}^d)$:

$$H^{\mathbf{T},\alpha\beta} \circ \gamma + \mathrm{Var}(B^\alpha) \circ \gamma + \mathrm{Var}(B^\beta) \circ \gamma + \widetilde{C}^\alpha \circ \gamma + \widetilde{C}^\beta \circ \gamma \prec F(\alpha, \beta)$$

(where the notation $\circ\gamma$ stands for the function of γ: $f \circ \gamma \equiv f(\gamma)$).

I_2) *(continuity condition)* for all $\alpha, \beta \in \Theta, t \geq 0, g \in C_2(\mathbf{R}^d)$ the functions $\gamma \longrightarrow H_t^{\mathbf{T},\alpha\beta} \circ \gamma$, $B_t^\alpha \circ \gamma$, $\widetilde{C}_t^\alpha \circ \gamma$, $g * \nu_t^\alpha \circ \gamma$ are continuous in Skorokhod topology,

I_3) *(large jumps condition)* for all $\alpha \in \Theta$, $L > 0$

$$\lim_{a\to+\infty} \sup_{\gamma\in D(\mathbf{R}^d)} \nu^\alpha(\gamma; [0, L] \times \{x : |x| > a\}) = 0.$$

Let $\widetilde{C}^{n,\alpha}$ stand for the modified second characteristic of the P_α^n-semimartingale X^n:

$$\widetilde{C}^{n,\alpha} = C^{n,\alpha} + ff^T * \nu^{n,\alpha} - \sum_{s\leq \cdot} \Delta B_s^{n,\alpha}(\Delta B_s^{n,\alpha})^T.$$

Next denote by $H^{n,\alpha\beta}$ the Hellinger process (of order 1/2) between P_α^n and P_β^n (see (Jacod and Shiryaev, 1987, Chapter IV)).

Finally, to avoid trivial complications we will assume that the restrictions of P_α^n on $\mathcal{F}_0^n$ coincide for all $\alpha \in \Theta$.

THEOREM 1. *Let* $\mathrm{I}_1-\mathrm{I}_3$ *hold as well as the following conditions (for all* $\alpha, \beta \in \Theta$ *and* $L > 0$*):*
a) *(convergence condition)*

$$\sup_{t\leq L} |B_t^{n,\alpha} - B_t^\alpha \circ X^n| \xrightarrow{P_\alpha^n} 0,$$

$$\sup_{t\leq L} |\widetilde{C}_t^{n,\alpha} - \widetilde{C}_t^\alpha \circ X^n| \xrightarrow{P_\alpha^n} 0,$$

$$\sup_{t\leq L} |g * \nu_t^{n,\alpha} - g * \nu_t^\alpha \circ X^n| \xrightarrow{P_\alpha^n} 0 \qquad \text{for all } g \in \mathbf{C}_2(\mathbf{R}^d),$$

$$\sup_{t\leq L} |H_t^{n,\alpha\beta} - H_t^{\mathbf{T},\alpha\beta} \circ X^n| \xrightarrow{P_\alpha^n} 0,$$

b) *the laws of* X_0^n *under* P_α^n *converge to some distribution* η *on* $\mathbf{R}^d$,
c) *(uniqueness condition): there is a unique probability measure* P_α *on* $(D(\mathbf{R}^d), \mathcal{D}(\mathbf{R}^d))$ *such that the canonical process* X *is a* P_α*-semimartingale on* $(D(\mathbf{R}^d), \mathcal{D}(\mathbf{R}^d), \mathbf{D}(\mathbf{R}^d), P_\alpha)$ *with the characteristics* $(B^\alpha, C^\alpha, \nu^\alpha)$ *and the initial distribution* η.
Then there are an experiment $\mathcal{E}$ *and a process* X *of the form of* (1) *such that*
(i) X *is a* P_α*-semimartingale with the characteristics* $(B^\alpha, C^\alpha, \nu^\alpha), H^{\mathbf{T},\alpha\beta}$ *is the Hellinger process (of order 1/2) between* P_α *and* P_β,
(ii) *for all* $\alpha, \beta_1, ..., \beta_{N'} \in \Theta$ *the laws of the* $N + d$*-dimensional processes* $(Z^{n,\beta_i}(\alpha), X^{n,j})_{i=1,...,N;\, j=1,...,d}$ *under* P_α^n *weakly converge (in the Skorokhod topology) to the law of the process* $(Z^{\beta_i}(\alpha), X^j)_{i=1,...,N;\, j=1,...,d}$ *under* P_α, *where* $Z^\beta(\alpha)$ *is the density process of* P_β *w.r.t.* P_α.

Remark. In §4 we shall prove that $Z^\beta(\alpha) = Z^{\mathbf{T},\beta}(\alpha)$, where $Z^{\mathbf{T},\beta}(\alpha)$ is the partial density process of P_α w.r.t. P_β relative to X (see (Jacod, 1987, 1989) and §4, formula (29)).
The proof of Theorem 1 will be given in §5. Sections 2–4 contain some necessary results. In §6 we apply Theorem 1 to the problem of the convergence of experiments generated by Markov step processes to an experiment generated by diffusion processes.

2. THE ALDOUS CRITERION FOR A SEQUENCE OF DENSITY PROCESSES

Let $\mathcal{E}^n = (\Omega^n, \mathcal{F}^n, \mathbf{F}^n, (P_\alpha^n : \alpha \in \Theta))$, $n \geq 1$, be a sequence of filtered experiments. In this section we suppose that the parameter set Θ is finite and we denote by $|\Theta|$ the number of its elements. We set $Q^n = |\Theta|^{-1} \sum_{\alpha\in\Theta} P_\alpha^n$ and denote by $Z^n = (z^{n,\alpha})_{\alpha\in\Theta}$ the $|\Theta|$-dimensional likelihood ratio process of the experiment $\mathcal{E}^n$, where $z^{n,\alpha} = (z_t^{n,\alpha})_{t\geq 0}$ is the density process of P_α^n w.r.t. Q^n. We also write $H^{n,\alpha\beta}$ for the Hellinger process (of order 1/2) between the measures P_α^n and P_β^n.

We begin with reminding of some notions. For $n \geq 1$, let $(\Omega^n, \mathcal{F}^n, \mathbf{F}^n, P^n)$ be a stochastic basis endowed with a d-dimensional adapted cadlag process $X^n = (X^{n,i})_{i=1,\dots,d}$ (cadlag means right-continuous with left-hand limits). One says that the sequence $X^n, n \geq 1$, is $(P^n - C)$-*tight*, if the laws of X^n under $P^n, n \geq 1$, are tight (in the Skorokhod topology) and all its limit points are the distributions of continuous processes. One says also that the sequence $X^n, n \geq 1$ satisfies the P^n-*Aldous condition* if

a) the sequence of laws of X_0^n under P^n is tight,
b) for all $N > 0$, $\varepsilon > 0$

$$\lim_{\delta\to 0} \limsup_{n\to\infty} \sup_{S,T\in\mathcal{T}_N^n, S\leq T\leq S+\delta} P^n\left(\|X_T^n - X_S^n\| > \varepsilon\right) = 0,$$

where $\mathcal{T}_N^n$ is the collection of all $\mathbf{F}^n$-stopping times, which are less than N, and $\|\cdot\|$ denotes the Euclidean norm in $\mathbf{R}^d$.

THEOREM 2. *Let for all $\alpha, \beta \in \Theta$ the sequence $H^{n,\alpha\beta}$, $n \geq 1$, be $(P_\alpha^n + P_\beta^n - C)$-tight. Then the sequence Z^n, $n \geq 1$, satisfies the Q^n-Aldous criterion.*

Proof. (a) We begin with the binary case, that is we suppose that the family $(P_\alpha^n : \alpha \in \Theta))$ consists of two probability measures P^n and P'^n. We denote by z^n, z'^n the density processes of P^n and P'^n w.r.t. $Q^n = (P^n + P'^n)/2$ respectively, and by H^n the Hellinger process (of order 1/2) between P^n and P'^n. In view of the equality $z^n + z'^n = 2$ (Q^n–a.s.) it is sufficient to check the Q^n-Aldous condition only for the sequence $z^n, n \geq 1$.
Let $[z^n, z^n]$ stand for the quadratic characteristic of the Q^n-martingale z^n :

$$[z^n, z^n] = \langle z^{n,c}, z^{n,c}\rangle + \sum_{s\leq \cdot}(\Delta z_s^n)^2.$$

Denote by $\langle z^n, z^n\rangle$ its Q^n-compensator. According to the proof of Theorem 4.13 in (Jacod and Shiryaev, 1987) the sequence $z^n, n \geq 1$, meets Q^n-Aldous criterion if the sequence $\langle z^n, z^n\rangle$, $n \geq 1$, is $(Q^n - C)$-tight. Set

$$\Gamma^n = \{z_-^n > 0\}, \quad \Gamma'^n = \{z'^n_- > 0\}, \quad \Gamma''^n = \Gamma^n \cap \Gamma'^n.$$

Since $z^n = I_{\Gamma^n} z^n, z'^n = I_{\Gamma'^n} z'^n$ and $z^n + z'^n = 2$, then

$$\langle z^n, z^n\rangle = \langle z'^n, z'^n\rangle = I_{\Gamma''^n} \cdot \langle z^n, z^n\rangle. \tag{2}$$

Next, on Γ''^n we introduce the increasing process

$$A^n = \frac{1}{8}\left(\frac{1}{z_-^n} + \frac{1}{z'^n_-}\right)^2 \cdot \langle z^{n,c}, z^{n,C}\rangle + \sum_{s\leq\cdot} \varphi\left(\frac{\Delta z_s^n}{z_{s-}^n}, \frac{\Delta z'^n_s}{z'^n_{s-}}\right),$$

where $\varphi(x, y) = (\sqrt{1+x} + \sqrt{1+y})^2/2$ for $x, y \geq -1$. Let us recall that the Q^n-compensator of A^n equals the Hellinger process H^n on Γ''^n (see (Jacod and Shiryaev, 1987, Chapter IV, 1.36)). Since $\frac{1}{z_-^n} + \frac{1}{z'^n_-} \geq 2$, it follows that $\langle z^{n,c}, z^{n,c}\rangle <$

$2H^n$ on Γ''^n, and from (2) and the condition of the theorem we deduce that the sequence

$$\langle z^{n,c}, z^{n,c}\rangle,\ n \geq 1, \text{ is } (Q^n - C)\text{-tight}. \tag{3}$$

For $n \geq 1,\ N > 0$, we define the following increasing processes on Γ''^n :

$$D^n(N) = \sum_{s\leq \cdot}(\Delta z_s^n)^2 I\Big(\frac{|\Delta z_s^n|}{z_{s-}^n} \leq N,\ \frac{|\Delta z_s'^n|}{z_{s-}'^n} \leq N\Big),$$

$$F^n(N) = 4\sum_{s\leq \cdot} I\Big(\frac{|\Delta z_s^n|}{z_{s-}^n} > N\Big),\quad F'^n(N) = 4\sum_{s\leq \cdot} I\Big(\frac{|\Delta z_s'^n|}{z_{s-}'^n} > N\Big)$$

and we denote by $(D^n(N))^{Q^n,p}, (F^n(N))^{Q^n,p}$ and $(F'^n(N))^{Q^n,p}$ their Q^n-compensators on Γ''^n. Since $|\Delta z^n| \leq 2$, we have on Γ''^n :

$$\sum_{s\leq \cdot}(\Delta z_s^n)^2 \prec D^n(N) + F^n(N) + F'^n(N).$$

In view of (2), we deduce from what was said above that the sequence of Q^n-compensators $(\sum_{s\leq \cdot}(\Delta z_s^n)^2)^{Q^n,p}, n \geq 1$, is $(Q^n - C)$-tight, if the following conditions hold:

$$\text{for } N > 0, \text{ the sequence } I_{\Gamma''^n}\cdot (D^n(N))^{Q^n,p}, n \geq 1, \text{ is } (Q^n - C)\text{-tight}, \tag{4}$$

for $\varepsilon > 0,\ L > 0$

$$\lim_{N\to\infty}\ \limsup_{n} Q^n\Big(I_{\Gamma''^n}\cdot (F^n(N))_L^{Q^n,p} \geq \varepsilon\Big) = 0, \tag{5}$$

$$\lim_{N\to\infty}\ \limsup_{n} Q^n\Big(I_{\Gamma''^n}\cdot (F'^n(N))_L^{Q^n,p} \geq \varepsilon\Big) = 0. \tag{6}$$

We begin with (4). For $N > 0$, there is a constant $K(N)$ such that for $-1 \leq x \leq N$, $-1 \leq y \leq N$

$$(x-y)^2 \leq K(N)\varphi(x,y).$$

Since $\left|\frac{\Delta z^n}{z_-^n} - \frac{\Delta z'^n}{z_-'^n}\right| = 2\frac{|\Delta z^n|}{z_-^n z_-'^n} \geq |\Delta z^n|$, it follows that $D^n(N) \prec K(N)A^n$ on Γ''^n and therefore $(D^n(N))^{Q^n,p} \prec K(N)H^n$ on Γ''^n. Now (4) follows from the condition of the theorem.

Let us pass to (5). According to the Lenglart–Rebolledo inequality (see (Jacod and Shiryaev, 1987, Liptser and Shiryaev, 1986))

$$Q^n\Big(I_{\Gamma''^n}\cdot (F^n(N))_L^{Q^n,p} \geq \varepsilon\Big)$$

$$\leq \lim_{\eta\to 0}\left\{\frac{1}{\varepsilon}\Big[\eta + E_{Q^n}(\sup_{s\leq L} I_{\{s\in\Gamma''^n\}}\Delta F_s^n(N))\Big] + Q^n\Big(I_{\Gamma''^n}\cdot F^n(N)_L \geq \eta\Big)\right\}$$

$$= \frac{4+\varepsilon}{\varepsilon} Q^n\Big(\sup_{s\leq L} I_{\Gamma''^n} \frac{|\Delta z_s^n|}{z_{s-}^n} \geq N\Big). \tag{7}$$

Since $z^n + z'^n = 2$, then for $N \geq 2$ and $|\Delta z^n|/z_{s-}^n \geq N$, we have $|\Delta z_s'^n|/z_{s-}'^n \leq N/(N-1) \leq 2$ and $\varphi/(\Delta z_s^n/z_{s-}^n, \Delta z_s'^n/z_{s-}'^n) \geq (\sqrt{N+1}-3)^2/2 = K(N)$. It follows that

$$Q^n\Big(\sup_{s\leq L} I_{\{s\in\Gamma''^n\}} \frac{|\Delta z^n|}{z_{s-}^n} \geq N\Big) \leq Q^n\Big(I_{\Gamma''^n}\cdot A^n \geq K(N)\Big). \tag{8}$$

Using the Lenglart–Rebolledo inequality, we get

$$Q^n\Big(I_{\Gamma''^n}\cdot A_L^n \geq K(N)\Big) \leq \frac{1}{\sqrt{K(N)}} + Q^n\Big(H_L^n \geq K(N)\Big). \tag{9}$$

By virtue of the theorem condition the right-hand side in (9) tends to zero as $N \to \infty$, $n \to \infty$. Now (5) follows from (7), (8) and (9).

The condition (6) may be verified in a similar way. So the sequence $(\sum_{s\leq\cdot}(\Delta z_s^n)^2, n \geq 1)^{Q^n,p}$, is $(Q^n - C)$-tight, and in view of (3) the sequence $\langle z^n, z^n\rangle$, $n \geq 1$, is $(Q^n - C)$-tight as well. The result follows.

b) (General case). Let $\alpha \in \Theta$. By virtue of part (a), the sequence $z^{n,\alpha}/(z^{n,\alpha} + z^{n,\beta})$, $n \geq 1$, meets the P_α^n-Aldous condition for each $\beta \in \Theta$. Moreover, setting $\tau^n = \inf\{t \geq 0 : z_t^{n,\alpha} \leq \varepsilon\}$ we deduce that

$$P_\alpha^n(\inf_{t\geq 0} z_t^{n,\alpha} \leq \varepsilon) \leq \int z_{\tau^n}^{n,a} I(z_{\tau^n}^{n,\alpha} \leq \varepsilon)dQ \leq \varepsilon.$$

Finally in view of $\sum_{\beta\in\Theta} z^{n,\beta} = |\Theta|$ and from what was said above we deduce step by step that the P_α^n-Aldous condition holds for the following sequences: $z^{n,\alpha}/z^{n,\beta}, n \geq 1$; $|\Theta|/z^{n,\alpha} = \sum_{\beta\in\Theta} z^{n,\beta}/z^{n,\alpha}, n \geq 1$; $z^{n,\alpha}, n \geq 1$; $z^{n,\beta} = z^{n,\alpha}z^{n,\beta}/z^{n,\alpha}, n \geq 1$; $Z^n = (z^{n,\beta})_{\beta\in\Theta}$, $n \geq 1$.

So the sequence Z^n, $n \geq 1$, meets the P_α^n-Aldous condition for each $\alpha \in \Theta$. Hence this sequence satisfies the Q^n-Aldous condition as well and the proof is completed.

3. THE IDENTIFICATION OF THE LIMITING EXPERIMENT

In this section we treat the following problem: if the sequence of density processes converges, will some prescribed process be the Hellinger process of the limiting experiment?

3.1. Assume that we are in the setting of the previous section. Moreover, we suppose that each experiment $\mathcal{E}^n = (\Omega^n, \mathcal{F}^n, \mathbf{F}^n, (P_\alpha^n : \alpha \in \Theta))$ is endowed with a d-dimensional $\mathbf{F}^n$-adapted process $X^n = (X^{n,i})_{i=1,\dots,d}$.

As a limiting model for the sequence $(\mathcal{E}^n, X^n), n \geq 1$, we consider the experiment $\mathcal{E} = (\Omega, \mathcal{F}, \mathbf{F}, (P_\alpha : \alpha \in \Theta))$ and the process $X = (X^i)_{i=1,\dots,d}$ such that $(\Omega, \mathcal{F}, \mathbf{F}) = (D(\mathbf{R}^{|\Theta|+d}), \mathcal{D}(\mathbf{R}^{|\Theta|+d}), \mathbf{D}(\mathbf{R}^{|\Theta|+d}))$ and (Z, X) is the canonical process, where $Z = (z^\alpha)_{\alpha\in\Theta}$ is the likelihood ratio process of $\mathcal{E}$, i.e., $z^\alpha = (z_t^\alpha)_{t\geq 0}$ is the density process of P_α w.r.t. $Q = (1/|\Theta|)\sum_{\alpha\in\Theta} P_\alpha$.

Let $a = (a_\alpha)_{\alpha\in\Theta}$ be a collection of non-negative numbers such that $\sum_{\alpha\in\Theta} a_\alpha = 1$. We suppose that on the filtered space $(D(\mathbf{R}^d), \mathcal{D}(\mathbf{R}^d), \mathbf{D}(\mathbf{R}^d))$ there is a predictable nondecreasing process $h(a) = (h_t(a))_{t\geq 0} \circ \gamma$, $\gamma \in D(\mathbf{R}^d)$. Then we set

$I(a) = \{\alpha \in \Theta : a_\alpha > 0\}$ and denote by $h^n(a) = (h_t^n(a))_{t\geq 0} \circ \gamma$ the Hellinger process of order a for the experiment $\mathcal{E}^n$.

THEOREM 3. *Let S be a dense subset of $\mathbf{R}_+$ such that $Q(\Delta X_t \neq 0$ or $\Delta Z_t \neq 0) = 0$ for $t \in S$, and let the following conditions hold:*
(i) $\mathcal{L}(Z^n, X^n | Q^n) \longrightarrow \mathcal{L}(Z, X | Q) = Q$,
(ii) *for all $t \in S$ and $\alpha \in I(a)$*

$$\mathcal{E}(-h_t^n(a)) - \mathcal{E}(-h_t(a)) \circ X^n \xrightarrow{P_\alpha^n} 0,$$

(iii) *for all $t \in S$ the map $D(\mathbf{R}^d) \longrightarrow \mathcal{E}(-h_t(a)) \circ \gamma$ is Skorokhod-continuous,*
(iv) *for each $t \in S$ there is a constant $C_t(a) > 0$ such that for all $\gamma \in D(\mathbf{R}^d)$*

$$\mathcal{E}(-h_t(a)) \circ \gamma \geq C_t(a).$$

Then $h(a)$ is the Hellinger process of order a for the experiment $\mathcal{E}$.

Proof. We set $Q^n(a) = |I(a)|^{-1} \sum_{\alpha \in I(a)} P_\alpha^n$, $Q(a) = |I(a)|^{-1} \sum_{\alpha \in I(a)} P_\alpha$ and we denote by $z^{n,\alpha}(a)\,(z^\alpha(a))$ the density process of $P_\alpha^n\,(P_\alpha)$ w.r.t. $Q^n(a)\,(Q(a))$. Then we set $Y^n(a) = \prod_{\alpha\in\Theta}(z^{n,\alpha}(a))^{a_\alpha}$ and $Y(a) = \prod_{\alpha\in\Theta}(z(a)^\alpha)^{a_\alpha}$. According to (Jacod, 1989b) it is sufficient to prove that the process $N(a) = Y(a)\mathcal{E}^{-1}(-h(a))$ is a Q-local martingale.
It is easily to see that (i) implies
(i′) $\mathcal{L}(Z^n, X^n | Q^n(a)) \longrightarrow \mathcal{L}(Z, X | Q(a))$.
We fix $t_0 \in S$. According to (Jacod and Shiryaev, 1987, Theorem 1.4) in view of (i′), (ii), (iv), the process $M = N_{\cdot\wedge t_0}(a)$ is a $Q(a)$-martingale, if there is a sequence of uniformly (in n) bounded $Q^n(a)$-martingales $M^n, n \geq 1$, such that for $t \in S$, $t \leq t_0$,

$$M_t^n - M_t \circ (Z^n, X^n) \xrightarrow{Q^n(a)} 0. \tag{10}$$

We fix $t_1 > t_0$, $t_1 \in S$, and we set $2\delta = t_1 - t_0$. Then, for $n \geq 1$, we define the predictable stopping time

$$\tau^n = \inf\left\{t > 0 : \mathcal{E}(-h_t^n(a)) \leq \frac{1}{2} C_{t_1}(a)\right\},$$

where the constant $C_{t_1}(a)$ is taken from (iv). From (ii) and (iv) it follows that for $\alpha \in I(a)$

$$\limsup_{n\to\infty} P_\alpha^n(\tau^n \leq t_1) = \limsup_{n\to\infty} P_\alpha^n\Big(\mathcal{E}(-h_{t_1}^n(a)) \leq \frac{1}{2}C_{t_1}(a)\Big)$$

$$\leq \limsup_{n\to\infty} P_\alpha^n\Big(|\mathcal{E}(-h_{t_1}^n(a)) - \mathcal{E}(-h_{t_1}(a)) \circ X^n| > \frac{1}{2}C_{t_1}(a)\Big) = 0. \tag{11}$$

Next, for $n \geq 1$, we define a stopping time σ^n such that $\sigma^n < \tau^n$ (Q^n-a.s.) and for $\alpha \in I(a)$

$$P_\alpha^n(\sigma^n < \tau^n - \delta) \leq \frac{1}{n}$$

(the existence of such σ^n follows from the predictability of τ^n).

Now from (11) we deduce for $\alpha \in I(a)$

$$\limsup_{n\to\infty} P^n_\alpha(\sigma^n \le t_0) = 0 \tag{12}$$

and from the definition of τ^n we obtain for $n \ge 1$

$$\mathcal{E}(-h^n(a))_{\cdot\wedge\sigma^n} \ge \frac{1}{2} C_{t_1}(a) \qquad (Q^n\text{-a.s.}). \tag{13}$$

According to (Jacod, 1989b) the process $M^n = N^n_{\cdot\wedge\sigma^n}(a) = Y^n(a)\cdot\mathcal{E}^{-1}(-h^n(a))$ is a $Q^n(a)$-martingale, and in view of (13) and from the inequality $Y^n(a) \le |I(a)|$, the processes $M^n, n \ge 1$, are uniformly bounded. Moreover, the following inequality holds (for $n \ge 1, \varepsilon > 0$, and $t \le t_0, t \in S$):

$$|I(a)|Q^n(a)\Big(|M^n_t - M_t \circ (Z^n, X^n)| > \varepsilon\Big)$$

$$= \sum_{\alpha\in I(a)} P^n_\alpha(|Y^n(a))_{t\wedge\sigma^n}\mathcal{E}^{-1}(-h^n(a))_{t\wedge\sigma^n}$$

$$-Y^n_t(a)\mathcal{E}^{-1}(-h_t(a))\circ X^n| > \varepsilon)) \le \sum_{\alpha\in I(a)} P^n_\alpha(\sigma^n \le t_0)$$

$$+ \sum_{\alpha\in I(a)} P^n_\alpha(|I(a)|\,|\mathcal{E}^{-1}(-h^n_t(a)) - \mathcal{E}^{-1}(-h_t(a))\circ X^n| > \varepsilon). \tag{14}$$

By virtue of (ii), (iv), and (12) the right-hand side vanishes as $n \to \infty$. So (10) holds for all $t_0 \in S$ and the proof is completed.

3.2. In the following proposition we exhibit more tractable conditions then the ones in the previous theorem.

THEOREM 4. *Let the process $h(a) = (h_t(a))_{t\ge0} \circ \gamma$ be continuous (i.e., all its paths are continuous functions) and let the following conditions be valid:*
(i_1) $\mathcal{L}(Z^n, X^n|Q^n) \longrightarrow \mathcal{L}(Z, X|Q) = Q$,
(ii_2) *for all $L \ge 0$ and $\alpha \in I(a)$*

$$\sup_{t\le L} |h^n_t(a) - h_t(a)\circ X^n| \xrightarrow{P^n_\alpha} 0,$$

(iii_3) *for each $t \ge 0$ the function $h_t(a) : D(\mathbf{R}^d) \longrightarrow h_t(a)\circ\gamma$ is Skorokhod-continuous,*
(iv_4) *for each $t \ge 0$ there is a constant $C'_t(a) < \infty$ such that for all $\gamma \in D(\mathbf{R}^d)$*

$$h_t(a)\circ\gamma \le C'_t(a).$$

Then $h(a)$ is the Hellinger process of order a for the experiment $\mathcal{E}$.

Proof. Let us check the conditions of Theorem 3. Since the process $h(a)$ is continuous, then $\mathcal{E}(-h(a)) = \exp(-h(a))$ and so conditions (iii) and (iii_1), (iv) and (iv_1) are equivalent.

Let $\alpha \in I(a)$. We fix $t > 0$ and we set

$$\omega_t^n(a) = \sup_{s\le t}\left|\frac{\log(1-\Delta h_s^n(a)) + \Delta h_s^n(a)}{\Delta h_s^n(a)}\right|.$$

Since the process $h(a)$ is continuous, (ii_1) implies

$$\sup_{s\le t}\left|\Delta h_s^n(a)\right| \xrightarrow{P_\alpha^n} 0$$

and so

$$\omega_t^n(a) \xrightarrow{P_\alpha^n} 0. \tag{15}$$

Moreover,

$$\begin{aligned} &\left|\mathcal{E}(-h_t^n(a)) - \mathcal{E}(-h_t(a))\circ X^n\right| \\ &\le \exp\left(-h_t^n(a)\right)\left|\prod_{s\le t}\left(1-\Delta h_s^n(a)\right)\exp\left(\Delta h_s^n(a)\right) - 1\right| \\ &+\left|\exp\left(-h_t^n(a)\right) - \exp\left(-h_t(a)\right)\circ X^n\right| \le \exp\left(-h_t^n(a)\right)\left|\exp\left(\omega_t^n(a)\right)h_t^n(a) - 1\right| \\ &+\left|\exp\left(-h_t^n(a)\right) - \exp\left(-h_t(a)\right)\circ X^n\right|. \end{aligned} \tag{16}$$

From (15) and (ii) the right-hand term in (16) vanishes as $n \to \infty$. Therefore condition (ii) of Theorem 3 is satisfied and the result follows.

4. THE CHARACTERIZATION OF THE DISTRIBUTIONS OF DENSITY PROCESSES

If $\mathcal{E} = (\Omega, \mathcal{F}, \mathbf{F}, (P, P'))$ is a filtered experiment and $X = (X_i)_{i=1,\dots,d}$ is a $\mathbf{F}$-adapted càdlàg process such that X is a P_α-semimartingale for all $\alpha \in \Theta$, then we can associate with $\mathcal{E}$ and X the family of predictable objects: Hellinger processes and semimartingale characteristics. In this section we treat the converse problem: we have predictable objects and we are going to characterize the model itself.

4.1. For simplicity we restrict our attention to the binary case. Let us consider a filtered experiment $\mathcal{E} = (\Omega, \mathcal{F}, \mathbf{F}, (P, P'))$ and a $\mathbf{F}$-adapted d-dimensional process X, which is a semimartingale under P and P'. We assume that $P = P'$ on $\mathcal{F}_0$. Let Q be a probability measure on $(\Omega, \mathcal{F})$ such that $P \underset{\text{loc}}{\ll} Q$, $P' \underset{\text{loc}}{\ll} Q$, X be a Q-semimartingale and $Q = P = P'$ on $\mathcal{F}_0$. We denote by z, z' the density processes of the measures P, P' w.r.t. Q and we set $\Gamma = \{z_- > 0\}$, $\Gamma' = \{z'_- > 0\}$, $\Gamma'' = \Gamma \cap \Gamma'$. Then we fix a truncation function f and we denote by (B, C, ν), $(\mathbf{B}', C', \nu')$, and $(\overline{B}, \overline{C}, \bar{\nu})$ the semimartingale characteristics of X relative to P, P', and Q respectively.

Following (Jacod, 1987), let us construct the processes Z^{T} and H^{T}, arising in the theory of partial likelihood. First the quadratic variation $\overline{C}$ of the continuous martingale part of the Q-semimartingale X admits the representation:

$$\overline{C} = c \cdot A \quad (Q\text{-a.s.}),$$

where (i) $A = (A_t)_{t\geq 0}$ is a nondecreasing continuous predictable process, (ii) $c = (c^{ij})_{i,j=1,\dots,d}$ is a predictable process with values in the set of $d \times d$ symmetric non-negative matrices.
Secondly by Girsanov's theorem (see (Jacod and Shiryaev, 1987; Liptser and Shiryaev, 1986))

$$\begin{cases} B = \overline{B} + c\beta\cdot A + f(x)(Y-1) * \nu, \\ C = \overline{C} = c\cdot A,\ \nu = Y\bar{\nu}, \quad Q\text{-a.s. on } \Gamma, \end{cases}$$

$$\begin{cases} B' = \overline{B} + c\beta'\cdot A + f(x)(Y'-1) * \bar{\nu}, \\ C = \overline{C} = c\cdot A, \nu' = Y'\bar{\nu}, \quad Q\text{-a.s. on } \Gamma', \end{cases}$$

where $\beta = (\beta^i)_{i=1,\dots,d}; \beta' = (\beta'^i)_{i=1,\dots,d}$ are d-dimensional predictable processes, Y, Y' are $\mathcal{P}\times\mathcal{B}(\mathbf{R}^d)$-measurable non-negative functions on $\Omega\times\mathbf{R}_+\times\mathbf{R}^d$ ($\mathcal{P}$ is the predictable σ-field).
Then, according to (Jacod and Shiryaev, 1987, Proposition III, 5.10) we can define on Γ (respectively Γ') a Q-local martingale N (respectively N') by

$$N = \beta^T\cdot X^c + \Big(Y - 1 + \frac{\widehat{Y} - \bar{a}}{1-\bar{a}} I(\bar{a} < 1)\Big) * (\mu - \bar{\nu}),$$

$$N' = \beta'^T\cdot X^c + \Big(Y' - 1 + \frac{\widehat{Y}' - \bar{a}}{1-\bar{a}} I(\bar{a} < 1)\Big) * (\mu - \bar{\nu}), \tag{17}$$

where $\bar{a}_t(\omega) = \bar{\nu}(\omega;\{t\}\times\mathbf{R}^d)$ and $\widehat{Y}_t(\omega) = \int_{\mathbf{R}^d} Y(\omega,t,x)\nu(\omega;\{t\}\times dx))$.
Next, we denote by v (respectively v') the Doleans–Dade exponential of N on Γ (respectively N' on Γ') and define

$$Z^{\mathbf{T}} = \begin{cases} v_t/v'_t, & t\in\Gamma'', \\ 0, & t\notin\Gamma''. \end{cases} \tag{18}$$

Then $\{v > 0\} \subset \{z > 0\}$ (Q–a.s.) and $Z^{\mathbf{T}}$ is a P-non-negative supermartingale (see (Jacod, 1987)).
Finally, we define the predictable non-decreasing process $H^{\mathbf{T}}$ on Γ'' by

$$H^{\mathbf{T}} = \frac{1}{8}(\beta-\beta')^T c(\beta-\beta')\cdot A + \frac{1}{2}\big(\sqrt{Y}-\sqrt{Y'}\big)^2 * \bar{\nu}$$

$$+\frac{1}{2}\sum_{s\leq\cdot}(\sqrt{1-a_s} - \sqrt{1-a'_s})^2. \tag{19}$$

According to (Jacod, 1990, §7), if the density process $Z = z'/z$ of P' w.r.t. P equals $Z^{\mathbf{T}}$, then the Hellinger process (of order 1/2) between P and P' equals $H^{\mathbf{T}}$ up to a Q-null set on Γ''. In the sequel we need the conversion of this result.

THEOREM 5. *Let $H^{\mathbf{T}}$ be the Hellinger process between P and P' on Γ''. Then $Z^{\mathbf{T}}$ is the density process of P' w.r.t. P.*

Remark. In the next section we give a direct construction of $Z^{\mathbf{T}}$ and $H^{\mathbf{T}}$ by means of the semimartingale characteristics of X relative to P and P'. So,

Theorem 5 contains the desired characterization result: if all the semimartingale characteristics are given and it is known that the Hellinger process equals $H^{\mathbf{T}}$, then the density process is well-defined: $Z = Z^{\mathbf{T}}$.

Proof of Theorem 5. According to (Jacod and Shiryaev, 1987, III 5.17) the density processes z and z' have the following form on Γ'':

$$z = 1 + z_- \cdot N + z_- \cdot \widetilde{N}, \quad z' = 1 + z'_- \cdot N' + z'_- \cdot \widetilde{N}',$$

where the processes N and N' are given by (17) and $\widetilde{N}$ and $\widetilde{N}'$ are Q-local martingales on Γ'' such that, up to Q-null set on Γ'',

$$\langle \widetilde{N}^c, X^c \rangle = 0, \quad M^Q_\mu[\Delta\widetilde{N} | \mathcal{P} \times \mathcal{B}(\mathbf{R}^d)] = 0,$$

$$\langle \widetilde{N}'^c, X^c \rangle = 0, \quad M^Q_\mu[\Delta\widetilde{N}' | \mathcal{P} \times \mathcal{B}(\mathbf{R}^d)] = 0, \tag{20}$$

where $M^Q_\mu(d\omega, dt, dx) = P(d\omega)\mu(d\omega, dx)$.

Set $M = \frac{1}{z_-} \cdot z = N + \widetilde{N}$, $M' = \frac{1}{z'} \cdot z' = N' + \widetilde{N}'$. Since Z and $Z^{\mathbf{T}}$ are P-nonnegative supermartingales, they are P-indistinguishable everywhere if they are P-indistinguishable on the predictable interval $\widetilde{\Gamma} = \{Z_- > 0\} \cap \{Z^{\mathbf{T}}_- > 0\}$ or, by virtue of Doleans–Dade formula, if the processes $K = \frac{1}{Z_-} \cdot Z$ and $L = (1/Z^{\mathbf{T}}_-) \cdot Z^{\mathbf{T}}$ are Q-indistinguishable on $\widetilde{\Gamma}$. Note that $P\{z_- > 0\} = 1$, so $\{Z_- > 0\} = \{z'_- > 0\} = \Gamma''$ (P-a.s.) and therefore $\widetilde{\Gamma} \subset \Gamma''$ (P-a.s.).

We denote by A and B the predictable nondecreasing processes from Doob–Meyer decomposition of the P-supermartingales K and L, respectively. The processes $1 + \Delta M = z/z_-$ and $1 + \Delta N = v/v_-$ are strictly positive (P-a.s.) and the following representations (P-a.s.) are valid on $\widetilde{\Gamma}$ (see (Jacod, 1989b, 1990)):

$$A = \Big(\sum_{s \le \cdot}(1 + \Delta M')I(1 + \Delta M = 0)\Big)^{Q,p}, \quad K^{P,c} = M'^c - M^c - \langle M'^c - M^c, M^c \rangle,$$

$$\Delta K = \frac{1 + \Delta M'}{1 + \Delta M} - 1, \quad B = \Big(\sum_{s \le \cdot}(1 + \Delta N')I(1 + \Delta N = 0)\Big)^{Q,p},$$

$$L^{P,c} = N'^c - N^c - \langle N'^c - N^c, N^c \rangle, \quad \Delta L = \frac{1 + \Delta N'}{1 + \Delta N} - 1,$$

where $K^{P,c}$ and $L^{P,c}$ are the continuous martingale parts of the P-supermartingales K and L, respectively; M^c, M'^c, N^c, and N'^c are the continuous martingale parts of Q-local martingales M, M', N, and N', and if S is a process with Q-locally intergrable variation, then $S^{Q,p}$ denotes its Q-compensator.

Let a random measure $p(\omega, dt)$ on $(\mathbf{R}_+, \mathcal{B}(\mathbf{R}_+))$ be defined by

$$p(\omega, F) = \sum_{s \in F} I(\bar{a}_s(\omega) > 0)I(\Delta X = 0), \quad F \in \mathcal{B}(\mathbf{R}_+).$$

We associate with it the Doleans–Dade measure

$$M^Q_p(d\omega, dt) = Q(d\omega)p(\omega, dt).$$

For the Q-compensator of p, we have (see (Liptser and Shiryaev, 1986, p.133))

$$q(\omega, F) = \sum_{s \in F} I(\bar{a}_s(\omega) > 0)(1 - \bar{a}_s(\omega)).$$

From (20) and the fact that $\widetilde{N}$ and $\widetilde{N}'$ are Q-local martingales on Γ'', it follows that on Γ'' we have (see (Liptser and Shiryaev, 1986, p. 134))

$$M_p^Q[\Delta \widetilde{N} | \mathcal{P}] = M_p^Q[\Delta \widetilde{N}' | \mathcal{P}] = 0. \tag{21}$$

Let us assume for the time being that the following equalities hold on Γ'':

$$M'^c - M^c = N'^c - N^c, \tag{22}$$

$$(1 + \Delta M)(1 + \Delta N') = (1 + \Delta M')(1 + \Delta N). \tag{23}$$

Since (20) implies $\langle N'^c - N^c, N^c \rangle = \langle N'^c - N^c, M^c \rangle$, we deduce from (22) that $L^{P,c} = K^{P,c}$ up to a P-null set on $\widetilde{\Gamma}$. Next, (23) implies $\Delta K = \Delta L$ (P-a.s.) on $\widetilde{\Gamma}$. Let us show that (23) also implies the P-indistinguishability of A and B. Since $\{\bar{a} = 1\} \cap \Gamma \subset \{a = 1\}$ and $\{\bar{a} = 1\} \cap \Gamma' \subset \{a' = 1\}$ Q-a.s., relation (17) yields

$$\Delta N = (Y - 1) I(\Delta X \neq 0) + \frac{\bar{a} - a}{1 - \bar{a}} I(\bar{a} < 1) I(\Delta X = 0),$$

$$\Delta N' = (Y' - 1) I(\Delta X \neq 0) + \frac{\bar{a} - a'}{1 - \bar{a}} I(\bar{a} < 1) I(\Delta X = 0). \tag{24}$$

Then the sets $\{1 + \Delta N = 0\} \cap \Gamma$ and $(\{Y = 0, \Delta X \neq 0\} \cup \{a = 1, \bar{a} < 1, \Delta X = 0\}) \cap \Gamma$ are Q-indistinguishable, and in view of (21) and (23), we obtain

$$A = \Big(\sum_{s \leq \cdot} (1 + \Delta M') I(1 + \Delta M = 0)\Big)^{Q,p} = M_\mu^Q \big[1 + \Delta M' | \mathcal{P} \times \mathcal{B}(\mathbf{R}^d)\big] I(Y = 0) * \bar{\nu}$$

$$+ M_p^Q \big[1 + \Delta M' | \mathcal{P}\big] I(a = 1) * q = M_\mu^Q \big[1 + \Delta N' | \mathcal{P} \times \mathcal{B}(\mathbf{R}^d)\big] I(Y = 0) * \bar{\nu}$$

$$+ M_p^Q \big[1 + \Delta N' | \mathcal{P}\big] I(a = 1) * q = \big(\sum_{s \leq \cdot} (1 + \Delta N') I(1 + \Delta N = 0)\big)^{Q,p} = B.$$

Thus, if (22) and (23) hold, then the processes A and B, $K^{P,c}$ and $L^{P,c}$, ΔK and ΔL, and hence K and L are P-indistinguishable on $\widetilde{\Gamma}$ and the result follows. So, to complete the proof we must show that (22) and (23) hold.

The Hellinger process for the experiment $\mathcal{E}$ has the following form on Γ'' (see (Jacod and Shiryaev, 1987, IV, Theorem 1.33)):

$$H = \frac{1}{8} \langle M^c - M'^c \rangle + \big(\sum_{s \leq \cdot} \varphi(\Delta M, \Delta M')\big)^{Q,p},$$

where $\varphi(x, y) = (\sqrt{1 + x} - \sqrt{1 + y})^2 / 2$ is a concave function on $\{x \geq -1, y \geq -1\}$.

Since $\langle \widetilde{N}'^c - N^c, X^c \rangle = 0$ and $N'^c - N^c = (\beta' - \beta)^T \cdot X^c$, we have

$$\langle M'^c - M^c \rangle = \langle N'^c - N^c \rangle + \langle \widetilde{N}'^c - \widetilde{N}^c \rangle \geq (\beta - \beta')^T c(\beta - \beta') \cdot A. \tag{25}$$

Furthermore, for $t \in \Gamma''$,

$$\Big(\sum_{s\leq \cdot} \varphi(\Delta M, \Delta M')\Big)^{Q,p} \geq (\varphi(\Delta M, \Delta M') * \mu)^{Q,p} + (\varphi(\Delta M, \Delta M') * p)^{Q,p}$$

$$= M_\mu^Q \Big[\varphi(\Delta M, \Delta M') | \mathcal{P} \times \mathcal{B}(\mathbf{R}^d)\Big] * \bar{\nu} + M_p^Q \Big[\varphi(\Delta M, \Delta M') | \mathcal{P}\Big] * q$$

$$\geq \varphi\Big(M^Q[\Delta M | \mathcal{P} \times \mathcal{B}(\mathbf{R}^d)], M^Q[\Delta M' | \mathcal{P} \times \mathcal{B}(\mathbf{R}^d)]\Big) * \bar{\nu}$$

$$+\varphi\Big(M^Q[\Delta M | \mathcal{P}], M^Q[\Delta M | \mathcal{P}]\Big) * q, \tag{26}$$

(at the last step we used Jensen's inequality and concavity of the function φ). Next (24), (20), and (21) imply that, on Γ'',

$$M_\mu^Q[\Delta M | \mathcal{P} \times \mathcal{B}(\mathbf{R}^d)] = Y - 1, \quad M_\mu^Q[\Delta M' | \mathcal{P} \times \mathcal{B}(\mathbf{R}^d)] = Y' - 1 \quad M_\mu^Q\text{-a.s.},$$

$$M_p^Q[\Delta M | \mathcal{P}] = \frac{\bar{a} - a}{1 - \bar{a}} I(\bar{a} < 1), \quad M_p^Q[\Delta M' | \mathcal{P}] = \frac{\bar{a} - a'}{1 - \bar{a}} I(\bar{a} < 1) \quad M_p^Q\text{-a.s.}$$

and consequently the inequality (26) takes the form

$$\Big(\sum_{s\leq \cdot} \varphi(\Delta M, \Delta M')\Big)^{Q,p} \geq \varphi(Y - 1, Y' - 1) * \bar{\nu} + \varphi\Big(\frac{\bar{a} - a}{1 - \bar{a}}, \frac{\bar{a} - a'}{1 - \bar{a}}\Big) I(\bar{a} < 1) * q$$

$$= \frac{1}{2}\Big(\sqrt{Y} - \sqrt{Y'}\Big)^2 + \frac{1}{2} \sum_{s\leq \cdot} \Big(\sqrt{1 - a_s} - \sqrt{1 - a'_s}\Big)^2. \tag{27}$$

On the other hand, according to the theorem condition, the Hellinger process H has the form (19), which implies that actually (25), (26), and (27) are equalities. Let us show that this implies (22) and (23).

First the equality in (25) implies that $\langle \widetilde{N}^c - \widetilde{N}'^c \rangle = 0$, so $\widetilde{N}^c - \widetilde{N}'^c \equiv 0$ Q-a.s. on Γ'' and (22) holds.

Second $\varphi(x, y) = 0$ if and only if $x = y$, and, for $u \in (0,1)$, the equality $\varphi(ux + (1-u)x', uy + (1-u)y') = u\varphi(x, y) + (1-u)\varphi(x', y')$ holds if and only if $(1+x)(1+y') = (1+x')(1+y)$. It follows that (26) is an equality if and only if

(i) $\Delta M = \Delta M'$ on the set $\{\Delta X = 0\} \cap \{\bar{a} = 0\} \cap \Gamma''$ (Q-a.s.). Note that on this set $\Delta N = \Delta N' = 0$;

(ii) $(1 + \Delta M) M_\mu^Q[1 + \Delta M' | \mathcal{P} \times \mathcal{B}(\mathbf{R}^d)] = (1 + \Delta M') M_\mu^Q[1 + \Delta M | \mathcal{P} \times \mathcal{B}(\mathbf{R}^d)]$ M_μ^Q-a.s. on Γ'' and therefore $(1 + \Delta M)(1 + \Delta N') = (1 + \Delta M')(1 + \Delta N)$ on the set $\{\Delta X \neq 0\} \cap \Gamma''$ (Q-a.s.)

(iii) $(1+\Delta M) M_p^Q[1+\Delta M' | \mathcal{P}] = (1+\Delta M') M_p^Q[1+\Delta M | \mathcal{P}]$ (M_p^Q -a.s.) on Γ'' and therefore $(1 + \Delta M)(1 + \Delta N') = (1 + \Delta M')(1 + \Delta N)$ on $\{\Delta X = 0\} \cap \{\bar{a} > 0\} \cup \Gamma''$ (Q-a.s.).

Thus, (23) holds, which is the desired result.

4.2. In this section, following to the (Jacod, 1990) (see also (Jacod and Shiryaev, 1987, §3)), we present another construction of the processes Z^{T} and H^{T} from (18) and (19) without using any additional measure Q.
First there are a continuous increasing predictable process A and two predictable processes c, c' with values in the set of $d \times d$ symmetric non-negative matrices, such that

$$C = c \cdot A, \quad C' = c' \cdot A \quad (P + P')\text{-a.s.}\ .$$

Next, let λ be a predictable random measure on $(\mathbf{R}_+ \times \mathbf{R}^d)$ such that $\nu + \nu' \ll \lambda$. Then there are non-negative predictable functions U and U' on $\Omega \times \mathbf{R}_+ \times \mathbf{R}^d$ such that

$$\nu = U\lambda, \quad \nu' = U'\lambda \quad (P + P')\text{-a.s.}\ .$$

Then we define the predictable set Σ and the predictable process $\widetilde{B}$ such that

$$\Sigma = \big\{(\omega, t) : |f(x)(U - U')| * \lambda_t(\omega) < \infty\big\},$$

$$\widetilde{B} = B_t - B'_t - f(x)(U - U') * \lambda_t(\omega), \ t \in \Sigma.$$

For $\widetilde{B}$, we have the decomposition

$$\widetilde{B} = c\widetilde{\beta} \cdot A + \widetilde{b} \cdot A + \widetilde{A} \quad (P + P')\text{-a.s. on}\, \Sigma,$$

where
(i) $\widetilde{\beta}, \widetilde{b}$ are predictable processes, $\widetilde{A}$ is a predictable process with finite variation on intervals strictly enclosed in Σ,
(ii) $d\widetilde{A}^i_t$ and dA_t are $(P + P')$-a.s. singular on Σ,
(iii) for all $(\omega, t) \in \Sigma$ the vector $\widetilde{b}_t(\omega)$ is orthogonal to the image of $\mathbf{R}^d$ under the linear map associated with $c_t(\omega)$.
Furthermore, define

$$\sigma = \inf\{t : |c_t - c'_t| + |\widetilde{A}_t + \widetilde{b} \cdot A_t| > 0\}, \ \Delta = \Sigma \cap [0, \sigma].$$

Since $\Gamma'' \subset \Delta$ $(P + P')$-a.s. (see (Jacod and Shiryaev, 1987, IV, Lemma 3.19)), the process H^{T} on Γ'' has the form

$$H^{\mathrm{T}} = \frac{1}{8}\widetilde{\beta}^T c\widetilde{\beta} I_\Delta \cdot A + \frac{1}{2}\Big(\sqrt{U} - \sqrt{U'}\Big)^2 I_\Delta * \lambda$$

$$+\frac{1}{2}\sum_{s\le\cdot}\Big(\sqrt{1 - a_s} - \sqrt{1 - a'_s}\Big)^2 I(s \in \Delta). \tag{28}$$

Set $\Delta' = \Delta \cap \{H^{\mathrm{T}}_- < +\infty\}$. According to (Jacod, 1987) the formulas

$$\bar{A} = U' I(U = 0) * \lambda + \sum_{s\le\cdot}(1 - a'_s) I(a_s = 1),$$

$$\overline{M}=\widetilde{\beta}^{T}\cdot X^{C}+\Big(\frac{U'}{U}I(U>0)-1+\frac{a'-a}{1-a}I(a<1)\Big)*(\mu-\nu),\ \overline{L}=\overline{M}-\overline{A}$$

define on Δ' an increasing predictable process $\overline{A}$, a P-local martingale $\overline{M}$ and a P-supermartingale $\overline{L}$ such that $\Delta\overline{L}\geq -1$. Moreover, $\Delta'\subset\Gamma''$ $(P+P')$-a.s. and the process $Z^{\mathbf{T}}$ has the form

$$Z^{\mathbf{T}}=\begin{cases}\mathcal{E}_t(\overline{L}) & \text{for } t\in\Delta',\\ 0 & \text{for } t\notin\Delta'.\end{cases} \tag{29}$$

4.3. The following result implied by Theorem 5 plays a key role in the proof of Theorem 1.

Let $\mathbf{F}_{+}^{X}=(\mathcal{F}_{t+}^{X})_{t\geq 0}$ be the natural flow of the σ-fields generated by X:

$$\mathcal{F}_{t+}^{X}=\bigcap_{\varepsilon>0}\big(\sigma(X_s,0\leq s\leq t+\varepsilon)\big).$$

COROLLARY 6. *Let $H^{\mathbf{T}}$ be the Hellinger process of the experiment $\mathcal{E}$ and let the characteristics (B,C,ν) and (B',C',ν') be $\mathbf{F}_{+}^{X}$-adapted. Then the density process Z of the measure P' w.r.t. P is $\mathbf{F}_{+}^{X,P}$-adapted.*

Proof. Replacing $(\Omega,\mathcal{F}_{\infty}^{X},\mathbf{F}_{+}^{X})$ on $(\Omega,\mathcal{F},\mathbf{F})$ in the constructions of §4.2, we deduce that the process $Z^{\mathbf{T}}$, defined in (28), is $\mathbf{F}_{+}^{X,P}$-adapted. Now the result follows from Theorem 5.

5. PROOF OF THEOREM 1

For simplicity, we assume that the parametric set Θ is finite. We denote by $|\Theta|$ the number of elements in Θ and for $n\geq 1$ we use the notation $Z^n=(z^{n,\alpha})_{\alpha\in\Theta}$ for the likelihood ratio process of the experiment $\mathcal{E}^n$, where $z^{n,\alpha}=(z_t^{n,\alpha})_{t\geq 0}$ is the density process of P_α^n w.r.t. $Q^n=(1/|\Theta|)\sum_{\alpha\in\Theta}P_\alpha^n$.

First let us show that the sequence of the $(|\Theta|+d)$-dimensional processes (Z^n,X^n), $n\geq 1$, is Q^n-tight in the Skorokhod topology. By virtue of Aldous theorem (see (Jacod and Shiryaev, 1987) or (Liptser and Shiryaev, 1986)), this is implied by the Q^n-Aldous condition for each of the sequences $Z^n, n\geq 1$, and $X^n, n\geq 1$. Note that, for $\alpha,\beta\in\Theta$, the claims (a) and I_1 imply $(P_\alpha^n+P_\beta^n-C)$-tightness of the sequence $H^{n,\alpha\beta}, n\geq 1$ (see Lemma 3.3 in (Jacod and Shiryaev, 1987)). Then from Theorem 2 of §2 it follows that the sequence $Z^n, n\geq 1$, satisfies Q^n-Aldous criterion. Furthermore, according to (Jacod and Shiryaev, 1987, §4) the claims a) and I_1 imply the P_α^n-Aldous condition for the sequence $X^n, n\geq 1$, for all $\alpha\in\Theta$, and hence the Q^n-Aldous condition as well.

Let a measure Q on $(D(\mathbf{R}^{|\Theta|+d}),\mathcal{D}(\mathbf{R}^{|\Theta|+d}))$ be the limit of the sequence of laws $(Z^{n'},X^{n'})$ under $Q^{n'}$, where (n') is a subsequence of (n). We denote by (Z,X) the canonical process on $(D(\mathbf{R}^{|\Theta|+d}))$. Since for $n\geq 1$ the process $Z^n=(z^{n,\alpha})_{\alpha\in\Theta}$ is a $|\Theta|$-dimensional Q-martingale such that $\sum_{\alpha\in\Theta}z^{n,\alpha}=|\Theta|$, $z^{n,\alpha}\geq 0$, and $z_0^{n,\alpha}=1$, the process $Z=(z^\alpha)_{\alpha\in\Theta}$ is a $|\Theta|$-dimensional Q-martingale on $(D(\mathbf{R}^{|\Theta|+d}),\mathcal{D}(\mathbf{R}^{|\Theta|+d}),\mathbf{D}(\mathbf{R}^{|\Theta|+d}))$ (see (Jacod and Shiryaev, 1987, Theorem 1.10)) such that $\sum_{\alpha\in\Theta}z^\alpha=|\Theta|$ and $z^\alpha\geq 0$, $z_0^\alpha=1$ for all $\alpha\in\Theta$. For

$\alpha \in \Theta$ we define the probability measure P_α on $(D(\mathbf{R}^{|\Theta|+d}), \mathcal{D}(\mathbf{R}^{|\Theta|+d}))$ by setting $P_\alpha(A) = \int I_A z_t^\alpha dQ, A \in \mathcal{D}_t(\mathbf{R}^{|\Theta|+d})$. It is easily seen that $\mathcal{L}(Z^{n'}, X^{n'}|Q^{n'}) \longrightarrow Q = \mathcal{L}(Z, X|Q)$ implies $\mathcal{L}(Z^{n'}, X^{n'}|P_\alpha^{n'}) \longrightarrow \mathcal{L}(Z, X|P_\alpha)$ for $\alpha \in \Theta$. By virtue of (Jacod and Shiryaev, 1987, Theorem 2.11) the conditions a), I_1, I_2 imply that X is a P_α-semimartingale and $(B^\alpha, C^\alpha, \nu^\alpha)$ are its characteristics. Then by virtue of Theorem 4 in §3 from a), I_1, and I_2 it follows that $H^{\mathbf{T},\alpha\beta}$ is the Hellinger process (of order 1/2) between measures P_α and P_β.

Now for all $\alpha, \beta \in \Theta$ the assertion of Corollary 6 from §4 holds for the binary experiment $(D(\mathbf{R}^{|\Theta|+d}), \mathcal{D}(\mathbf{R}^{|\Theta|+d}), \mathbf{D}(\mathbf{R}^{|\Theta|+d}), (P_\alpha, P_\beta))$ and therefore the density process $Z^\alpha(\beta)$ is $\mathbf{F}_+^X$-adapted. This implies that $Z^\alpha(\beta)$ is the density process of $\mathcal{P}$ w.r.t. P_β^X of the experiment $(D(\mathbf{R}^d), \mathcal{D}(\mathbf{R}^d), \mathbf{D}(\mathbf{R}^d), (P_\alpha^X : \alpha \in \Theta))$ with the canonical process X, where $P_\alpha^X = \mathcal{L}(X|P_\alpha)$. It follows that $\mathcal{L}(Z^X, X|Q^X) = \mathcal{L}(Z, X|Q)$, where $Q^X = |\Theta|^{-1} \sum_{\alpha \in \Theta} P_\alpha^X$, $Z^X = (z^{X,\alpha})_{\alpha \in \Theta}$ and $z^{X,\alpha}$ is the density process of P_α^X w.r.t Q^X.

Thus, all the limiting points of the sequence $\mathcal{L}(Z^n, X^n|Q^n), n \geq 1$, have the form $\mathcal{L}(Z, X|Q)$, where Z is the likelihood ratio process of the experiments $\mathcal{E} = (\Omega, \mathcal{F}, \mathbf{F}, (P_\alpha : \alpha \in \Theta))$ with the canonical process X. Moreover, $H^{\mathbf{T},\alpha\beta}$ is the Hellinger process between P_α and P_β, and $(B^\alpha, C^\alpha, \nu^\alpha)$ is the P_α-characteristics of X. By virtue of c) each measure P_α is uniquely determined by the semimartingale characteristics and so is Q. Hence, $\mathcal{L}(Z^n, X^n|Q^n) \longrightarrow \mathcal{L}(Z, X|Q)$ and the result follows.

6. EXAMPLES

6.1. *Example* 1 (the convergence of experiments generated by Markov step processes to an experiment generated by diffusion processes). We suppose that, for each $n \geq 1$, there is an experiment $\mathcal{E}^n = (\Omega^n, \mathcal{F}^n, \mathbf{F}^n, (P_\alpha^n : \alpha \in \Theta))$ such that the canonical process X is a Markov step process under P_α^n with the generator $A^{n,\alpha} g(x) = \int K^{n,\alpha}(x, dy)[g(x+y) - g(x)]$, where $K^{n,\alpha}$ is a σ-finite measure on $\mathbf{R}^d$ and $X_0 = 0$. We assume that $K^{n,\alpha}(x, dy) = U^{n,\alpha}(x, dy)\lambda^n(x, dy)$, where λ^n is a σ-finite measure on $\mathbf{R}^d$ and, for $\alpha, \beta \in \Theta$, we set:

$$b^{n,\alpha}(x) = \int yI\Big(|y| < 1\Big)K^{n,\alpha}(x, dy), \quad c^{n,\alpha}(x) = \int yy^T I\Big(|y| < 1\Big)K^{n,\alpha}(x, dy),$$

$$h^{n,\alpha\beta}(x) = \frac{1}{2}\int \Big(\sqrt{U^{n,\alpha}} - \sqrt{U^{n,\beta}}\Big)^2 \lambda^n(x, dy).$$

According to (Jacod and Shiryaev, 1987), the process X is a P_α^n-semimartingale and its characteristics with respect to the truncation function $f(y) = yI(|y| < 1)$ are

$$B_t^{n,\alpha}(X) = \int_0^t b^{n,\alpha}(X_s)ds, \quad C^{n,\alpha} \equiv 0, \quad \widetilde{C}_t^{n,\alpha}(X) = \int_0^t c^{n,\alpha}(X_s)ds,$$

$$\nu^{n,\alpha}(X_t; dt \times dy) = K^{n,\alpha}(X_t, dy)dt.$$

Moreover, according to (Jacod and Shiryaev, 1987, Chapter IX, §4b), the Hellinger process between P_α^n and P_β^n is

$$H_t^{n,\alpha\beta}(X) = \int_0^t h^{n,\alpha\beta}(X_s)ds.$$

As a limiting model we consider the experiment $\mathcal{E} = (D(\mathbf{R}^d), \mathcal{D}(\mathbf{R}^d), \mathbf{D}(\mathbf{R}^d), (P_\alpha : \alpha \in \Theta))$ with the canonical process X such that P_α is a weak solution of the diffusion equation

$$dX_t = b^\alpha(X_t)dt + \sigma(X_t)dW_t, \quad X_0 = 0, \tag{30}$$

where $b^\alpha : \mathbf{R}^d \longrightarrow \mathbf{R}^d$ and $\sigma : \mathbf{R}^d \longrightarrow \mathbf{R}^{d\times d}$ are Borel functions.
Then we set $c = \sigma\sigma^T$ and, for $\alpha \in \Theta$, define $\tilde{b}^\alpha$ as the solution of $c\tilde{b}^\alpha = b^\alpha$. Finally, for $\alpha, \beta \in \Theta$, we set

$$h^{\mathbf{T},\alpha\beta}(x) = \frac{1}{8}(\tilde{b}^\alpha - \tilde{b}^\beta)^T c(\tilde{b}^\alpha - \tilde{b}^\beta)(x).$$

THEOREM 7. *Assume that, for $\alpha, \beta \in \Theta, L > 0, \varepsilon > 0$,*
(i) $b^{n,\alpha} \longrightarrow b^\alpha$, $c^{n,\alpha} \longrightarrow c^\alpha$, $h^{n,\alpha\beta} \longrightarrow h^{\mathbf{T},\alpha\beta}$, $n \to \infty$, *uniformly on compact sets,*
(ii) $\sup_{x\in\mathbf{R}^d}\{|b^\alpha(x)| + |c(x)| + |h^{\mathbf{T},\alpha\beta}(x)|\} < +\infty$, *where* $|\cdot|$ *denotes the corresponding norm,*
(iii) *the functions* $b^\alpha, c, h^{\mathbf{T},\alpha\beta}$ *are continuous,*
(iv) *the equation* (30) *has a unique weak solution.*
Then, for all $\alpha, \beta_1, \dots, \beta_N \in \Theta$, *the laws of* $(Z^{n,\beta_i}(\alpha), X^{n,j})_{i=1,\dots,N;\ j=1,\dots,d}$ *under* P_α^n *converge (in the Skorokhod topology) to the law of* $(Z^{\beta_i}(\alpha), X^j)_{i=1,\dots,N;\ j=1,\dots,d}$ *under* P_α, *where* $Z^{n,\beta}(\alpha)$ $(Z^\beta(\alpha))$ *denotes the density process of* P_β^n (P_β) *w.r.t.* P_α^n (P_α).

Proof. We verify the validity of the conditions of Theorem 1. The implications (ii) $\Rightarrow I_1$, (iii) $\Rightarrow I_2$, and (iv) $\Rightarrow$ c) are obvious. The conditions I_3 and b) are trivially met. So to complete the proof we must only verify (a).
According to (Jacod and Shiryaev, 1987, Chapter IX, Theorem 4.8) our conditions imply that, for all $\alpha \in \Theta$ and $t \geq 0$,

$$\lim_{N\to\infty} \limsup_{n\to\infty} P_\alpha^n(\sup_{s\leq t} |X_s| > N) = 0.$$

Now a) follows from (i) and the proof is completed.

6.2. If all the Hellinger processes are deterministic, then they determine the laws of density processes. As is shown by Coquet (Coquet, 1990) this statement does no longer hold for the experiments generated by point processes. Now we present a corresponding example for the case of diffusion processes.

Example 2. Let us consider the experiments $\mathcal{E} = (D(R), \mathcal{D}(R), \mathbf{D}(R), (P, P'))$ such that

(i) the canonical process X is the Wiener process under the measure P,
(ii) for $a \in (0,1)$ the Hellinger process of order a has the form

$$h_t(a)(X) = \frac{a(1-a)}{2} \int_0^t d^2(X_s)ds, \qquad d(x) \geq 0, \tag{31}$$

(iii) P' is a weak solution of the diffusion equation

$$dX_t = b(X_t)dt + dW_t, \tag{32}$$

where W is Wiener process and $|b(x)| = d(x)$, $x \in \mathbf{R}$.
Let us show that, by varying the sign of $b(x)$ and thereby obtaining the same family of the Hellinger processes, we can get different laws for the corresponding density processes.
To this end, we set for $n \geq 1$

$$b_1^n(x) = \begin{cases} 1 & \text{for } x \leq 1, \\ n & \text{for } x > 1, \end{cases} \quad b_2^n(x) = \begin{cases} -1 & \text{for } x \leq 1, \\ n & \text{for } x > 1. \end{cases}$$

For $i = 1,2$ we denote by P_i^n the distribution in (33) corresponding to b_i^n and by Z_i^n the density process of $P_i'^n$ w.r.t. P.
Set $\tau = \inf\{t : X_t > 1\}$ and note that, for i =1, 2,

$$P_i^n|\mathcal{F}_\tau = P_i^1|\mathcal{F}_\tau. \tag{33}$$

From the expression for b_i^n it follows that, for all $\varepsilon > 0, N > 0$,

$$\limsup_{n\to\infty} P_i'^n(X_{\tau+\varepsilon}^n < N) = 0$$

so the sequences $P_i'^n|\mathcal{F}_{\tau+\varepsilon}$ and $P^n = P|\mathcal{F}_{\tau+\varepsilon}$ are completely separated, which implies (see (Jacod and Shiryaev, 1987, Theorem 1.20)) that, for all $\varepsilon > 0, N > 0$,

$$\limsup_{n\to\infty} P_i'^n(\sup_{s\leq\tau+\varepsilon} Z_{i,s}^n < N) = 0. \tag{34}$$

Fix $t > 0$. By virtue of (33) and (34),

$$\limsup_{n\to\infty} P_i'^n(\sup_{s\leq t} Z_{i,s}^n < N, t > \tau)$$

$$\leq \lim_{\varepsilon\to 0} \{\limsup_{n\to\infty} P_i'^n(\sup_{s\leq\tau+\varepsilon} Z_{i,s}^n < N) + P_i'^1(\tau < t \leq \tau + \varepsilon)\} = 0.$$

In view of (33), we have

$$\limsup_{N\to\infty} \limsup_{n\to\infty} P_i'^n(\sup_{s\leq t} Z_{i,s}^n < N)$$

$$= \limsup_{N\to\infty} P_i'^1(\sup_{s\leq t} Z_{i,s}^n < N, t \leq \tau) = P_i'^1(t \leq \tau).$$

Since

$$P_1'^1(t \le \tau) = P_1'^1(\sup_{s \le t} X_s < 1) = P(\sup_{s \le t}(X_s + s) < 1)$$
$$< P\big(\sup_{s \le t}(X_s - s) < 1\big) < P_2'^1(t \le \tau),$$

for some $n_0 \ge 1, N \ge 1$ we obtain

$$P_1'^{n_0}(\sup_{s \le t} Z_{1,s}^{n_0} < N) < P_2'^{n_0}(\sup_{s \le t} Z_{2,s}^{n_0} < N).$$

Thus, the laws of $Z_1^{n_0}$ and $Z_2^{n_0}$ under P are different, which is the desired result.

Acknowledgment
I am indebted to Professor A.N. Shiryaev for the ideas underlying this paper and for helpful discussions.

REFERENCES

Coquet, F. (1990). *Processus de Hellinger des Processus Ponctuels.* Preprint, Univ. de Rennes Press, Rennes.

Greenwood, P. and Shiryaev, A. N. (1985). *Contiguity and the Statistical Invariance Principle. Stochastics Monographs.* V. 1. Gordon and Breach, London.

Jacod, J. (1987). Partial likelihood processes and asymptotic normality. *Stoch. Proc. Appl.* **26**, 47–71.

Jacod, J. (1989a). Convergence of filtered statistical models and Hellinger processes. *Stoch. Proc. Appl.* **32**, 47–68.

Jacod, J. (1989b). Filtered statistical models and processes. *Stoch. Proc. Appl.* **32**, 3–45.

Jacod, J. (1990). Sur le processus de vraisemblence partielle. *Ann. Inst. Henri Poincaré (B)* **26**, 2–20.

Jacod, J. and Shiryaev, A. N. (1987). *Limit Theorems for Stochastic Processes.* Springer-Verlag, Berlin, New York.

Liptser, R. S. and Shiryaev, A. N. (1986). *Theory of Martingales.* Nauka, Moscow (in Russian).

Memin, J. (1985). Theoremes limites functionnels pour les processus de vraisemblance (cadre asymptotiqment a-normal). In: *Seminaire de Probabilites de Rennes, Lecture Notes in Mathematics*, Springer-Verlag, New York.

Vostrikova, L. (1988). On the weak convergence of likelihood ratio processes of general statistical parametric models. *Stochastics* **23**, 277–298.

Frontiers in Pure and Applied Probab. 1, pp. 165 - 182
H. Niemi et al. (Eds.)
1993 TVP/VSP

MARTINGALE APPROACH TO THE PROCEDURES OF STOCHASTIC APPROXIMATION

A. V. MELNIKOV and A. E. RODKINA
Steklov Mathematical Institute, Vavilov Street 42, 117966 Moscow GSP-1, Russia
Voronezh Civil Engineering Institute, 20th October Street 86, 394006 Voronezh, Russia

ABSTRACT

Multidimensional Robbins–Monro and Kiefer–Wolfowitz stochastic approximation procedures are considered from the point of view of the theory of stochastic differential equations with respect to semimartingales. The main result is (a.s.)-convergence of the procedures if coefficients of corresponding equations belong to some Lyapunov class of functions.

1. INTRODUCTION AND STATEMENT OF THE PROBLEM

In this article the approach to the procedures of stochastic approximation (Robbins–Monro and Kiefer–Wolfowitz procedures (Robbins and Monro, 1951; Kiefer and Wolfowitz, 1952; Nevelson and Khasminskii, 1972)) based on the theory of stochastic equations with respect to the semimartingales (Gikhman and Skorokhod, 1982; Liptser and Shiryaev, 1986), which is sufficiently investigated at present, will be systematically developed. The idea of this approach was stated in (Melnikov, 1988; 1989), where features of asymptotic behaviour (convergence a.s., asymptotic normality) of one-dimensional stochastic procedures generalizing classical ones in continuous and discrete time were investigated from this point of view. In this article this approach is extended to a wider class of procedures (multidimensional procedures, more complicated "noise" etc.); in addition, the connection with stochastic equations is traced more accurately.

Let $(\Omega, F, (F_t)_{t\geq 0}, \mathbf{P})$ be a standard stochastic basis for which all the cadlag processes used below will be considered (right-continuous with finite left limits). Let $\mathcal{M}_{\text{loc}}(\mathbf{R}^k)$ $(\mathcal{M}^2_{\text{loc}}, \mathcal{M}^c_{\text{loc}})$ be the class of local (local square integrable, continuous martingales, starting from zero and having values in $\mathbf{R}^k$, $k \geq 1)$; $\mathcal{A}_{\text{loc}}$ $(\mathcal{A}^+_{\text{loc}})$ be the class of one-dimensional (increasing) processes starting from zero and having the locally integrable variation, and let $\mathcal{P}$ be the class of F_t-predictable processes. Consider two classes od stochastic equations which we call respectively, Robbins–Monro and Kiefer–Wolfowitz procedures.

Robbins–Monro procedure is defined by the stochastic equation

$$x_t = x_0 + \int_0^t \gamma_s R(x_{s-})\, da_s + \int_0^t \gamma_s \sigma(s, x_{s-})\, dm_s \tag{1}$$

for an approximation of the unique root $\Theta = 0$ of the equation $R(\Theta) = 0$, where $R\colon \mathbf{R}^d \to \mathbf{R}^d$, $d \geq 1$, is some continuous function.
Kiefer–Wolfowitz procedures is defined by more complicated equation

$$x_t = x_0 + \int_0^t \gamma_s \nabla_c J(x_{s-})\, da_s + \int_0^t \gamma_s c_s^{-1} \sigma(s, x_{s-})\, dm_s. \tag{2}$$

It is a natural approximation for the point Θ which is the unique argmin (argmax) of the convex function $J\colon \mathbf{R}^d \to \mathbf{R}^1$, $d \geq 1$, where the kth component of the vector $\nabla_c J$ is defined by $(\nabla_c J(u))_k = (2c_t)^{-1}[J(u + c_t i_k) - J(u - c_t i_k)]$, i_k being the kth vector in $\mathbf{R}^d$.
Let $m \in \mathcal{M}^2_{\text{loc}}(\mathbf{R}^k)$, $a \in \mathcal{A}^+_{\text{loc}} \cap \mathcal{P}$, positive processes γ, c belong to $\mathcal{P}$, and let Δ^m_t be the matrix of the quadratic characteristics $\langle m^i, m^j \rangle$ of the components m^i and m^j, $i, j = 1, \ldots, k$, of the martingale m, $\langle m \rangle_t = \operatorname{tr} \Delta^m_t$. For $m \in \mathcal{M}_{\text{loc}}(\mathbf{R}^k)$ let $m = m^c + m^d$ be the decomposition in the continuous and pure discontinues martingale components, $\Box^m_t$ be the matrix of quadratic characteristics $[m^i, m^j]$, $i, j = 1, \ldots, k$, of the components m^i and m^j of the martingale m, $[m]_t = \operatorname{tr} \Box^m_t$, $\widetilde{\Box}^m_t$ be the compensator of $\Box^m_t$.
We denote by $\{\omega\colon z_t(\omega) \to\}$ the set of all $\omega \in \Omega$, for which a finite $\lim z_t(\omega)$ exists besides for $x, y \in \mathbf{R}^d$ and a matrix B denote by (x, y) the scalar product of x and y, $\|x\| = (x, x)^{1/2}$, $\|B\|$, $\operatorname{tr} B$ are the norm and the trace of B.

2. PRELIMINARY RESULTS

Let $e_0 = 0$, $e_m = \exp e_{m-1}$, $m = 1, 2, \ldots$, $L(u) = L_\varepsilon(u) = L_{\varepsilon,m}(u) = \text{const} \times \prod_{i=0}^m \log_i(u + e_m + \varepsilon)$, $u, \varepsilon > 0$, $\log_0(x) = x$, $\log_i(x) = \log_{i-1}(\log x)$, $i = 1, 2, \ldots, x > 0$. The class of such functions is denoted by $\mathcal{L}$.

LEMMA 2.1. 1) *Every function $L_{\varepsilon,m} \in \mathcal{L}$ is correctly defined on $[0, \infty)$, positive, nondecreasing continuously differentiable and is such that*

$$\int_0^x L^{-1}(u)\, du \to \infty \Leftrightarrow x \to \infty.$$

2) *For every $\delta > 0$ there exists $\varepsilon > 0$ such that the corresponding function $L_{\varepsilon,m} \in \mathcal{L}$ for $u > \delta$ satisfies the inequality*

$$L_{\varepsilon,m}(u) - 2uL'_{\varepsilon,m}(u) < -L_{\varepsilon,m}(u).$$

3) *For a function $L_{\varepsilon,m}(u) \in \mathcal{L}$ there exist constants $H = H(\varepsilon, m) > 0$ and $H_1 = H_1(\varepsilon, m) > 0$ such that, for $u \geq 0$,*

$$uL'_{\varepsilon,m}(u) \leq HL_{\varepsilon,m}(u), \qquad u\big(L'_{\varepsilon,m}(u^2)\big)^{1/2} \leq H_1\big(L_{\varepsilon,m}(u^2)\big)^{1/2}.$$

Proof. 1) The first part is obvious from the constructions of the function L.
2) For $L_{\varepsilon,m}(u) \in \mathcal{L}$ we have $L_{\varepsilon,m}(u) - 2uL'_{\varepsilon,m}(u) = L_{\varepsilon,m}(u) - 2u[(u + e_m + \varepsilon)^{-1}L_{\varepsilon,m}(u) + \ldots + (u + e_m + \varepsilon)^{-1}\log^{-1}_{m-1}(u + e_m + \varepsilon)L_{\varepsilon,m}(u)] = L_{\varepsilon,m}(u) - 2(u + e_m + \varepsilon - e_m - \varepsilon)(u + e_m + \varepsilon)^{-1}L_{\varepsilon,m}(u) - 2u((u + e_m + \varepsilon)^{-1}\log^{-1}_{m-1}(u + e_m + \varepsilon)L_{\varepsilon,m}(u) + \ldots) = -L_{\varepsilon,m}(u) + 2(e_m + \varepsilon)(u + e_m + \varepsilon)^{-1}L_{\varepsilon,m}(u) - 2u(\ldots) = -L_{\varepsilon,m}(u) + \varphi(u)$. Note now that $\lim_{\varepsilon\to 0}\varphi_\varepsilon(0) = 0$ and, for $u > 0$, $\varphi'_\varepsilon(u) = \text{const}\,\{2(e_m + \varepsilon)[(u + e_m + \varepsilon)^{-1}\log_2(u + e_m + \varepsilon)\ldots\log_{m-1}(u + e_m + \varepsilon) + \ldots + (u + e_m + \varepsilon)^{-1}] - 2[\log_2(u + e_m + \varepsilon)\ldots\log_{m-1}(u + e_m + \varepsilon) + \ldots + 1] - 2u[(u + e_m + \varepsilon)^{-1}\log^{-1}(u + e_m + \varepsilon)\log_3(u + e_m + \varepsilon)\ldots\log_{m-1}(u + e_m + \varepsilon) + \ldots + (u + e_m + \varepsilon)^{-1}\log^{-1}(u + e_m + \varepsilon)\log^{-1}_{m-1}(u + e_m + \varepsilon)]\} <$ $\text{const}\,2u((u + e_m + \varepsilon)^{-1}(e_m + \varepsilon) - 1)[\log_2(u + e_m + \varepsilon)\ldots\log_{m-1}(u + e_m + \varepsilon) + \ldots + 1] =$ $\text{const}\,2(-u)(u + e_m + \varepsilon)[\log_2(u + e_m + \varepsilon)\ldots\log_{m-1}(u + e_m + \varepsilon) + \ldots + 1] < 0$. Therefore, for every $\delta > 0$ there exists $\varepsilon > 0$ such that

$$L_{\varepsilon,m}(u) - 2uL'_{\varepsilon,m}(u) \le -L_{\varepsilon,m}(u)$$

for $u > \delta$.
3) For $L_{\varepsilon,m} \in \mathcal{L}$ we have, for $u > 0$, $uL'_{\varepsilon,m}(u) = [u(u + e_m + \varepsilon)^{-1} + \ldots + u(u + e_m + \varepsilon)^{-1}\ldots\log^{-1}_{m-1}(u + e_m + \varepsilon)]L_{m,\varepsilon}(u + e_m + \varepsilon)$, $\log_{i-1}(u + e_m + \varepsilon) > \log_{i-1}(e_m + \varepsilon) = e^{m-i-1} \ge 1$, $i \le m - 1$, $\log_{m-1}(u + e_m + \varepsilon) > \log_{m-1}(e_m + \varepsilon) = \alpha_{m,\varepsilon} \to 0$ $(\varepsilon \to 0)$. Therefore $u(u + e_m + \varepsilon)^{-1} + \ldots + u(u + e_m + \varepsilon)^{-1}\ldots\log^{-1}_{m-1}(u + e_m + \varepsilon) \le m + \alpha^{-1}_{m,\varepsilon} = H(\varepsilon, m)$, and the proof of the lemma is completed.

LEMMA 2.2. *Let $Z = Z_0 + A + M$ be a non-negative semimartingale, $\mathbf{E}Z_0 < \infty$, $A \in \mathcal{A}_{\text{loc}} \cap \mathcal{P}$, $M \in \mathcal{M}_{\text{loc}}(\mathbf{R}^1)$, $A \le A^1 - A^2$, A^1, $A^2 \in A^+_{\text{loc}} \cap \mathcal{P}$ and, for every predictable stopping time (s.t.) σ, $\mathbf{E}\Delta A^1_\sigma I_{\{\omega:\,\sigma<\infty\}} < \infty$.*
Then $\{\omega\colon A^1_\infty < \infty\} \subseteq \{Z \to\} \cap \{A^2_\infty < \infty\}$ a.s.

Proof. Consider the following representation of Z:

$$Z_t = Z_0 + A^1_t - (A^1_t - A_t) + M_t.$$

According to Theorem 7 (Liptser and Shiryaev, 1986, p. 115)

$$\{A^1_\infty < \infty\} \subseteq \{Z_t \to\} \cap \{(A^1 - A)_\infty < \infty\} \quad \text{(a.s.)}.$$

Note now that $A^2_t \le -(A^1_t - A_t)$ (a.s.) and the necessary statement follows immediately.

LEMMA 2.3. *For derivatives of the twice continuously differentiable function*

$$W(x) = W_\varepsilon(x) = W_{\varepsilon,m}(x) = \int_0^{\|x\|^2} L^{-1}_{\varepsilon,m}(y)\,dy$$

the following bounds take place

$$\frac{1}{2}\frac{\partial^2 W(x)}{\partial x_i^2} \le L^{-1}(\|x\|^2), \quad i = 1, \ldots, d,$$

$$\frac{1}{2}\frac{\partial^2 W(x)}{\partial x_i \partial x_j} \leq H(m,\varepsilon)L^{-1}(\|x\|^2), \quad i,j=1,\dots,d.$$

In particular, for $\varepsilon = e_{m+1} - e_m$ the constant $H(m,\varepsilon)$ equals $m+1$.

Proof. It is easy to see that for $i = 1,\dots,d$

$$\frac{\partial W(x)}{\partial x_i} = 2x_i L^{-1}(\|x\|^2),$$

$$\begin{aligned}\frac{1}{2}\frac{\partial^2 W(x)}{\partial x_i^2} &= \Big(L(\|x\|^2) - 2x_i^2 L'(\|x\|^2)\Big)L^{-2}(\|x\|^2) \\ &\leq L(\|x\|^2)L^{-2}(\|x\|^2) = L^{-1}(\|x\|^2).\end{aligned}$$

Furthermore, by Lemma 2.1 we have for $i,j = 1,\dots,d$

$$\begin{aligned}\frac{1}{2}\frac{\partial^2 W(x)}{\partial x_i^2} &= -2x_i x_j L'(\|x\|^2)L^{-2}(\|x\|^2) \\ &\leq (x_i^2 + x_j^2)L'(\|x\|^2)L^{-2}(\|x\|^2) \\ &\leq \|x\|^2 L'(\|x\|^2)L^{-2}(\|x\|^2) \leq H(m,\varepsilon)L(\|x\|^2)L^{-2}(\|x\|^2) \\ &\leq H(m,\varepsilon)L^{-1}(\|x\|^2).\end{aligned}$$

The lemma is proved.

LEMMA 2.4. *Let the random variable x_0 be such that $\mathbf{E}\|x_0\| < \infty$. Then, for every $\varepsilon > 0$, $m = 0,1,\dots$, $\mathbf{E}W_{\varepsilon,m}(x_0) < \infty$.*

Proof. The direct calculation gives

$$\begin{aligned}\mathbf{E}W(x_0) &= \int_0^{\|x_0\|^2} L_{\varepsilon,m}^{-1}(y)\,dy \\ &= \mathbf{E}\int_0^{\|x_0\|^2} \text{const}\,\big[(y+e_m+\varepsilon)\log(u+e_m+\varepsilon)\dots\big]^{-1}dy \\ &= \mathbf{E}\int_0^{\|x_0\|^2} \text{const}\,\big[(y+e_m+\varepsilon)\log(u+e_m+\varepsilon)\dots\big]^{-1}d(y+e_m+\varepsilon) \\ &= \text{const}\,\mathbf{E}\log_m(u+e_m+\varepsilon)\Big|_0^{\|x_0\|^2} \\ &\leq \text{const}\,\Big[\mathbf{E}(\|x_0\|^2+e_m+\varepsilon) - \log_m(e_m+\varepsilon)\Big] \\ &\leq \text{const}\,(\mathbf{E}\|x_0\|^2 + 1).\end{aligned}$$

The lemma is proved.

3. CONVERGENCE OF PROCEDURES (1)–(2)

Represent procedure (1) in a form more suitable for us:

$$x_t = x_0 + \int_0^t R(x_{s-})\, da_s(\gamma) + \int_0^t \sigma(s,\, x_{s-})\, dm_s(\gamma), \tag{1'}$$

where

$$a_t(\gamma) = \int_0^t \gamma_s\, da_s m_t(\gamma) = \int_0^t \gamma_s\, dm_s.$$

Consider the process $v \in \mathcal{A}^+_{\mathrm{loc}} \cap \mathcal{P}$ such that $a_t(\gamma) \ll v_t$ and $\Delta^m_t(\gamma) \ll v_t$. Denote the corresponding scalar and matrix densities by $\alpha_t(\gamma)$ and $\beta_t(\gamma)$, respectively. Taking this into account, we obtain the following general theorem about the convergence (a.s.) of the procedure of type (1)–(1′).

THEOREM 3.1. *Let x be the regular solution of equation (1′) and let the following assumptions be fulfilled:*
1) $a_\infty(\gamma) = \infty$ (a.s.);
2) $\big(u,\, R(u)\big) < 0$ *for every* $u \in \mathbf{R}^d_0 = \mathbf{R}^d \setminus \{0\}$;
3) $\|R(u)\|^2 \le L_{\varepsilon,m}\big(\|u\|^2\big)$, $u \in \mathbf{R}^d$;
4) *there exists a predictable function l such that for every $t > 0$, $u \in \mathbf{R}^d$*

$$\operatorname{tr}\sigma(t,u)\Delta^m_t\sigma^*(t,u) \le l_t L_{\varepsilon,m}\big(\|u\|^2\big) \quad \text{(a.s.)},$$

$$\int_0^\infty l_s\, d\langle m(\gamma)\rangle_s < \infty, \qquad \int_0^\infty \Delta a_s(\gamma)\, da^d_s(\gamma) < \infty,$$

$$\mathbf{E}\sup_t l_t \Delta\langle m(\gamma)\rangle_t < \infty, \qquad \mathbf{E}\sup_t \Delta^2 a_t(\gamma) < \infty.$$

Then $x_t \to 0$ (a.s.) as $t \to \infty$.

Proof. We apply Itô formula (Gikhman and Skorokhod, 1982; Liptser and Shiryaev, 1986) to $W_{\varepsilon,m}(x_t) = W(x_t)$ and obtain

$$\begin{aligned} W(x_t) = {} & W(x_0) + 2\int_0^t L^{-1}\big(\|x_{s-}\|^2\big) x^*_{s-} R(x_{s-})\, da_s(\gamma) \\ & + \frac{1}{2}\operatorname{tr}\int_0^t \Big(\frac{\partial^2}{\partial x^2} W(x_{s-})\Big)\sigma(s,\, x_s)\, d\Delta^{m^c(\gamma)} \sigma^*_s(x_{s-}) \\ & + 2\int_0^t L^{-1}\big(\|x_{s-}\|^2\big) x^*_{s-}\sigma(s,\, x_{s-})\, dm_s(\gamma) \\ & + \sum_{s\le t}\Big[W(x_{s+}) - W(x_{s-}) - \big(\operatorname{grad} W(x_{s-}),\, \Delta x_s\big)\Big]. \end{aligned} \tag{3}$$

Consider the sth term of the sum. The following two cases are possible: either $\|x_s\|^2 \geq \|x_{s-}\|^2$ or $\|x_s\|^2 < \|x_{s-}\|^2$.
In the first case, since $L(x)$ is a non-negative and nondecreasing function, we have

$$\begin{aligned}
&W(x_s) - W(x_{s-}) - (\operatorname{grad} W(x_{s-}),\, \Delta x_s) \\
&= \int_0^{\|x_s\|^2} L_{\varepsilon,m}^{-1}(y)\,dy - \int_0^{\|x_{s-}\|^2} L_{\varepsilon,m}^{-1}(y)\,dy - (\operatorname{grad} W(x_{s-}),\, \Delta x_s) \\
&\leq L^{-1}(\|x_{s-}\|^2)\left[\|x_s\|^2 - \|x_{s-}\|^2\right] - 2L^{-1}(\|x_{s-}\|^2)(x_{s-},\, \Delta x_s) \\
&= L^{-1}(\|x_{s-}\|^2)(\Delta x_s,\, \Delta x_s).
\end{aligned}$$

The second case is analysed in a similar way. Return now to the general formula (3). Taking into account that $\operatorname{tr} AB \leq \|A\| \operatorname{tr} B$ for a symmetric matrix A and a symmetric positive definite matrix B, we obtain, by Lemma 2.4,

$$\begin{aligned}
W(x_t) \leq W(x_0) &+ 2\int_0^t L^{-1}(\|x_{s-}\|^2)x_{s-}^* R(x_{s-})\,da_s(\gamma) \\
&+ H\int_0^t L^{-1}(\|x_{s-}\|^2) \operatorname{tr} \sigma(s,\, x_{s-})\,d\Delta_s^{m^c(\gamma)} \sigma^*(s,\, x_{s-}) \\
&+ 2\int_0^t L^{-1}(\|x_{s-}\|^2)x_{s-}^* \sigma(s,\, x_{s-})\,dm_s(\gamma) \\
&+ \int_0^t L^{-1}(\|x_{s-}\|^2) R^*(x_{s-})R(x_{s-})\Delta a_s(\gamma)\,da_s^d(\gamma) \\
&+ \int_0^t L^{-1}(\|x_{s-}\|^2)\Delta a_s(\gamma)\big(dm_s^d(\gamma)\big)^* \sigma^*(s,\, x_{s-})R(x_{s-}) \\
&+ \int_0^t L^{-1}(\|x_{s-}\|^2)\Delta a_s(\gamma) R^*(x_{s-})\sigma(s,\, x_{s-})\,dm_s^d(\gamma) \\
&+ \int_0^t L^{-1}(\|x_{s-}\|^2) \operatorname{tr} \sigma^*(s,\, x_{s-})\,d\square_s^{m^d(\gamma)} \sigma(s,\, x_{s-}).
\end{aligned} \tag{4}$$

Furthermore, note that

$$\begin{aligned}
m_t^\sigma = &\int_0^t L^{-1}(\|x_{s-}\|^2) \operatorname{tr} \sigma^*(s,\, x_{s-})\,d\square_s^{m^d(\gamma)} \sigma(s,\, x_{s-}) \\
&- \int_0^t L^{-1}(\|x_{s-}\|^2) \operatorname{tr} \sigma^*(s,\, x_{s-})\,d\Delta_s^{m^d(\gamma)} \sigma(s,\, x_{s-}) \in \mathcal{M}_{\mathrm{loc}}(\mathbf{R}^d).
\end{aligned}$$

Therefore, adding and subtracting in (4) the integral

$$\int_0^t L^{-1}(\|x_{s-}\|^2)\,\mathrm{tr}\,\sigma^*(s,\,x_{s-})\,d\Delta_s^{m^d(\gamma)}\sigma(s,\,x_{s-}),$$

we obtain

$$\begin{aligned}
Z_t = W(x_t) \le W(x_0) &+ 2\int_0^t L^{-1}(\|x_{s-}\|^2)x_{s-}^* R(x_{s-})\,da_s(\gamma)\\
&+ \Bigg[H\int_0^t L^{-1}(\|x_{s-}\|^2)\,\mathrm{tr}\,\sigma(s,\,x_{s-})\,d\Delta_s^{m^c(\gamma)}\sigma^*(s,\,x_{s-})\\
&+ \int_0^t L^{-1}(\|x_{s-}\|^2)R^*(x_{s-})R(x_{s-})\Delta a_s(\gamma)\,da_s^d(\gamma)\Bigg]\\
&+ \Bigg\{2\int_0^t L^{-1}(\|x_{s-}\|^2)x_{s-}^*\sigma(s,\,x_{s-})\,dm_s(\gamma)\\
&+ \int_0^t L^{-1}(\|x_{s-}\|^2)\Delta a_s(\gamma)\big(dm_s^d(\gamma)\big)^*\sigma^*(s,\,x_{s-})R(x_{s-})\\
&+ \int_0^t L^{-1}(\|x_{s-}\|^2)\Delta a_s(\gamma)R^*(x_{s-})\sigma(s,\,x_{s-})\,dm_s^d(\gamma) + m_t^\sigma\Bigg\}\\
&= z_0 - A_t^2 + A_t^1 + M_t,
\end{aligned} \tag{5}$$

where $z_0 = W(x_0)$, A^2 is the second term on the right-hand side, A^1 is the term in square brackets, and M is the term in curly braces.
By Lemma 2.3, we have $\mathbf{E}Z_0 = \mathbf{E}W(x_0) < \infty$ and, evidently, $A^1,\ A^2 \in \mathcal{A}_{\mathrm{loc}}^+ \cap \mathcal{P}$, $M \in \mathcal{M}_{\mathrm{loc}}(\mathbf{R}^1)$.
Furthermore, for the predictable s.t. τ, we have, by conditions 3) and 4) of the theorem,

$$\begin{aligned}
\mathbf{E}\Delta A_\tau^l I_{\{\tau<\infty\}} &= \mathbf{E}\{HL^{-1}(\|x_{\tau-}\|^2)l_\tau\Delta\langle m^d(\gamma)\rangle_\tau\, L(\|x_\tau\|^2)\\
&\quad + L^{-1}(\|x_{\tau-}\|^2)L(\|x_{\tau-}\|^2)\Delta_\tau^2 a(\gamma)\}I_{\{\tau<\infty\}}\\
&\le H\mathbf{E}l_\tau\Delta\langle m^d(\gamma)\rangle_\tau + \mathbf{E}\Delta_\tau^2 a(\gamma)I_{\{\tau<\infty\}}\\
&\le H\mathbf{E}\sup_t l_t\Delta\langle m^d(\gamma)\rangle_t + \mathbf{E}\sup_t \Delta_t^2 a(\gamma) < \infty.
\end{aligned}$$

Therefore, we can apply Lemma 2.2 to the non-negative semimartingale Z and as a result we obtain that (a.s.)

$$\{\omega\colon\ A_\infty^1 < \infty\} \subseteq \{\omega\colon\ Z \to\} \cap \{\omega\colon\ A_\infty^2 < \infty\}. \tag{6}$$

By the assumption of the theorem

$$A^1_\infty \leq H \int_0^\infty l_s \, d\langle m(\gamma)\rangle_s + \int_0^\infty \Delta a_s(\gamma) \, da^d_s(\gamma) < \infty \quad \text{(a.s.)}.$$

This implies $\mathbf{P}\{\omega\colon\ W(x_t) \rightarrow\} = 1$ and therefore $\mathbf{P}\{\omega\colon\ \|x_t\|^2 \rightarrow\} = 1$. Suppose that $\|x_t\|^2 \rightarrow b(\omega)$ as $t \rightarrow \infty$ and that the set $\Omega_1 = \{\omega\colon\ b(\omega) \neq 0\}$ has a positive probability $\mathbf{P}(\Omega_1) = \alpha_1$.
Then we assert that

$$\mathbf{P}\Big\{\omega \in \Omega_1\colon\ \liminf_t \big(- x_{t-}, R(x_{t-})\big) > 0\Big\} = \alpha_1. \tag{7}$$

To prove (7) let us denote

$$\Omega_2 = \Big\{\omega \in \Omega_1\colon\ \liminf_t \big(- x_{t-}, R(x_{t-})\big) = 0\Big\}$$

and suppose that $\mathbf{P}(\Omega_2) = \alpha_2 < \alpha_1$. For a fixed $\omega \in \Omega_2$ choose a sequence of the real numbers $t_n = t_n(\omega) \rightarrow \infty$ such that $\big(- x_{t_n-}, R(x_{t_n-})\big)(\omega) \rightarrow \infty$ $(n \rightarrow \infty)$. Furthermore, since $\|x_t\|^2$ is convergent as $t \rightarrow \infty$ the sequence $\{x_{t_n(\omega)-}\}$ is bounded and therefore some its subsequence $x_{t'_n(\omega)}$ converges to $\xi(\omega)$. Consider the nontrivial case $\xi(\omega) \neq 0$. Then by virtue of the continuity of the scalar product and the function R we have, as $t'_n \rightarrow \infty$,

$$\big(- x_{t'_n(\omega)-}, R(-x_{t'_n(\omega)-})\big) \rightarrow \big(- \xi, R(\xi)\big)$$

on Ω_2 and therefore the equality $\big(- \xi, R(\xi)\big) = 0$ which is true is impossible by virtue of the assumptions of the theorem. Thus the relation (7) is established. Set $\eta(\omega) = \liminf_t \big(- x_{t-}, R(-x_{t-})\big) > 0$, $\omega \in \Omega_1$. There exists $T = T(\omega)$ such that, for every $T \geq T(\omega)$, $\omega \in \Omega_1$,

$$-L^{-1}\big(\|x_{t-}\|^2\big)\big(x_{t-}, R(x_{t-})\big) > L^{-1}(0)\eta(\omega) > 0.$$

Then, by assumption 1) of the theorem, we have on Ω_1

$$\int_{T(\omega)}^\infty \gamma_s L^{-1}\big(\|x_{s-}\|^2\big)\big(- x_{s-}, R(x_{s-})\big) \, da_s(\omega)$$

$$\geq L^{-1}(0)\eta(\omega) \int_{T(\omega)}^\infty \gamma_s \, da_s(\omega) = \infty.$$

It follows from here that $\mathbf{P}\{\omega\colon\ A^2_\infty(\omega) < \infty\} < 1$. This contradiction to (6) proves the theorem.

Remark 3.1. If we set $\varepsilon = e_m - e_{m-1}$, then $H = m + 1$ and condition (4) of the theorem may be weakened in the following way: there exists $\delta \in (0,1)$ such that

$$2\gamma_t(1-\delta)\big(u,R(u)\big)a_t+(m+1)\gamma_t^2\mathrm{tr}\,\sigma^*(t,u)\beta_t\sigma(t,u)\le l_tL(\|u\|^2)$$

$(dV\times dP)$-a.s.
The first term is negative, and, therefore, tr $\sigma_t^*\beta_t\sigma_t$ increases with respect to u quicker than $L(\|u\|^2)$.

Remark 3.2 (see (Melnikov and Rodkina, 1992)). In the case of continuous model (1)–(1′), assumption 3) of the theorem is superfluous. In addition, assumption 2) may be substituted by a weaker one that is the existence of a simmetric positive definite matrix C such that $\big(Cu,R(u)\big)<0$ for all $u\ne 0$.
Now we turn to the study of Kiefer–Wolfovitz procedure (2) in the following form

$$x_t=x_0+\int\limits_0^t\gamma_s\nabla_cJ(x_{s-})\,da_s+\int\limits_0^t\sigma(s,x_s)\,dm_s(c^{-1}\gamma). \tag{2'}$$

The next theorem shows, what assumption it is sufficient to impose on the grows of $\sigma(t,x)$ and on $J(x)$ and grad $J(x)$ also. To ensure the convergence (a.s.) of this procedure under "almost classical" assumptions on the behaviour at infinity of predictable processes $a_t(\gamma)$ and $\langle m(c^{-1}\gamma)\rangle_t$.
For convinience, denote $\nabla J(x)=\mathrm{grad}\,J(x)$ and let $\omega(u),u\ge 0$, be some continuous increasing function of the module of continuity type such that $\omega(0)=0$.

THEOREM 3.2. *Let x be a solution of equation (2) (procedure (2′)) with* $\mathbf{E}\|x_0\|^2<\infty$ *and let*
1) $a_\infty(\gamma)=\infty$ (a.s.);
2) $\big(u,\nabla J(u)\big)<0$ *for all* $u\in\mathbf{R}_0^d$;
3) $\|\nabla J(u)-\nabla J(y)\|\le R^{-1}L_{\varepsilon,m}(R^2)\omega(\|x-y\|)$, $\|\nabla J(x)\|^2\le L_{\varepsilon,m}(R^2)$, $L_{\varepsilon,m}\in\mathcal{L}$ *for some* $\delta_1\in(0,1]$, *for any* $R>0$ *and* $\delta\le\delta_1$, *and for all* x,y *such that* $\|x-y\|\le\delta$, $\|x\|$, $\|y\|\le R+\delta$;
4) *there exists a predictable function* l_t *such that* $\mathrm{tr}\,\sigma(t,u)\Delta_t^m\times\sigma^*(t,u)\le l_tL_{\varepsilon,m}(\|u\|^2)$;
5)

$$\int\limits_0^\infty\omega(c_s)\,da_s(\gamma)+\int\limits_0^\infty l_s\,d\langle m(c^{-1}\gamma)\rangle_s+\int\limits_0^\infty\Delta a_s(\gamma)\,da_s^d(\gamma)<\infty\quad\text{(a.s.)},$$

$$\mathbf{E}\big\{\sup_t l_t\Delta\langle m(c^{-1}\gamma)\rangle_t+\sup_t\Delta^2a_s(\gamma)+\sup_t\Delta a_t\omega(c_t)\big\}<\infty.$$

Then $x_t\to\infty$ (a.s.) *as* $t\to\infty$.

Proof. By analogy with the proof of Theorem 3.1, applying the Itô formula to $W(x_t)=W_{\varepsilon,m}(x_t)$ and using the conditions of Theorem 3.2, we have

$$W(x_t)\le W(x_0)+4\int\limits_0^tL^{-1}(\|x_{s-}\|^2)x_{s-}^*\nabla J(x_{s-})\,da_s(\gamma)$$

$$+2\int\limits_0^tL^{-1}(\|x_{s-}\|^2)x_{s-}^*\big[\nabla_cJ(x_{s-})-\nabla J(x_{s-})\big]\,da_s(\gamma)$$

$$+ \operatorname{const} \int_0^t L^{-1}\left(\|x_{s-}\|^2\right) \operatorname{tr} \sigma(s, x_s) \Delta_s^m \sigma^*(s, x_{s-})\, d\left\langle m(c^{-1}\gamma)\right\rangle$$

$$+ \int_0^t L^{-1}\left(\|x_{s-}\|^2\right) \|\nabla a_s(\gamma)\, da_s^d(\gamma) + M_t = Z_0 + \sum_{i=1}^4 I_i + M_t, \qquad (8)$$

where $M \in \mathcal{M}_{\text{loc}}(\mathbf{R}^1)$. Without loss of generality, we assume that $c_t < \delta_1/2$ (a.s) for all $t \geq 0$. Furthermore, according to the average theorem, for every $k = 1, \ldots, d$, there exists $\theta_s^k \in [x_{-s}^k - c_s, x_{s-}^k + c_s]$ such that

$$J(x_{s-} + c_s i_k) - J(x_{s-} - c_s i_k) = \frac{\partial}{\partial x_k} J(x_{s-}^1, \ldots, x_{s-}^{k-1}, \theta_s^k, x_{s-}^{k+1}, \ldots, x_{s-}^d) 2c_s. \quad (9)$$

Denote the right-hand side of (9) by $J'_{x_k}(\theta_s^k)$. Using (9) and assumption 3), we obtain

$$\left\|\nabla_c J(x_{s-}) - \nabla J(x_{s-})\right\| = \left(\sum_{k=1}^d \left| J'_{x_k}(\theta_s^k) - J'_{x_k}(x_{s-}) \right|^2 \right)^{1/2}$$

$$\leq \sup_{\substack{u,v:\, \|u-v\| \leq c_s; \\ \|u\|, \|v\| \leq \|x_{s-}\| + c_s}} \left(\sum_{k=1}^d \left| J'_{x_k}(u) - J'_{x_k}(v) \right|^2 \right)^{1/2}$$

$$\leq \|x_{s-}\|^{-1} L\left(\|x_{s-}\|^2\right) \omega(c_s). \qquad (10)$$

Analogously,

$$\left\|\nabla_c J(x_{s-})\right\| \leq \sup_{u:\, \|u\| \leq \|x_{s-}\| + c_s} \left\|\nabla J(u)\right\| \leq L^{1/2}\left(\|x_s\|^2\right). \qquad (11)$$

Now, using estimates (10)–(11) and the assumptions of Theorem 3.2, we obtain

$$I_2 \leq 2 \int_0^t L^{-1}\left(\|x_{s-}\|^2\right) \|x_{s-}\| \left\|\nabla_c J(x_{s-}) - \nabla J(x_{s-})\right\| da_s(\gamma)$$

$$\leq \operatorname{const} \int_0^t \omega(c_s)\, da_s(\gamma),$$

$$I_3 \leq \operatorname{const} \int_0^t l_s d\left\langle m(c^{-1}\gamma)\right\rangle_s, \qquad I_4 \leq \int_0^t \Delta a_s(\gamma)\, da_s^d(\gamma). \qquad (12)$$

Denoting the sum of the right-hand sides of inequalities (12) by A_t^1 and I_1 by A_t^2, we get from relation (8) that

$$Z_t \leq Z_0^1 + A_t^1 - A_t^2 + M_t.$$

Assumption 5) makes the Lemma 2 applicable. Reasonings needed to prove the theorem repeat, actually, the proof of Theorem 3.1. Theorem 3.2 is proved.

Remark 3.3. The assumption 3) and the first condition in 5) for model (2′) with a continuous process a and a twice continuously differentiable function $J(x)$ can be substituted by the following ones

$$\left\| \frac{\partial^2 J}{\partial x^2}(x) - \frac{\partial^2 J}{\partial x^2}(y) \right\| \leq R^{-1} L(R^2)\omega(\|x-y\|), \qquad \int_0^\infty \omega(c_s)c_s\, da_s(\gamma) < \infty \quad \text{(a.s.)}.$$

In this case, indeed, according to the average theorem, we have

$$\left\| \nabla_c J(x_{s-}) - \nabla J(x_{s-}) \right\| \leq \operatorname{const} c_s \|x_{s-}\|^{-1} L(\|x_{s-}\|^2)\omega(c_s).$$

This allows us to estimate I_2 in (12) by

$$\int_0^t \omega(c_s)c_s\, da_s(\gamma).$$

Remark 3.4. As in model (1), we can relax assumptions on the growth of $\sigma(t,u)$ (see condition 4)) for continuous Kiefer–Wolfowitz model (2) by the folowing ones: for $\varepsilon = e_m - e_{m-1}$ there exist $\delta \in (0,1)$ and $l \in \mathcal{P}$, $\int_0^\infty l_s\, dV_s < \infty$, such that $2\gamma_t(1-\delta)(u, \nabla J(u))\alpha_t + (m+1)\gamma_t^2 \operatorname{tr} \sigma(t,u)\beta_t\sigma^*(t,u) \leq l_t L_{\varepsilon,m}(\|u\|^2)$ $(dV \times dP)$-(a.s).

4. LINE STOCHASTIC INTEGRALS AND PROCEDURES OF STOCHASTIC APPROXIMATION

Simulation of procedures of stochastic approximation as solutions of stochastic equations with line stochastic integrals is, in our opinion, a more general and natural approach to studies of these procedures. The purpose of this paragraph is to justify this fact for continuous procedures; the discrete case of this approach is well-known. But, in fact, the connection with the general theory of stochastic line integration has never been used. In addition, we prove some standard existence and uniqueness theorems for such equations (see (Gikhman and Skorokhod, 1982)). Consider the stochastic equation

$$x_t = x_0 + \int_0^t a_s(x_s, da_s) + \int_0^t b_s(x_s, ds), \quad t > 0, \tag{13}$$

in $\mathbf{R}^d$, where $a_s(u,h) = a(s+h,u) - a(s,u)$, $b_s(u,h) = b(s+h,u) - b(s,u)$, $h > 0$, $a(s,u)$ is a continuous process of bounded variation with values in $\mathbf{R}^d$ for a fixed $u \in \mathbf{R}^d$, $d \geq 1$, $b(\cdot,u) \in \mathcal{M}^c_{\text{loc}}(\mathbf{R}^d)$.

All notions concerning the theory of line stochastic integrals see in (Gikhman and Skorokhod, 1982). Let $B_t(u,v)$ be the quadratic matrix characteristic of martingales $b(t,u)$ and $b(t,v)$ and let $B_t(u,v,h) = B_{t+h}(u,v) - B_t(u,v)$ for $h > 0$.

Furthermore, let N be a continuous convex nondecreasing non-negative function defined on $[0,\infty]$ such that $\int\limits_{0+}^{\infty} N^{-1}(u)\,du = \infty$.

THEOREM 4.1. *Let the following assumptions be fulfilled:*
1) *there exists a continuous process* $\lambda \in \mathcal{A}^+_{\mathrm{loc}} \cap \mathcal{P}$ *such that*

$$\|a_t(u,h)a_t(v,h)\| \le (\lambda(t+h)-\lambda(t))N^{1/2}(\|u-v\|^2),$$
$$\|B_t(u,v,h)-B_t(u,u,h)\| \le (\lambda(t+h)-\lambda(t))N(\|u-v\|^2)$$

for all $t>0$, u, $v \in \mathbf{R}^d$, $h>0$ (a.s.);
2) $\sup_{u:\,\|u\|\le n}\{\|a_t(u,h)\|, \|B_t(u,u,h)\|\} \le \varphi(n)(\lambda(t+h)-\lambda(t))$, *where* $\varphi\colon [0,\infty) \to [0,\infty)$ *is a locally bounded function,* $h>0$;
3) *there exist* $0<l \in \mathcal{P}$, $\int\limits_0^{\infty} l_s\,d\lambda_s < \infty$ (a.s.), $L_{\varepsilon,m} \in \mathcal{L}(\varepsilon = e_m - e_{m-1})$ *such that*

$$2(u, a_t(u,h)) + (m+1)\,\mathrm{tr}\,B_t(u,u,h) \le l_t L_{\varepsilon,m}(\|u\|^2)(\lambda(t+h)-\lambda(t))$$

for $t>0$, $u \in \mathbf{R}^d$ (a.s.).
Then there exists a unique solution of equation (13).

To prove the theorem, we need the following lemma.

LEMMA 4.1. *Let assumption* 1) *be fulfilled and let*

$$\|a_t(u,h)\| \le (\lambda(t+h)-\lambda(t))(1+\|u\|),$$
$$\mathrm{tr}\,B_t(u,u,h) \le (\lambda(t+h)-\lambda(t))(1+\|u\|^2) \tag{14}$$

for $t>0$, $u \in \mathbf{R}^d$ (a.s.).
Then there exists a unique solution of equation (13).

Proof. Consider the successive approximations

$$(x_{n+1})_t = x_0 + \int_0^t a_s((x_n)_s, ds) + \int_0^t b_s((x_n)_s, ds), \quad (x_0)_t \equiv x_0. \tag{15}$$

Denote the Banach space of F_t-adopted processes Y_t with values in $\mathbf{R}^d$ such that

$$\|Y\|_{B_T} = \left(\mathbf{R}\sup_{s\le T}\|Y_S\|^2\right)^{1/2} < \infty$$

for a fixed $T>0$ by B_T.
Let $\eta_u = \inf(t\colon \lambda(t) > u)$. Then, in accordance with properties of stochastic line integrals and conditions (14)–(15), we have

$$\mathbf{E}\|x_{n+1}\|^2_{\eta_t} \le H_1\left((t+1)^2 + (t+1)\int_0^t \mathbf{E}\|x_n\|^2_{\eta_s}\,ds\right).$$

Furthermore, using the Gronwall lemma (Liptser and Shiryaev, 1986), from this inequality we obtain

$$\mathbf{E}\|x_n\|^2_{\eta_T} \le H_2(T), \qquad \mathbf{E}\|x_n - x_0\|^2_{\eta_T} \le H_\xi(T),$$

where H_1, H_2, H_3 are some constants. Now let us show that

$$\mathbf{E}\|x_{m+n} - x_m\|^2_{\eta_t} \le Z_m(t)$$

for $t < T$, where

$$Z_{m+1}(t) = 2(T + 8d)\int_0^T N\big(Z_m(s)\big)\,ds, \qquad Z_0(t) \equiv H_3(T).$$

Note the important property (see, for example, (Rodkina, 1984)) of the functions Z_m:

$$\sup_{t\le T} Z_m(t) \to 0 \quad (m \to \infty). \tag{16}$$

Furthermore, using induction on m, condition (1), and the Doob inequality, we have

$$\mathbf{E}\|x_{n+m+1} - x + m + 1\|^2_{\eta_t}$$
$$\le 2\left\{\mathbf{E}\sup_{n\le t}\left\|\int_0^{\eta_n} N^{1/2}\Big(\|(x_{m+n})_s - (x_n)_s\|^2\Big)\,d\lambda(s)\right\|^2 + 4\mathbf{E}\,\mathrm{tr}\,B^\eta(t)\right\},$$

where $B^\eta(t)$ is the quadratic characteristic of the martingale

$$\int_0^{\eta_t} b_s\big((x_{m+n})_s, ds\big) - \int_0^{\eta_t} b_s\big((x_m)_s, ds\big).$$

In addition, according to (Gikhman and Skorokhod, 1982),

$$\mathrm{tr}\,B^\eta(t) \le 2d\int_0^{\eta_t} N\Big(\|(x_{m+n})_s - (x_m)_s\|^2\Big)\,d\lambda(s) \quad \text{(a.s.)}.$$

Therefore, using the connection between $\lambda(t)$ and η_t and the inductive hypothesis, we have

$$\mathbf{E}\|x_{n+m+1} - x_{m+1}\|^2_{\eta_t} \le 2\left\{\mathbf{E}\left|\int_0^{\eta_t} N^{1/2}(\|x_{m+n} - x_m\|^2_s)\,d\lambda(s)\right|^2\right.$$
$$\left.+ 8d\mathbf{E}\int_0^{\eta_t} N(\|x_{m+n} - x_m\|^2_s)\,d\lambda(s)\right\} \le 2\left\{\mathbf{E}\left|\int_0^{t} N^{1/2}(\|x_{m+n} - x_m\|^2_{\eta_s})\,ds\right|^2\right.$$

$$+ 8d\mathbf{E}\int_0^t N\big(\|x_{m+n} - x_m\|^2_{\eta_s}\big)\,ds\Big\}$$

$$\le 2\Big\{\mathbf{E}t\int_0^t N\big(\|x_{m+n} - x_m\|^2_{\eta_s}\big)\,ds + 8d\mathbf{E}\int_0^t N\big(\|x_{m+n} - x_m\|^2_{\eta_s}\big)\,ds\Big\}$$

$$\le 2(T + 8d)\int_0^t N(Z_m(s))\,ds = Z_{m+1}(t).$$

Hence (16) implies the fundamentality of $(x_n)_{\eta_t}$ in the space B_T. It is not difficult to show that its limit $y^* \in B_T$ and $x_t = y^*_{\lambda_t}$ is the unique solution of equation (13) on $[0, \eta_T]$. Taking into account that $\eta_t \to \infty$ as $t \to \infty$, it is possible to construct the solution on $[0, \infty)$. The lemma is proved.

Proof of Theorem 4.1. Let

$$a^n_s(u, h) = \begin{cases} a_s(u, h), & \|u\| \le n, \\ a_s(n\frac{u}{\|u\|}, h), & \|u\| > n, \end{cases}$$

$$b^n_s(u, h) = \begin{cases} b_s(u, h), & \|u\| \le n, \\ b_s(n\frac{u}{\|u\|}, h), & \|u\| > n, \end{cases}$$

for $n = 1, 2, \ldots$.
Consider equation (13) with the coefficients a^n and b^n instead of a and b, respectively. Then, according to Lemma 4.1, this equation has a unique solution x^n on $[0, \infty)$. Consider the a.s. increasing sequence

$$\tau_n = \inf\{t > 0\colon \|x^n(t)\| > n\}.$$

It is obvious that $x_t = x^n_t$ is a solution of (13) on the stochastic interval $[0, \tau_n[$. Let us show that

$$\mathbf{P}\{\omega\colon \tau = \lim_n \tau_n = \infty\} = 1. \tag{17}$$

Suppose that $\mathbf{P}\{\omega\colon \tau(\omega) < \infty\} > 0$. Consider the function

$$W(u) = \int_0^{\|u\|^2} L^{-1}(u)\,du \quad \text{and s.t.} \quad \theta_m = \inf\Big\{t\colon \int_0^t l_s\,d\lambda(s) > m\Big\}.$$

Then, using the estimates from the lemma and the assumptions of the theorem, by the Itô formula we have

$$\mathbf{E}W(x_{t\wedge\tau_n\wedge\theta_m}) \le W(x_0) + \mathbf{E}\Big(\int_0^{t\wedge\tau_n\wedge\theta_m} L^{-1}(\|x_s\|^2)2x^*_s a_s(x_s, ds)$$

$$+ (m+1)\,\mathrm{tr}\,B_s(x_s, x_s, ds)\Big)$$

$$\le \mathbf{E}W(x_0) + \mathbf{E}\int_0^{t\wedge\tau_n\wedge\theta_m} l_s L^{-1}(\|x_s\|^2)L(\|x_s\|^2)\,d\lambda(s)$$

$$\le \mathbf{E}W(x_0) + m.$$

Note that $\inf_{\|u\|>n} W(u) \to \infty \ (n \to \infty)$ and

$$\mathbf{E}W(x_{t\wedge\tau_n\wedge\theta_m}) \geq \int\limits_{(\tau_n<t\wedge\theta_m)} W(x_{t\wedge\tau_n\wedge\theta_m})\,d\mathbf{P} \geq \inf_{\|u\|>n} W(u)\mathbf{P}\{\omega\colon \tau_n < t < \theta_m\}.$$

Thus,

$$\begin{aligned}\mathbf{P}\{\tau_n < t \wedge \theta_m\} &\leq \big(\inf_{\|u\|>n} W(u)\big)^{-1}\mathbf{E}W(x_{t\wedge\tau_n\wedge\theta_m}) \\ &\leq \big(\inf_{\|u\|>n} W(u)\big)^{-1}\mathbf{E}W(x_0) + m \longrightarrow 0 \quad \text{as} \quad n \to \infty\end{aligned}$$

for a fixed m. Taking into account that $\theta_m \to \infty \ (m \to \infty)$, we obtain (17) and, therefore, the solution is constructed on the interval $[0, \infty)$. The theorem is proved.

Now let us consider the stochastic (continuous) Robbins–Monro and Kiefer–Volfowitz procedures

$$x_t = x_0 + \int\limits_0^t \gamma_s R(x_s)\,d\lambda(s) + \int\limits_0^t \gamma_s b_s(x_s,\, ds), \tag{18}$$

$$x_t = x_0 + \int\limits_0^t \gamma_s \nabla_c J(x_s)\,d\lambda(s)_s + \int\limits_0^t \gamma_s c_s^{-1} b_s(x_s,\, ds), \tag{19}$$

where the functions R, J, γ, c, ∇_c and the random field b are the same as above. In addition, the stochastic integral $\int_0^t \gamma_s b_s(x_s,\, ds)$ requires a formal definition. It is possible to do this in the standard way for a function γ such that $\int_0^t \gamma_s^2\,d\lambda(s) < \infty$. Therefore, the trajectories of the stochastic procedures are supposed to be constructed as random curves in random fields which are induced by the functions R, J and "errors" of b.

THEOREM 4.2. *Let x be the solution of* (18) *and let the following assumptions be fulfilled:*
1) $\lambda_\infty(\gamma) = \infty$ (a.s.);
2) $(u, \mathbf{R}(u)) < 0$ *for all* $u \in \mathbf{R}_0^d$;
3) *there exist* $L \in \mathcal{L}$ *and a predictable function* l_t *such that* $\int_0^\infty l_s\,d\lambda(s) < \infty$ (a.s.) *and*

$$\gamma_t^2 \operatorname{tr} B_t(u^2, u, h) \leq l_t L(\|u\|^2)\big(\lambda(t+h) - \lambda(t)\big)$$

for all $u \in \mathbf{R}^d$.
Then $x_t \to 0$ (a.s.) *as* $t \to \infty$.

Proof. As in the proof of Theorem 4.1, we have

$$\begin{aligned}W(x_t) \leq W(x_0) &+ 2\int\limits_0^t \gamma_s L^{-1}(\|x_s\|^2)(x_s, R(x_s))\,d\lambda(s) \\ &+ (m+1)\int\limits_0^t \gamma_s^2 L^{-1}(\|x_s\|^2)\operatorname{tr} B_s(x_s, x_s, ds) + \text{martingale} \\ = W(x_0) &- A_t^2 + A_t^1 + \text{martingale}.\end{aligned}$$

As it was mentioned earlier, according to Lemma 2.2 we have $\{A^1_\infty < \infty\} \subseteq \{W(x_t) \to\} \cap \{A^2_\infty < \infty\}$ (a.s.). Using this key relation, we complete the proof of the theorem in a standard way.

THEOREM 4.3. *Let x be the solution of* (19) *and let following assumptions be fulfilled*:
1) $\lambda_\infty(\gamma) = \infty$ (a.s.);
2) $(u, \nabla J(u)) < 0$ *for all* $u \in \mathbf{R}^d_0$;
3) *there exists* $L = L_{m,\varepsilon} \in \mathcal{L}$ *and a predictable function* l *such that* $\int_0^\infty l_s \, d\lambda(s) < \infty$ (a.s.), $\gamma_t^2 c_t^{-1} \operatorname{tr} B_t(u,u,h) \le l_t L(\|u\|^2)(\lambda(t+h) - \lambda(t))$ *for all* $u \in \mathbf{R}^d$;
4) *there exists* $\delta_0 > 0$ *such that*

$$\|\nabla J(u) - \nabla J(v)\| \le R^{-1} L(R^2)\, \omega(\|u - v\|),$$

$$\int_0^\infty \gamma_s \omega(c_s) d\lambda(s) < \infty \quad \text{a.s.}$$

for all $\delta \le \delta_0$ *and* u, $v \in \mathbf{R}^d$ *such that* $\|u\|$, $\|v\| \le R$, $\|u - v\| < \delta$.
Then $x_t \to 0$ (a.s.) *as* $t \to \infty$.

Proof. By the Itô formula we have

$$\begin{aligned} W(x_t) \le{}& W(x_0) + \int_0^t \gamma_s L^{-1}(\|x_s\|^2)(x_s, \nabla J(x_s))\, d\lambda(s) \\ &+ 2\int_0^t \gamma_s L^{-1}(\|x_s\|^2)(x_s, \nabla_c J(x_s) - \nabla J(x_s))\, d\lambda(s) \\ &+ (m+1)\int_0^t \gamma_s^2 c_s^{-2} L^{-1}(\|x_s\|^2) \operatorname{tr} B_s(x_s, x_s, ds) + \text{martingale} \\ ={}& W(x_0) - A_t^2 + A_t^1 + A_t^3 + \text{martingale}. \end{aligned}$$

By condition 4) we have (a.s.) $A^1_\infty \le \text{const} \cdot \int_0^\infty \gamma_s l_s d\lambda(s) < \infty$, $A_t^2 < 0$, and $A^3_\infty \le \text{const} \cdot \int_0^\infty \gamma_s \omega(c_s) d\lambda(s) < \infty$. Then Lemma 2.2 combined with these relations implies the convergence of $W(x_t)$ (a.s.) to a finite limit as $t \to \infty$. The rest of the proof is standard.

Remark 4.1. Only continuous procedures simulating "noises" with the help of stochastic line integrals were considered in this article. Clearly, it is not difficult to investigate quasi-continuous processes. Finally, the natural consideration of the general case containing discrete and continuous models of stochastic approximation needs an essential generalization of classical results of the theory of stochastic line integrals (see (Gikhman and Skorokhod, 1982)) that may be the subject of a separate investigation.

5. CONCLUDING REMARKS

Fist of all, observe that the continuons (diffusion) model of a stochastic approximation is imbedded directly in general model (1)–(2): for that purpose, has to put $a_t \equiv t$ and $m_t = W_t$ (the standard Wiener process). As we can see from the examples, the obtained results for procedures (1) and (2) are the nontrivial generalizations of the corresponding theorems for diffusion procedures (cf. (Nevelson and Khasminskii, 1972)). If we consider, for example, the discrete Robbins–Monro procedure (with the stochastic basis $(\Omega, F, (F_n)_{n=1,2,\ldots}, \mathbf{P})$, $x_{n+1} = x_n + \gamma_n y_{n+1}$, $n \geq 0$, where $y_{n+1} = R(x_n) + \varepsilon_{n+1}$ is the observation and ε_{n+1} is its error (martingale difference with $\mathbf{E}(\varepsilon_{n+1}^2 \mid F_n) \equiv \delta^2$), then its embedding in model (1) may be constructed in the following way: $a_t = [t]$ (the integer part of t), $m_t = \sum_{k=1}^{[t]} \varepsilon_k$ and so on.

Remark 5.1 (see (Melnikov and Rodkina, 1992)). If $d = k = 1$ in continuous model (1)–(2), then it is possible to relax the assumptions on the growth of the functions R, J, and σ. Namely, the convergence of procedures (1) and (2) is provided under conditions 1)–3) of the corresponding multi-dimensional theorems and the following assumption: there exist $\delta_0 > 0$ and a predictable non-negative function l, $\int_0^\infty l_s dV_s < \infty$ (a.s.), such that

$$2u\gamma_t R(u)\alpha_t + \gamma_t^2 \beta_t \sigma^2(t,u) \leq l_t,$$
$$2u\gamma_t \nabla_c J(u)\alpha_t + \gamma_t^2 c_t^{-1} \beta_t \sigma^2(t,u) \leq l_t$$

for $|u| \leq \delta_0 (dV \times dP)$ (a.s.). It should be noted that different variants of condition 2) of the theorem are modifications of the well-known in the theory of stochastic equations monotone condition (see (Gyongy and Krylov, 1980)).

Remark 5.2. Consider the linear one-dimensional procedure (1)

$$x_t = x_0 - \int_0^t \beta(x_{s-} - \theta)\gamma_s da_s - \int_0^t \gamma_s dm_s, \quad \beta > 0.$$

Setting $\varepsilon_t = 1 - \beta \int_0^t \gamma_s \varepsilon_{s-} da_s$, $\varepsilon_t^{-1} = 1 + \beta \int_0^t \gamma_s \varepsilon_s^{-1} da_s$ and using the Itô formula, we have

$$\varepsilon_t^{-1}(x_t - \theta) = (x_0 - \theta) - \int_0^t \gamma_s \varepsilon_s^{-1}(\beta)\, dm_s = x_0^\theta + M_t. \tag{20}$$

Note that the local square integrable martingale M has the square characteristic

$$\langle M \rangle_t = \int_0^t \gamma_s^2 \varepsilon_s^{-2}\, d\langle m \rangle_s.$$

Dividing both sides of (20) by $(2\langle M \rangle_t \log\log\langle M \rangle_t)^{-1/2}$ we obtain

$$\varepsilon_t^{-1}(x_t - \theta)\big(2\langle M \rangle_t \log\log\langle M \rangle_t\big)^{1/2} = x_0^\theta \big(2\langle M \rangle_t \log\log\langle M \rangle_t\big)^{-1/2}$$
$$- M_t\big(2\langle M \rangle_t \log\log\langle M \rangle_t\big)^{-1/2}. \tag{21}$$

Relation (21) allows one to obtain results of the iterated logarithm law type. For the procedures considered one can deduce them from the corresponding results for martingales (see (Lepingle, 1977)). Thus, for the continuous model with $\gamma^{-1} = (1 + a_t)^{-1}$, $a_t = \langle m \rangle_t \to \infty$ $(t \to \infty)$ we obtain (compare with (Gaposhkin and Krasulina, 1974))

$$\limsup_t \frac{|x_t - \theta|}{(\langle m \rangle \log\log \langle m \rangle)_t^{1/2}} = (2\beta - 1)^{1/2}.$$

The same relation (20) shows that it is possible to obtain the result on asymptotic normality of $\varepsilon_t^{-1/2}(x_t - \theta)$ as a corollary of martingale limit theorems (see (Lipser and Shiryaev, 1986; Melnikov, 1989)).

In conclusion, note that it was not our purpose in this article to cover all the variety of problems of a stochastic approximation which are sufficiently represented in many books and monographs, the common feature of which is to study continuous and discrete models separately (see, e.g. (Gaines, 1988; Duflo, 1990; Korostelev, 1984)).

REFERENCES

Caines, P.E. (1988). *Linear Stochastic System*. John Willey, New York.

Duflo, M. (1990). *Methodes Recursives Aleatoires*. Masson, Paris.

Gaposhkin, B.F. and Krasulina, T.P. (1974). On the law of iterated logarithm in processes of stochastic approximation. *Theory Probab. Appl.* **19**, 879–885.

Gikhman, I.I. and Skorokhod, A.V. (1982). *Stochastic Differential Equations and Its Applications*. Naukova Dumka, Kiev (in Russian).

Gyongy, I. and Krylov, N.V. (1980). On stochastic equations with respect to semimartingales. *Stochastics* **4**, 1–22.

Kiefer, J. and Wolfowitz, J. (1952). Stochastic estimation of the maximum of a regression function. *Ann. Math. Statist.* **23**, 462–466.

Korostelev, A.P. (1984). *Stochastic Recursion Procedures*. Nauka, Moscow (in Russian).

Lepingle, D. (1977). Sur le comportments asymptotiques des martingaleslocales *Lect. Notes Math.* **649**, 148–161.

Liptser, R.S. and Shiryaev, A.N. (1986). *Theory of Martingales*. Nauka, Moscow (in Russian).

Melnikov, A.V. (1988). Procedures of stochastic approximation in statistics of processes of martingale-type. *Russian Math. Surveys* **43**, 216–216 (in Russian).

Melnikov, A.V. (1989). Procedures of stochastic approximation for semimartingales. In: *Statistics and Control of Random Processes*, Nauka, Moscow, pp. 147–156 (in Russian).

Melnikov, A.V. and Rodkina, A.E. (1992). Consistent statistical estimation in semimartingale models of stochastic approximation. *Ann. Acad. Sci. Fennicae. Ser. A.I. Mathematica* **17**, 85–92.

Nevelson, M.B. and Khasminskii, R.Z. (1972). *Stochastic Approximation and Recursion Estimation*. Nauka, Moscow (in Russian).

Robbins, H. and Monro, S. (1951). A stochastic approximation method. *Ann. Math. Statist.* **22**, 400–407.

Rodkina, A.E. (1984). On existence and uniqueness of solutions of stochastic differential equations with the hereditary. *Stochastics* **12**, 187–200.

Frontiers in Pure and Applied Probab. 1, pp. 183 – 196
H. Niemi et al. (Eds.)
1993 TVP/VSP

ON A GENERAL CLASS OF STOCHASTIC APPROXIMATION ALGORITHMS

A. V. MELNIKOV, A. E. RODKINA, and E. VALKEILA

Steklov Mathematical Institute, Vavilov Street 42, 117966 Moscow GSP-1, Russia

Voronezh Civil Engineering Institute, 20th October Street 86, 394006 Voronezh, Russia

Computer Center of University of Helsinki, Teollisuusk. 23, SF-00510 Helsinki, Finland

ABSTRACT

This paper is devoted to a general class of stochastic approximation algorithms of the Robbins–Monro type. The results about (a.s.)-convergence, $\mathbf{L}_2$-convergence and asymptotic normality of these algorithms are presented.

1. INTRODUCTION

The theory of stochastic approximation (see (Nevelson and Khasminskii, 1972)) deals with the problem of how to construct a (stochastic) sequence θ_n or a (stochastic) process θ_t converging to the unique root θ^* of the regression equation

$$R(\theta^*) = 0. \tag{1.1}$$

If the errors of observation of the function R have a martingale structure (i.e., the errors are sequences of independent random variables (i.r.v.'s), martingale differences, Wiener process, etc.), then one can describe the corresponding stochastic algorithm as a strong solution of a stochastic differential equation with respect to semimartingales (see (Liptser and Shiryaev, 1986; Gikhman and Skorokhod, 1982))

$$d\theta_t = -\gamma_t R(\theta_{t-}) da_t - \gamma_t dm_t, \tag{1.2}$$

where the initial point θ_0 is a finite r.v., a_t is an increasing predictable process, m_t is a local square-integrable martingale with quadratic characteristic $\langle m_t \rangle$, γ_t is a non-negative predictable process.

Fixing a standard stochastic basis $(\Omega, F, (F_t)_{t \geq 0}, \mathbf{P})$ (see (Liptser and Shiryaev, 1986; Gikhman and Skorokhod, 1982)) for all our references from the general theory of stochastic processes) we shall consider our model (1.2) on it. Also we assume that the function R is a real continuous function and all the processes are cadlag (right continuous with finite left limits).

Model (1.2) combines both discrete time and pure continuous time models which were treated in (Nevelson and Khasminskii, 1972) separately. Our approach, first proposed in (Melnikov, 1988, 1989), is based on the general theory of martingales, which makes it possible to treat the two cases in a unified way. To investigate the asymptotic properties of θ_t (a.s. convergence, $\mathbf{L}_2$-convergence and asymptotic normality), it is advantageous to use the so-called Dolean exponent.
Note that this paper is close by related to the paper of Melnikov and Rodkina in this volume.

2. ON PROPERTIES OF DOLEAN'S EXPONENT

We give some properties of Dolean's exponent (see (Liptser and Shiryaev, 1986; Gikhman and Skorokhod, 1982)) when it is defined with respect to an increasing (decreasing) function. They are useful here but some of them are interesting in more general aspects.
In what follows, A and B are increasing processes with $A_0 = B_0 = 0$. Dolean's exponent $\varepsilon(A)$ is the unique solution of the equation

$$Z_t = 1 + \int_0^t Z_{s-} dA_s. \tag{2.1}$$

We have $\varepsilon_t(A) = e^{A_t} \prod_{s \leq t} (1 + \Delta A_s) e^{-\Delta A_s}$.
Denote by $\Delta A_t = A_t - A_{t-}$ the jump of A at t and by $[A, B]$ the increasing function $[A, B]_t = \sum_{s \leq t} \Delta A_s \Delta B_s$ and put $[A] = [A, A]$. The notation $(f \cdot A)_t$ means the integral $\int_0^t f_s dA_s$ and $(f_- \cdot A)_t$ means the integral $\int_0^t f_{s-} dA_s$. Recall the following properties of $\varepsilon(A)$ (see, e.g. (Liptser and Shiryaev, 1986)). The function $\varepsilon^{-1}(A)$ satisfies the equation

$$\varepsilon_t^{-1}(A) = 1 - \int_0^t \varepsilon_s^{-1}(A) dA_s. \tag{2.2}$$

Note that $\varepsilon_t^{-1}(A) = \varepsilon_t(-1/(1 + \Delta A) \cdot A)$ and hence $\varepsilon_t^{-1}(A)$ satisfies (2.1) with respect to the decreasing function $(-1/(1 + \Delta A) \cdot A)_t$. The following formula is called Yor's formula

$$\varepsilon(A)\varepsilon(B) = \varepsilon(A + B + [A, B]). \tag{2.3}$$

If $A_t \leq B_t$ for all $t \geq 0$, we write $A \leq B$.

LEMMA 2.1. *If $A \leq B$, then $\varepsilon(A) \leq \varepsilon(B)$.*

Proof. By Yor's formula we have

$$\varepsilon_t(A)\varepsilon_t^{-1}(B) = \varepsilon_t \left(\frac{1}{1 + \Delta B} \cdot (A - B) \right) \leq 1,$$

and the assertion follows.

Henceforth α and β are positive real numbers.

LEMMA 2.2. *If $\beta\Delta A < 1$, then*

$$\varepsilon_t^{-1}(-\beta A) = \varepsilon_t\left(\beta A + \frac{\beta^2}{1-\beta\Delta A}\cdot[A]\right)$$

and

$$\varepsilon_t^{-1}(-\beta A)\varepsilon_t(-\alpha A) = \varepsilon_t\left((\beta-\alpha)A + \frac{\beta(\beta-\alpha)}{1-\beta\Delta A}\cdot[A]\right) = \varepsilon_t\left(\frac{\beta-\alpha}{1-\beta\Delta A}\cdot A\right).$$

If $\beta\Delta A < -1$, then

$$\varepsilon_t^{-1}(\beta A) = \varepsilon_t\left(-\beta A + \frac{\beta^2}{1-\beta\Delta A)}\cdot[A]\right)$$

and

$$\varepsilon_t^{-1}(\beta A)\varepsilon_t(\alpha A) = \varepsilon_t\left((\alpha-\beta)A + \frac{\beta(\beta-\alpha)}{1+\beta\Delta A)}\cdot[A]\right) = \varepsilon_t\left(\frac{\alpha-\beta}{1-\beta\Delta A}\cdot A\right).$$

Proof. We only prove the first statement. Using Yor's formula (2.3), we get

$$\varepsilon_t\left(\beta A + \sum\frac{\beta^2(\Delta A)^2}{(1-\beta\Delta A)}\right)\varepsilon_t(-\beta A) = \varepsilon_t\Bigg(0\cdot A + \sum\frac{\beta^2(\Delta A)^2}{(1-\beta\Delta A)}$$
$$-\sum\beta\Delta A\left(\beta\Delta A + \frac{\beta^2(\Delta A)^2}{(1-\beta\Delta A)}\right)\Bigg) = \varepsilon_t(0) = 1.$$

Hence

$$\varepsilon_t^{-1}(-\beta A)\varepsilon_t(-\alpha A) = \varepsilon_t\Bigg((\beta-\alpha)A + \sum\frac{\beta(\beta-\alpha)(\Delta A)^2}{(1-\beta\Delta A)}$$

$$-\alpha\sum\Delta A\left(\beta\Delta A + \frac{\beta^2(\Delta A)^2}{(1-\beta\Delta A)}\right)\Bigg) = \varepsilon_t\left((\beta-\alpha)A + \sum\frac{\beta(\beta-\alpha)(\Delta A)^2}{(1-\beta\Delta A)}\right)$$
$$= \varepsilon_t\left((\beta-\alpha)A + \frac{\beta(\beta-\alpha)}{1-\beta\Delta A}\cdot[A]\right).$$

Denote $A_t^d = \sum\limits_{s\le t}\Delta A_s$ and $A_t^c = A_t - A_t^d$.

We can rewrite now

$$(\beta-\alpha)A + \sum\frac{\beta(\beta-\alpha)(\Delta A)^2}{(1-\beta\Delta A)} = (\beta-\alpha)A_t^c + (\beta-\alpha)\sum\Delta A_s$$
$$+\sum\frac{\beta(\beta-\alpha)(\Delta A_s)^2}{(1-\beta\Delta A_s)} = (\beta-\alpha)A_t^c + \sum_{s\le t}(\beta-\alpha)\Delta A_s\left(1+\frac{\beta\Delta A_s}{1-\beta\Delta A_s}\right)$$

$$= (\beta-\alpha)A_t^c + (\beta-\alpha)\sum_{s\le t}\frac{\Delta A_s}{1-\beta\Delta A_s} = \left(\frac{(\beta-\alpha)}{1-\beta\Delta A}\cdot A\right)_t$$

and the lemma follows.

COROLLARY 2.1.

$$\varepsilon_t^{-1}(-\beta A) \ge \varepsilon_t(\beta A),$$

$\varepsilon_t^{-1}(-\beta A)\varepsilon_t(-\alpha A)$ is greater or less than $\varepsilon_t(\beta-\alpha)A$ according to whether $\beta-\alpha>0$ or $\beta-\alpha<0$; $\varepsilon_t^{-1}(\beta A)\varepsilon_t(\alpha A)$ is greater or less than $\varepsilon_t((\alpha-\beta)A)$ according whether to $\beta-\alpha>0$ or $\beta-\alpha<0$.

Next we compare $\varepsilon_t(\alpha A)$ and $\varepsilon_t^\alpha(A)$.

LEMMA 2.3. *1) $\varepsilon_t^\alpha(A)\le\varepsilon_t(\alpha A)$ if $\alpha\in(0,1)$,*
2) $\varepsilon_t^\alpha(A)\ge\varepsilon_t(\alpha A)$ if $\alpha\in(1,\infty)$,
3) $\varepsilon_t^\alpha(A)\le\xi\varepsilon_t(\alpha A)$ if $|\Delta A|\le\mu$, where ξ and μ are some finite (random) variables.

Proof. Recall that

$$\varepsilon_t(A) = e^{A_t^c}\prod_{s\le t}(1+\Delta A_s). \tag{2.4}$$

From the representation $(1+x)^\alpha = (1+\alpha x)[(1+x)^\alpha(1+\alpha x)^{-1}] = (1+\alpha x)g_\alpha(x)$, $x>0$, we have

$$g'_\alpha(x) = \frac{\alpha(1+x)^{\alpha-1}}{(1+\alpha x)^2}(\alpha-1)x.$$

It is clear that $g'_\alpha(x)<0$, if $\alpha\in(0,1)$, so $g'_\alpha(x)$ is a decreasing function and therefore $(1+x)^\alpha\le(1+\alpha x)g_\alpha(0)=1+\alpha x$. The statement 1) follows from this and (2.4). In the case $\alpha\in(1,\infty)$, we have $g'_\alpha(x)>0$, so g_α is increasing and therefore $(1+x)^\alpha\ge(1+\alpha x)g_\alpha(0)=1+\alpha x$. Now, applying (2.4), we get the second statement. In the third case we just note $g_\alpha(x)$ is bounded above by $\xi=(\mu+1)^\alpha$. The lemma is proved.

Remark 2.1. Assume that $\Delta A<1$. We have then in the previous lemma that $\varepsilon(-\alpha A)\ge\varepsilon^\alpha(-A)$ for $\alpha>1$ and $\varepsilon(-\alpha A)\le\varepsilon^\alpha(-A)$ for $0<\alpha<1$.
In the case $\alpha=1/2$ there are exact upper and lower bounds for $\varepsilon^\alpha(\pm A)$.

LEMMA 2.4. *For $\alpha=1/2$ we have*

$$\varepsilon\left(\frac{1}{2+\Delta A}\cdot A\right)\le\varepsilon^{1/2}(A)\le\varepsilon(A/2) \tag{2.5}$$

and

$$\varepsilon(-A/2)\le\varepsilon^{1/2}(-A)\le\varepsilon\left(-\frac{1}{2-\Delta A}\cdot A\right). \tag{2.6}$$

Proof. If $B>0$ is a function of bounded variation, then $dB^{1/2}=\left(B_-^{1/2}+B^{1/2}\right)^{-1}dB$. Using this and (2.1), we get

$$\varepsilon^{1/2}(A) = \varepsilon\left(\frac{1}{1+(1+\Delta A)^{1/2}} \cdot A\right), \tag{2.7}$$

Since

$$\left(\frac{1}{2+\Delta A} \cdot A\right)_t \le \left(\frac{1}{1+(1+\Delta A)^{1/2}} \cdot A\right)_t \le \frac{1}{2}A_t.$$

We obtain (2.5) from Lemma 2.1 and (2.7). The proof of (2.6) is similar.

LEMMA 2.5 (see (Melnikov, 1989)). *We have*

$$\varepsilon_t(A) \to \infty \Leftrightarrow A_t \to \infty.$$

as $t \to \infty$.

Proof. Note that for $x \ge -1$ $e^x \ge 1 + x \ge 0$, hence

$$\varepsilon_t(A) = e^{A_t} \prod_{s \le t} e^{-\Delta A}(1+\Delta A_s) \le e^{A_t}.$$

So we got the direct implication. To verify the converse implication, denote $A^1 = 1_{\{\Delta A \ge 1\}} \cdot A$, $A^2 = A - A^1$ and put $S_t^1 = \sum_{s \le t} 1_{\{\Delta A \ge 1\}}$, where 1_C is the indicator function of a set C.
The following inequality is obvious:

$$\left(\frac{1}{1+\Delta A} \cdot A\right)_t \ge \frac{1}{2}(A^2 + S_2^1)_t. \tag{2.8}$$

If $A_t \to \infty$, then the right-hand side of (2.8) also tends to infinity since either A^2 or S^1 tends to infinity.
We also have

$$\varepsilon^{-1}(A) = \varepsilon\left(-\frac{1}{1+\Delta A} \cdot A\right) \le \exp\{-1/2(A^2 + S^1)\},$$

which implies the desired converse implication.

COROLLARY 2.2. *If* $A_t \to \infty$, *then, for any* $k > 0$, $\varepsilon_t(-A)A_t^k \to 0$ *as* $t \to \infty$.

We need to know the asymptotical behaviour for $t \to \infty$ of the following expression

$$E_t = \varepsilon_t^{1/2}(-\beta A) \int_0^t \varepsilon_s^{-1}(-\beta A)\varepsilon_{s-}(-\alpha A)dA_s.$$

This behaviour is described by the following lemma.

LEMMA 2.6. *Suppose that* $\beta \Delta A < 1$ *and* $\beta \ge \alpha > 0$. *Then we have*

$$\varepsilon_t^{1/2}(-\beta A)\left(\int_0^t \varepsilon_s^{-1}(-\beta A)\varepsilon_{s-}(-\alpha A)dA_s + (\beta-\alpha)^{-1}\right) \le (\beta-\alpha)^{-1}$$

$$\times \varepsilon_t \left(\frac{\beta - 2\alpha}{(1 - \beta\Delta A)(2 - \beta\Delta A)} \cdot A + \frac{\alpha^2}{(1 - \beta\Delta A)(2 - \beta\Delta A)} \cdot [A, A] \right). \tag{2.9}$$

In particular, if $A_t \to \infty$, $[A, A]_\infty < \infty$, and $\beta < 2\alpha$, *then* $E_t \to 0$.

Proof. We note that, by Lemma 2.2 and 2.4, the left-hand side of (2.9) is less than

$$F_t = \varepsilon_t \left(\frac{\beta - \alpha}{(1 - \beta\Delta A)} \cdot A \right) \cdot \varepsilon_t \left(-\frac{\beta}{(2 - \beta\Delta A)} \cdot A \right) (\beta - \alpha)^{-1}$$

and, by Yor's formula,

$$(\beta - \alpha) F_t \leq \varepsilon_t \left(\frac{\beta - 2\alpha}{(1 - \beta\Delta A)(2 - \beta\Delta A)} \cdot A + \frac{\alpha^2}{(1 - \beta\Delta A)(2 - \beta\Delta A)} \cdot [A, A] \right).$$

This implies (2.9).

3. GENERAL DESCRIPTION OF THE APPROACH TO ALGORITHM (1.2)

Let us denote by $\theta_t \to \theta^* = 0$ (a.s.) the almost sure convergence, by $\theta_t \xrightarrow{\mathbf{L}_2} \theta^* = 0$ the $\mathbf{L}_2$-convergence and by $(\xrightarrow{D})$ the convergence in distribution. Our conditions can be divided into two groups. On the one hand we formulate the existence-uniqueness and stability conditions for the solution (algorithm) θ_t in terms of the function R. On the other hand we need to choose a normalizing process γ which would guarantee stability properties of θ_t. As a result, our basic stability condition combines both of them and is a natural generalization of the well-known ones (see Gyongy and Krylov, 1980; Hall and Heyde, 1980).
Our first type conditions are

$$R(x)x > 0 \quad \text{for all} \quad x \neq 0; \tag{3.1}$$

$$k_1|x| \leq |R(x)| \leq k_2|x| \quad \text{for all} \quad x \in \mathbf{R} \tag{3.2}$$

where k_1 and k_2 are positive numbers;

$$R(x) = \beta x + V(x) \quad \text{and} \quad V(x) = o(x^2) \quad \text{as} \quad x \to 0. \tag{3.3}$$

Let k_t be a non-negative predictable process integrable w.r.t. on every finite interval $(0, t]$.
One of the second type conditions is

$$\varepsilon_t(-ka) \to 0 \quad \text{(a.s.) as} \quad t \to \infty, \tag{3.4}$$

where $\varepsilon_t(-ka)$ is Dolean's exponent. Recall that (3.4) is equivalent to $k \cdot a_t \to \infty$ (a.s.) as $t \to \infty$.
Applying the Ito formula to $\varepsilon_t^{-1}(-ka)\theta_t^2$, we get (see (Liptser and Shiryaev, 1986; Gikhman and Skorokhod, 1982; Gyogy and Krylov, 1980))

$$Z_t = \varepsilon_t^{-1}(-k \cdot a)\theta_t^2 = \theta_0^2 + \int_0^t \varepsilon_s^{-1} d\theta_s^2 + \int_0^t \theta_{s-}^2 d\varepsilon_s^{-1}$$

$$= \theta_0^2 + \int_0^t \varepsilon_s^{-1}\{ -2\theta_{s-}R(\theta_{s-})\gamma_s da_s + \Delta a_s \gamma_s^2 R^2(\theta_{s-})da_s + k_s\theta_{s-}^2 da_s$$

$$+\gamma_s^2 d\langle m\rangle_s\} + \int_0^t \varepsilon_s^{-1}\{ -2\theta_{s-}\gamma_s dm_s + \gamma_s^2 d([m]_s - \langle m\rangle_s) + 2\Delta a_s \gamma_s^2 R^2(\theta_{s-})dm_s\}$$

$$= \theta_0^2 + \int_0^t \gamma_s^2 \varepsilon_s^{-1} d\langle m\rangle + \int_0^t \varepsilon_s^{-1}\{k_s\theta_{s-}^2 + 2\Delta a_s R^2(\theta_{s-}) - 2\gamma_s\theta_{s-}R(\theta_{s-})\}da$$

$$+\gamma_s^2 d\langle m\rangle_s\} + \int_0^t \varepsilon_s^{-1}\{ -2\theta_{s-}\gamma_s dm_s + \gamma_s^2 d([m]_s - \langle m\rangle_s) + 2\Delta a_s \gamma_s^2 R^2(\theta_{s-})dm_s\}$$

$$= \theta_0^2 + \int_0^t \gamma_s^2 \varepsilon_s^{-1} d\langle m\rangle_s + \int_0^t \varepsilon_s^{-1}\{k_s\theta_{s-}^2 + \Delta a_s R^2(\theta_{s-}) - 2\gamma_s\theta_{s-}R(\theta_{s-})\}da_s$$

$$+ \text{martingale} = Z_0 + A_t' + A_t^2 + M_t. \tag{3.5}$$

Here we used the usual notation from the theory of martingales. If $m_t = m_t^c + m_t^d$ is the decomposition of m into the pure continuous part m^c and the pure discontinuous part m^d, then $[m]_t = \langle m^c\rangle_t + \sum_{s\le t}(\Delta m_s)^2$ and $m_t^d = \int_0^t \int_{\mathbf{R}^1\setminus\{0\}} x d(\mu - \nu)$, where μ is the jump measure of m and ν is its compensator (see (Liptser and Shiryaev, 1986)).
Yet another condition of the second type is

$$A_\infty^1 = \int_0^\infty \gamma_s^2 \varepsilon_s^{-1}(-k \cdot a)d\langle m\rangle_s < \infty \quad \text{(a.s.)}. \tag{3.6}$$

Also we assume below that all the jumps of A_t^1 are *bounded*. The fixed stability condition is

$$k_s x^2 + \Delta a_s \gamma_s^2 R^2(x) \le 2\gamma_s^2 x R(x) \quad \text{(a.s.)} \tag{3.7}$$

for all $x \in \mathbf{R}^1$ and $s \in [0, \infty)$.

In view of (3.7), the increasing process $A_t^2 \leq 0$. So applying the theorem on the sets of (a.s.)-convergence of semimartingales (see (Liptser and Shiryaev, 1986)) and taking into account (3.6), we see that Z_t converges (a.s.) to a finite (random) limit as $t \to \infty$. Moreover, $\theta_t^2 \to 0$ (a.s.) $(t \to \infty)$ in view of (3.4). Thus, we have proved the following

THEOREM 3.1 (see (Melnikov, 1989; Melnikov and Rodkina, 1992)). *Let conditions* (3.1), (3.2), (3.4), (3.6), *and* (3.7) *be fulfilled. Then* $\theta_t \to \theta^*$ (a.s.) *as* $t \to \infty$.

Assume now that $\varepsilon_t(-k \cdot a)$ is *deterministic*. Let us take the expectation in (3.5). We have, by (3.7),

$$\mathbf{E}\theta_t^2 \leq \left(\mathbf{E}\theta_0^2 + \mathbf{E}\int_0^\infty \gamma_s^2 \varepsilon_s^{-1}(-k \cdot a)d\langle m\rangle_s\right)\varepsilon_t(-k \cdot a). \tag{3.8}$$

If we assume now that

$$\mathbf{E}\theta_0^2 < \infty \quad \text{and} \quad \mathbf{E}\int_0^\infty \gamma_s^2 \varepsilon_s^{-1}(-k \cdot a)d\langle m\rangle_s < \infty, \tag{3.9}$$

then we get $\mathbf{E}\theta_t^2 \to 0$ $(t \to 0)$ which is the statement of the next theorem.

THEOREM 3.2. *Let conditions* (3.1), (3.2), (3.4), (3.6), (3.7) *and* (3.9) *be fulfilled. Then* $\theta_t \xrightarrow{\mathbf{L}_2} \theta^*$ *as* $t \to \infty$.

Now we shall try to state a variant of the central limit theorem for θ_t. To formulate the corresponding result concerning the asymptotical normality of θ_t, we impose condition (3.3) which is usual for this statement (see (Nevelson and Khasminskii, 1972) and (Hall and Heyde, 1980)). Then, using Ito's formula we have

$$\varepsilon_t^{-1}(-\beta\gamma \cdot a)\theta_t = \theta_0 + \int_0^t \gamma_s \varepsilon_s^{-1}(-\beta\gamma \cdot a)V(\theta_{s-})da_s$$

$$- \int_0^t \gamma_s \varepsilon_s^{-1}(-\beta\gamma \cdot a)dm_s. \tag{3.10}$$

Multiplying (3.10) by $\varepsilon_t^{1/2}(-\beta\gamma \cdot a)$, we get

$$\varepsilon_t^{-1/2}(-\beta\gamma \cdot a)\theta_t = \varepsilon_t^{1/2}(-\beta\gamma \cdot a)\theta_0 + \varepsilon_t^{1/2}(-\beta\gamma \cdot a)\int_0^t \gamma_s \varepsilon_s^{-1}(-\beta\gamma \cdot a)V(\theta_{s-})da_s$$

$$-\varepsilon_t^{1/2}(-\beta\gamma \cdot a)\int_0^t \gamma_s \varepsilon_s^{-1}(-\beta\gamma \cdot a)V(\theta_{s-})dm_s. \tag{3.11}$$

Assume (instead of (3.4)) that

$$\varepsilon_t(-\beta\gamma\cdot a)\to 0 \quad \text{as} \quad t\to\infty. \tag{3.12}$$

In this case the first term in (3.11) tends to 0 as $t\to\infty$. Now, we shall find a condition to handle the second term in (3.11) using the inequality (3.8).

LEMMA 3.1. *For any $\varepsilon>0$ and $\delta>0$, there exists a deterministic $T>0$ such that*

$$\mathbf{P}\{\omega\colon \sup_{t\geq T}\theta_t^2<\delta\}\geq 1-\varepsilon.$$

Proof. We have seen that $\varepsilon_t^{-1/2}(-k\cdot a)\theta_t^2$ has a finite limit and therefore $\varepsilon_t^{-1/2}(-k\cdot a)\theta_t^2\leq\eta<\infty$ (a.s.) or $\theta_t^2\leq\eta\varepsilon_t(-k\cdot a)$ (a.s.) for all $t\geq 0$. From this and (3.4) it follows that

$$\sup_{t\geq T}\theta_t^2\leq\sup_{t\geq T}\eta\varepsilon_t(-k\cdot a)=\eta\varepsilon_T(-k\cdot a)$$

and

$$\mathbf{P}\{\omega\colon \sup_{t\geq T}\theta_t^2<\delta\}=1-\mathbf{P}\{\omega\colon \sup_{t\geq T}\theta_t^2\geq\delta\}$$

$$\geq 1-\mathbf{P}\{\omega\colon \eta\varepsilon_T(-k\cdot a)\geq\delta\}\geq 1-\mathbf{P}\{\omega\colon \eta\geq\delta\varepsilon_T^{-1}(-k\cdot a)\}\geq 1-\varepsilon$$

for some large T.

Let us choose a fixed moment T from the Lemma 3.1, where δ is such that $|V(x)|\leq c|x|^2$ for $|x|<\delta$. Then we can write the second term in (3.11) as

$$\varepsilon_t^{1/2}(-\beta\gamma\cdot a)\int_0^t\gamma_s\varepsilon_s^{-1}(-\beta\gamma\cdot a)V(\theta_{s-})da_s$$

$$=\varepsilon_t^{1/2}(-\beta\gamma\cdot a)\int_0^T\gamma_s\varepsilon_s^{-1}(-\beta\gamma\cdot a)V(\theta_{s-})da_s$$

$$+\varepsilon_t^{1/2}(-\beta\gamma\cdot a)\int_T^t\gamma_s\varepsilon_s^{-1}(-\beta\gamma\cdot a)V(\theta_{s-})da_s. \tag{3.13}$$

The first term in (3.13) tends to 0 by (3.12). Let us estimate the second term. Following (Nevelson and Khasminskii, 1972, p.176), for T, ε, δ defined as specified above we have, by the Chebyshev inequality, (3.3), and (3.8),

$$\mathbf{P}\{\omega\colon |\varepsilon_t^{1/2}(-\beta\gamma\cdot a)\int_T^t\gamma_s\varepsilon_s^{-1}(-\beta\gamma\cdot a)V(\theta_{s-})da_s|>\varepsilon\}$$

$$\leq \varepsilon + \mathbf{P}\{\omega\colon |\varepsilon_t^{1/2}(-\beta\gamma\cdot a)\int_T^t \gamma_s\varepsilon_s^{-1}(-\beta\gamma\cdot a)V(\theta_{s-})da_s| > \varepsilon,\ \sup_{t\geq T}\theta_t^2 < \delta\}$$

$$\leq \varepsilon + \mathbf{P}\{\omega\colon \varepsilon_t^{1/2}(-\beta\gamma\cdot a)\int_T^t \gamma_s\varepsilon_s^{-1}|V(\theta_{s-})|da > \varepsilon,\ \sup_{t\geq T}\theta_t^2 < \delta\}$$

$$\leq \varepsilon + \varepsilon^{-1}\mathbf{E}\varepsilon_t^{1/2}\int_T^t \gamma_s\varepsilon_s^{-1}|V(\theta_{s-})|da_s > \varepsilon,\ \sup_{t\geq T}\theta_t^2 < \delta\}$$

$$\leq \varepsilon + \varepsilon^{-1}c_1\varepsilon_t^{1/2}(-\beta\gamma\cdot a)\int_T^t \gamma_s\varepsilon_s^{-1}(-\beta\gamma\cdot a)\mathbf{E}\theta_{s-}^2 da_s$$

$$\leq \varepsilon^{-1}c_2\varepsilon_t^{1/2}(-\beta\gamma\cdot a)\int_T^t \gamma_s\varepsilon_s^{-1}(-\beta\gamma\cdot a)\varepsilon_s(-k\cdot a)da_s. \tag{3.14}$$

Assume that as

$$\varepsilon_t^{1/2}(-\beta\gamma\cdot a)\int_T^t \gamma_s\varepsilon_s^{-1}(-\beta\gamma\cdot a)\varepsilon_s(-k_s\cdot a)da_s \to 0 \tag{3.15}$$

$t\to\infty$. In this case the left-hand side of (3.14) can be made smaller than const$\cdot\varepsilon$ for all large t.

Turning back to (3.11), we consider the last term which is the martingale $M_t = \int_0^t \gamma_s\varepsilon_s^{-1}(-\beta\gamma\cdot a)dm_s$ divided by the infinitely increasing function $A_t = \varepsilon_t^{-1/2}(-\beta\gamma\cdot a)$.

Thus, we are led to the following result on the asymptotic normality of normalized martingales. Assume that N_t is a square-integrable (local) martingale with the compensator $\langle N\rangle$ and the jump measure μ^N and its compensator ν^N (see (Liptser and Shiryaev, 1986) and the previous section).

LEMMA 3.2. *Suppose there exists an infinitely increasing function g_t such that*

$$g_T^{-2}\langle N\rangle_T \xrightarrow{\mathbf{P}} 1 \ (\textit{in probability}), \tag{3.16}$$

$$g_T^{-2}\int_0^T\int_{|x|>\varepsilon g_T} x^2 d\nu^N \xrightarrow{\mathbf{P}} 0 \text{ as } T\to\infty,\quad \varepsilon>0. \tag{3.17}$$

Then $g_T^{-1}N_T \xrightarrow{D} N(0,1)$, where $N(0,1)$ is the normal distribution with parameters 0 and 1.

Proof. Select a sequence of real numbers b_n such that $b_n\to\infty$ as $n\to\infty$. Put $N_t^n = g_{b_n}^{-1}N_{b_nt}$ and $F_t^n = F_{b_nt}$. It is clear that $\{N_t^n, F_t^n\}_{t\geq 0,\, n=1,2,\ldots}$ is a sequence of martingales with $\langle N^n\rangle_t = \langle N\rangle_{b_nt}$.

Then from (3.16)–(3.17) we have $\langle N^n\rangle_1 \xrightarrow{\mathbf{P}} 1$ and

$$\int_0^1 \int_{|x|>\varepsilon} x^2 d\nu^{N^n} = g_{b_n}^{-2} \int_0^{b_n} \int_{|x|>\varepsilon g_{b_n}} x^2 d\nu \xrightarrow{\mathbf{P}} 0 \quad \text{as} \quad n \to \infty,$$

so the lemma follows from Theorem 5.5.4 of (Liptser and Shiryaev, 1986).

Now we are able to formulate the corresponding result on asymptotic normality of algorithm (1.2). For $t \to \infty$, consider the conditions

$$\varepsilon_t(-\beta\gamma \cdot a) \int_0^t \gamma_s^2 \varepsilon_s^{-2}(-\beta\gamma \cdot a) d\langle m\rangle_s \xrightarrow{\mathbf{P}} 1, \tag{3.18}$$

$$\varepsilon_t(-\beta\gamma \cdot a) \int_0^t \int_{|x|>\epsilon\varepsilon_t} x^2 \gamma_s^2 \varepsilon_s^{-2}(-\beta\gamma \cdot a) d\nu \xrightarrow{\mathbf{P}} 0. \tag{3.19}$$

Our next result is a direct corollary of Lemma 3.2.

THEOREM 3.3. *Assume that conditions* (3.1)–(3.4), (3.6)–(3.9), (3.12), (3.15), (3.18), *and* (3.19) *are satisfied. Then*

$$\varepsilon_t^{-1/2}(-\beta\gamma \cdot a)(\theta_t - \theta^*) \xrightarrow{D} N(0,1) \quad \text{as} \quad t \to \infty.$$

Remark 3.1. Conditions (3.1)–(3.3) are classical (see (Nevelson and Khasminskii, 1972) and (Hall and Heyde, 1980)). As stated in Section 2, conditions (3.4) and (3.12) are close to the classical condition $\int_0^\infty \gamma_s da_s = \infty$ (a.s.). Conditions (3.6) and (3.9) are close to the classical assumption $\int_0^\infty \gamma_s^2 da_s < \infty$ (a.s.). Conditions (3.18) and (3.19) stem from the theory of limit theorems for martingales (see (Liptser and Shiryaev, 1986) and (Hall and Heyde, 1980)).

Remark 3.2. The "monotonicity" condition (3.7) (compare with (Gyongy and Krylov, 1980) for pure continuous and discrete models (1.2) looks like the so-called stability condition from the classical theory (Nevelson and Khasminskii, 1972): if $k_t = k/t$, $t > 0$ or $k_n = k/n$, $n = 1, 2, \ldots$, then (3.7) reduces to $2\beta \geq k$.

Remark 3.3. The following question arises: is it possible to choose (simultaneously) constants k_1, k_2 (see (3.2)) and k_t from (3.7)? The answer is affirmative. If we take $k_t = k\gamma_t$ and assume that $\Delta a_t \gamma_t \leq c$, then $k \leq 2k_1 - ck_2^2$ and therefore the correlation between k_1 and k_2 is $2k_1 - ck_2^2 > 0$.

4. ALGORITHM (1.2) WITH GAUSSIAN ERRORS

In this section, we apply the method of the investigation of algorithm (1.2) above described in the case where a_t is deterministic, $\gamma_t = \alpha(a_t + 1)^{-1}$, $\alpha > 0$, and m_t is a Gaussian martingale (hence $\langle m\rangle_t$ is deterministic, see (Liptser and Shiryaev, 1986).

Assume that our basic conditions (3.1)–(3.3) for the real function $R(x)$ are satisfied and according to Remark 3.3 we can take $k_t = \beta\gamma_t$, $\beta > 0$. Throughout this section we suppose that

$$a_t \to \infty \quad (t \to \infty) \tag{4.1}$$

and

$$[\gamma \cdot a,\ \gamma \cdot a]_\infty = \alpha^2 \sum_s \left(\frac{\Delta a_s}{1 + a_s}\right)^2 < \infty. \tag{4.2}$$

Using (4.1)–(4.2) by straightforward simulations we get

$$\int_0^\infty \gamma_s da_s = \infty, \quad \int_0^\infty \gamma_s^2 da_s < \infty,$$

and

$$\varepsilon_t(-\beta\gamma \cdot a) = (1 + a_t)^{-\alpha\beta} \exp\Big\{\sum_{s \le t} {}^{\alpha\beta}\Big\}, \tag{4.3}$$

where

$$\sum_{s \le t} {}^{\alpha\beta} = \sum_{s \le t} \Big[(\log(1 - \alpha\beta\Delta a_s(1 + a_s)^{-1}) + \alpha\beta\Delta a_s(1 + a_s)^{-1})$$

$$+\alpha\beta\left(\log(1 - \alpha\beta\Delta a_s(1 + a_{s-})^{-1}) + \alpha\beta\Delta a_s(1 + a_{s-})^{-1}\right)\Big]$$

$$= \sum_{s \le t} \Big[\left(\log(1 - \alpha\beta\Delta a_s(1 + a_s)^{-1} + \alpha\beta\Delta a_s(1 + a_s)^{-1}\right)\Big] + O\Big(\left(\Delta a_s(1 + a_s)^{-1}\right)^2\Big),$$

$$\varepsilon_t(-\beta\gamma \cdot a) = O\Big((1 + a_t)^{-\alpha\beta} \exp\{\text{const} \sum_{s \le t} (\Delta a_s(1 + a_s)^{-1})^2\}\Big)$$

$$= O((1 + a_t)^{-\alpha\beta}). \tag{4.4}$$

We can now give specific conditions for (a.s.)-convergence, $\mathbf{L}_2$-convergence and the asymptotic normality of θ_t.
In view of (4.4), assumption (3.6) reduces to

$$\int_0^\infty \gamma_s^2 \varepsilon_s(-\beta\gamma \cdot a) d\langle m\rangle_s \le \text{const} \int_0^\infty (1 + a_s)^{\alpha\beta - 2} d\langle m\rangle_s < \infty. \tag{4.5}$$

Note that in the usual case $\langle m\rangle_t = a_t$ we get the inequality $\alpha\beta < 1$. In our special situation, the stability condition (3.2) can be simplified to become

$$\beta + \alpha k_2^2 \Delta a_t (1 + a_t)^{-1} \le 2k_1 \quad \text{or} \quad \beta + \alpha k_2^2 \le 2k_1. \tag{4.6}$$

As a result, we obtain

THEOREM 4.1. *Suppose conditions* (3.1)–(3.3) *and* (4.5)–(4.6) *are satisfied. Then* $\theta_t \to \theta^*$ *(a.s) and* $\theta_t \to \theta^*$ *in* $\mathbf{L}_2(\Omega, F, \mathbf{P})$ *as* $t \to \infty$.

Let us analyse the asymptotic normality. Multiplying (3.10) by $\varepsilon_t^{1-(2\alpha\beta)^{-1}} \times (-\beta\gamma \cdot a)$, we have

$$\varepsilon_t^{-(2\alpha\beta)^{-1}}\theta_t = \theta_0\varepsilon_t^{1-(2\alpha\beta)^{-1}} + \varepsilon_t^{1-(2\alpha\beta)^{-1}} \int_0^t \gamma_s\varepsilon_s^{-1}V(\theta_{s-})da_s$$

$$-\varepsilon_t^{1-(2\alpha\beta)^{-1}} \int_0^t \gamma_s\varepsilon_s^{-1}dm_s. \tag{4.7}$$

Assume that $2\alpha\beta > 1$. Then, in view of (4.1)–(4.4) (see also §2), $\varepsilon_t^{-(2\alpha\beta)^{-1}} \to 0$ $(t \to 0)$ and the corresponding analog of (3.15) holds: $\varepsilon_t^{-(2\alpha\beta)^{-1}} \log(a_t + 1) = \text{const}\,((1 + a_t)^{1-(2\alpha\beta)^{-1}}) \log(1 + a_t) \to 0$ as $t \to \infty$. So the first and the second terms in (4.7) both tend to 0 (in probability) as $t \to \infty$.
It is clear that the third term in (4.7) has a normal distribution. Let us calculate its variance D_m. We have (for some $T > 0$)

$$D_m = \varepsilon_t^{2(1-(2\alpha\beta)^{-1})} \int_0^t \gamma^2\varepsilon_s^{-2}d\langle m\rangle_s$$

$$= \varepsilon_t^{2(1-(2\alpha\beta)^{-1})} \int_0^T \gamma^2\varepsilon_s^{-2}d\langle m\rangle_s + \varepsilon_t^{2(1-(2\alpha\beta)^{-1})} \int_T^t \gamma^2\varepsilon_s^{-2}d\langle m\rangle_s. \tag{4.8}$$

The first term in (4.8) tends to 0. But the second term, as a straightforward calculation shows, in the case $\langle m\rangle_t = a_t$ tends to

$$\alpha(2\alpha\beta - 1)^{-1} \exp\Big\{ -\frac{1}{\alpha\beta} \sum_{s<\infty} {}^{\alpha\beta}\Big\}.$$

As a result, we get

$$\varepsilon_t^{-(2\alpha\beta)^{-1}}(\theta_t - \theta^*) \xrightarrow{D} N(0, \alpha(2\alpha\beta - 1)^{-1} \exp\Big\{ -\frac{1}{\alpha\beta} \sum {}^{\alpha\beta}\Big\}. \tag{4.9}$$

On the other hand, using (4.3), we have

$$\varepsilon_t^{-(2\alpha\beta)^{-1}} = (1 + a_t)^{1/2} \exp\Big\{ -\frac{1}{2\alpha\beta} \sum_{s\le t} {}^{\alpha\beta}\Big\}.$$

Combining the last equality and (4.9), we obtain the following result.

THEOREM 4.2. *Assume that* $1 < 2\alpha\beta < 2$, $\langle m\rangle_t = a_t$, *and let* (3.1)–(3.3) *and* (4.6) *be satisfied. Then*

$$(a_t+1)^{1/2}(\theta_t-\theta^*)\xrightarrow{D}N\left(0,\frac{\alpha}{2\alpha\beta-1}\right)\quad \text{as}\quad t\to\infty.$$

Remark 4.1. Applying the above approach, one can state similar asymptotic results for the more general algorithm

$$d\theta_t=-\gamma_t R(\theta_{t-})da_t-\gamma_t G(\theta_{t-})dm_t,$$

where G is a Lipshitzian function.
In particular,

$$(a_t+1)^{1/2}(\theta_t-\theta^*)\xrightarrow{D}N\left(0,\frac{\alpha G_0}{2\alpha\beta-1}\right)\quad \text{as}\quad t\to\infty$$

under the condition of the previous theorem, where $G_0=\lim_{x\to\theta^*}G(x)>0$.
Our asymptotic normality result with Gaussian errors is valid for both the diffusion and the discrete time models of stochastic approximation:

$$t^{1/2}(\theta_t-\theta^*)\xrightarrow{D}N\left(0,\frac{\alpha G_0}{2\alpha\beta-1}\right)\quad \text{as}\quad t\to\infty$$

and

$$n^{1/2}(\theta_n-\theta^*)\xrightarrow{D}N\left(0,\frac{\alpha G_0}{2\alpha\beta-1}\right)\quad \text{as}\quad n\to\infty.$$

Recall that earlier these last results were proved separately (see, e.g., (Nevelson and Khasminskii, 1972)).

REFERENCES

Gikhman, I.I. and Skorokhod, A.V. (1982). *Stochastic Differential Equations and Its Applications.* Naukova Dumka, Kiev (in Russian).

Gyongy, I. and Krylov, N.V. (1980). On stochastic equations with respect to semimartingales. *Stochastics* **4**, 1–22.

Hall, P. and Heyde, C.C. (1980). *Martingale Limit Theory and Applications.* Academic Press, New York, London.

Liptser, R. Sh. and Shiryaev, A.N. (1986). *Theory of Martingales.* Nauka, Moscow (in Russian).

Melnikov, A.V. (1988). Procedures of stochastic approximation in statistics of processes of martingales type. *Russian Math. Surveys.* **43**, 215–216 (in Russian).

Melnikov, A.V. (1989). Procedures of stochastic approximation for semimartingales. In: *Statistics and Control of Stochastic Processes,* Nauka, Moscow, pp. 147–156 (in Russian).

Melnikov, A.V. and Rodkina, A.E. (1992). Martingale approach to the procedures of stochastic approximation. In: *Frontiers in Pure and Applied Probability. Vol.* 1. Ed. by H. Niemi et al., TVP/VSP, Moscow/Utrecht, pp. 165–182.

Nevelson, M.B. and Khasminskii, R.Z. (1972). *Stochastic Approximation and Recursion Estimation.* Nauka, Moscow (in Russian).

Frontiers in Pure and Appl. Probab. 1, pp. 197 - 205
H. Niemi et al. (Eds.)
1993 TVP/VSP

STOCHASTIC APPROXIMATIONS IN THE MAXIMUM LIKELIHOOD INFERENCE FOR MARKOV RANDOM FIELDS

A. PENTTINEN
University of Jyväskylä, P.O. Box 35, SF-40351 Jyväskylä, Finland

ABSTRACT

The likelihood function of a Markov random field (MRF) model contains a factor depending on the model parameters which is practically intractable. This is a drawback in the maximum likelihood inference. Stochastic approximation is the only general approach applicable in parameter estimation through the maximum likelihood method. In this paper different approaches, most of which have been introduced initially for Gibbs point processes, are presented and applied in the parameter estimation for a simple one-parametric MRF model. Extensions are discussed and an example is also given.

1. INTRODUCTION

Markov random fields (MRFs) appear to be a flexible family of statistical models for data with dependent observations. Since first presented for the audience of statisticians by Besag (Besag, 1974), these models intended originally for describing ferromagnetic phenomena in physics played an important role in spatial statistics and image analysis (e.g., (Ripley, 1981; Besag, 1986; Besag et al., 1991)) and, recently, sampled from the exponential family in the analysis of dependent observations (Geyer and Thompson, 1992).

Any statistician when applying MRFs in his or her statistical problems with dependent data is confronted by the unfamiliarity of these models. The difficulty lies in the partition function, an intractable factor in the likelihood function of the MRF which makes the parameter estimation using the maximum likelihood method difficult. This problem is usually avoided in data analysis by adopting the pseudo-likelihood estimation method (Besag, 1975). If one is interested in applying the maximum likelihood method, then the only general procedure to proceed is the approximation of the likelihood function through stochastic approximation, the method based on stochastic simulation.

The solution of the likelihood equation for Gibbs point processes through stochastic approximation has been introduced by Penttinen (Penttinen, 1984), see also (Ripley, 1988). Recently, Moyeed and Baddeley (Moyeed and Baddeley, 1991) have modified the classical Robbins–Monro method for the same purpose. So far Geyer and Thompson (Geyer and Thompson, 1992) were the first to consider stochastic

approximations for MRFs in connection with the exponential family of models that are not tied with spatial data.
In this paper the three existing stochastic approximation methods developed mainly for Gibbs point processes are recalled and applied to the likelihood inference for MRFs.

2. THE STATISTICAL MODEL AND PARAMETER ESTIMATION

Let us consider a finite square lattice E with sites denoted by t. The neighbourhood of a site t is denoted by ∂t. Usually ∂t is the four or eight nearest sites of t (call them the N_1 and $N_{\sqrt{2}}$ -systems, respectively). In order to keep the notations as simple as possible, we consider the special case of binary site variables. Let us consider sites with two possibilities: the values 1/2 (black) and −1/2 (white). Let X be a measurable mapping from a probability space to the set $\mathbf{X} = \{-1/2, 1/2\}^E$. The projection of X to site t is called a site variable and is denoted by X_t. We assume that the distribution of X is of the form

$$\mathbf{P}_\theta\{X = x\} = p(x;\theta) = Z(\theta)^{-1} e^{-U(x;\theta)}, \quad x \in \mathbf{X}.$$

Let us consider the following special case called the Ising model (a special case of the Potts model):

$$U(x;\theta) = -\sum_{s,t\in\partial t}\sum \theta\|x_s - x_t\| = -Y\theta,$$

where Y is the number of neighbouring pairs of different value (colour). The model is thought to be conditioned by the number of (1/2)'s in E. (This assumption will be relaxed at the end of the paper). Let us note that Y is sufficient for θ; so, the data X can be reduced to $Y = Y(X) =: Y_{obs}$ without loss of information. In parameter estimation two alternative methods are of use. These are the maximum pseudo-likelihood method (MPL) and the maximum likelihood method (ML).

Pseudo-likelihood. The maximum pseudo-likelihood method (Besag, 1975) is based on the optimization of the pseudo-likelihood function

$$\prod_{t\in E} p(x_t \mid x_s, s \neq t),$$

the product of the conditional probabilities $\mathbf{P}\{X_t = x_t \mid X_s = x_s, s \neq t\}$ called also the local characteristics. The benefit of this method is that it is easy to apply it, standard program packages can be used, and some asymptotics are known. Despite of its merits and popularity, the pseudo-likelihood method seems to be rather weak or even useless when interactions are strong ((Särkkä, 1990), see also the discussion to (Besag et al., 1991)). One should note that even then the maximum likelihood method works.

Maximum likelihood. As our model belongs to the exponential family of distributions, the existence and uniqueness of the ML-estimator follows from the general theory. The ML-estimator is obtained through the maximization of the log-likelihood function

$$l(\theta; x) = -\log Z(\theta) + Y\theta, \tag{1}$$

where

$$Z(\theta) = \sum_{x \in \mathbf{X}} e^{\theta Y(x)},$$

the partition function is unknown even when E is of a moderate size. The optimization of (1) is, therefore, impossible. The purpose of this paper is to describe stochastic approximation methods and apply them in the ML inference for model (1) and its generalizations.

3. MONTE CARLO MAXIMUM LIKELIHOOD METHOD

3.1. *Stochastic approximation*

Sampson in Encyclopedia of Statistical Sciences (Sampson, 1988) defines stochastic approximation as follows: *Stochastic approximation refers to a class of iterative procedures that estimate the optima or roots of functions that are only observable up to some statistical noise. Many of these procedures can be viewed as stochastic counterparts of iterative numerical analysis procedures for optimizing known evaluateable functions.*

In this connection, the ML-estimator $\hat{\theta}$ is the root of the likelihood equation

$$-\frac{d}{d\theta} \log Z(\theta) + Y = 0. \tag{2}$$

The noise comes from some stochastic system which makes the distribution of the statistical model under consideration a stationary distribution. The two most common simulation techniques are the Metropolis method (Metropolis et al., 1953) and the Gibbs sampler (Geman and Geman, 1984). Both of them are Markov chains with $\mathbf{X}$ as their state space, and the transition probabilities are constructed with the help of the distribution $p(x; \theta)$. If the model is "clustered", the method proposed by Swendsen and Wang (Swendsen and Wang, 1987) is preferable. These algorithms operate locally and are, therefore, easy to apply.

The key idea of the stochastic ML-estimation lies in the following two steps:

1) Present the unknown part of (2) in the form $\mathbf{E}_\theta h(X)$ with some measurable function h on $\mathbf{X}$ and

2) approximate it consistently (with increase of N) by

$$\frac{1}{N} \sum_{i=1}^{N} h(X^{(i)})$$

calculated from simulations $X^{(1)}, ..., X^{(N)}$ under the model assumption.

At least three choices of the function h exist depending on what we want to approximate. These lead to different approximations and to different algorithms also.

3.2. *Monte Carlo likelihood function*

The first method is based on direct approximation of the partition function $Z(\theta)$. Let us first note that

$$Z(\theta) = Z(0)\mathbf{E}_0 e^{\theta Y}. \tag{3}$$

Hence, $Z(\theta)$ can be estimated from simulations under the distribution $p(x;\theta)$ which corresponds to independent X_t's. This approximation works when the interactions are very weak only (Penttinen, 1984). Another possibility is to present $Z^{-1}(\theta)$ as the expectation $Z^{-1}(\theta) = \mathbf{E}_\theta e^{-\theta Y}$ which is approximated by $1/N\sum_{i=1}^N e^{-\theta Y_i}$, where $X^{(1)},...,X^{(N)}$ are simulations from $p(x;\theta)$, not necessarily independent, and $Y_i = Y(X^{(i)})$. This leads to the approximation of the likelihood but does not provide us with a reasonable estimation equation.

The variability of the estimator based on (3) can be reduced by applying the importance sampling. In this connection, let θ_0 be a fixed parameter value which we call "pivot parameter". Then

$$\begin{aligned} Z(\theta) = Z(0)\mathbf{E}_0 e^{\theta Y} &= Z(0)\mathbf{E}_{\theta_0}[e^{\theta Y(X)}p(X;0)/p(X;\theta_0)] \\ &= Z(\theta_0)\mathbf{E}_{\theta_0}[e^{(\theta-\theta_0)Y(X)}]. \end{aligned}$$

Let $X^{(1)},...,X^{(N)}$ be simulations (not necessarily independent) under the model assumption $p(x;\theta_0)$ and let $Y_i = Y(X^{(i)})$ be a sufficient statistic giving the approximation

$$\mathbf{E}_\theta[e^{(\theta-\theta_0)Y(X)}] \simeq \frac{1}{N}\sum_{i=1}^N e^{(\theta-\theta_0)Y_i}. \tag{4}$$

Now, likelihood equation (2) is approximated by

$$\frac{1}{N}\sum_{i=1}^N (Y_i - Y_{obs})e^{(\theta-\theta_0)Y_i} = 0. \tag{5}$$

This problem can be solved numerically, e.g., by using the Newton method.

The idea of importance sampling is due to Ripley (Ripley, 1988) but in a different and less efficient form. This principle gives approximation (4) which has its origins in (Ogata and Tanemura, 1981). Estimation equation (5) is due to (Geyer and Thompson, 1992). They also propose a stepwise algorithm: when an approximative solution $\hat{\theta}_1$ has been achieved using the pivot parameter θ_0, then the value $\hat{\theta}_1$ is applied as a new pivot parameter.

3.3. *Stochastic Newton-Raphson algorithm*

Let us start with the equation, well-known in the theory of exponential family of statistical models,

$$\frac{d}{d\theta}\log Z(\theta) = \mathbf{E}_\theta Y(X).$$

The approximation of $d/d\theta(\log Z(\theta))$ is $1/N\sum_{i=1}^N Y_i$, the sample mean of the Y_i's results from the simulations $X^{(i)}$ from the distribution $p(x;\theta)$. This approximation is due to (Ogata and Tanemura, 1984).

This reasoning yields the stochastic approximation of likelihood equation (2) of the form

$$-(\overline{Y}_\theta - Y_{obs}) = 0, \tag{6}$$

where $\overline{Y}_\theta = 1/N \sum_{i=1}^N Y(X^{(i)})$ from simulations $X^{(1)},...,X^{(N)}$ under $p(x;\theta)$. Equation (6) does not contain the parameter to be found. One possibility for bypassing this problem is to approximate the second derivative of the log-likelihood function:

$$\frac{d^2}{d\theta^2} l(\theta;x) = \mathbf{E}_\theta[Y^2] - [\mathbf{E}_\theta Y]^2$$

$$\simeq \frac{1}{N}\sum_{i=1}^N Y_i^2 - [\frac{1}{N}\sum_{i=1}^N Y_i]^2 =: S_{Y,\theta}^2.$$

This brings us to the following stepwise procedure called the stochastic Newton–Raphson algorithm:
1) Set the initial value $\hat{\theta}^{(0)}$.
2) $\hat{\theta}^{(k+1)} = \hat{\theta}^{(k)} - S_{Y,\theta}^{-2}(\hat{Y}_{\hat{\theta}^{(k)}} - Y_{obs})$.
This algorithm is due to Penttinen (Penttinen, 1984).

3.4. *The Robbins-Monro method*
The Robbins-Monro method is a general stochastic approximation method for solving equations like

$$H(\theta) = \varphi,$$

where the function H is unknown, see e.g. (Wasan, 1969). Assume, further, that there exists a statistic V such that $\mathbf{E}_\theta V = H(\theta)$ and i.i.d. simulations $V_1,...,V_N$ from the common distribution $p_V(v;\theta)$ are available. Here we approximate the unknown $H(\theta) = \mathbf{E}_\theta V$ by the average $\overline{V}_\theta = 1/N \sum_{i=1}^N V_i$. The algorithm is now:
1) Take the initial value $\hat{\theta}^{(0)}$.
2) $\hat{\theta}^{(k+1)} = \hat{\theta}^{(k)} - a_i[\overline{V}_{\hat{\theta}^{(k)}} - \varphi]$, where $\{a_i\}$ is a decreasing sequence of positive real numbers, $\sum_i a_i = \infty$ and $\sum_i a_i^2 < \infty$. The sequences $a_i = 1/i$ and $a_i = A/i$ are in common use.
In our case the application is rather straightforward: choose $H(\theta) = d/d\theta(\log Z(\theta))$ and $\varphi = Y_{obs}$. Therefore, $\overline{V}_\theta = 1/N \sum_{i=1}^N Y(X^{(i)})$, where $X^{(1)},...,X^{(N)}$ are i.i.d. simulations under $p(x;\theta)$ (results from the Gibbs sampler).
Something is known about the precision of the method. The following Central Limit Theorem version helps in applications: under some conditions (see (Hall and Heyde, 1980))

$$\sqrt{N}(\hat{\theta}^{(N)} - \hat{\theta}) \to N(0, A^2\sigma^2/(2A\mu - 1)),$$

where $\mu = H'(\hat{\theta})$, $\sigma^2 = \mathbf{D}(\overline{V}_{\hat{\theta}})$, $A > 1/(2\mu)$. One should note that a clever choice of A can reduce the time needed for convergence.
We shall comment on extensions and further possibilities of the three approximation methods in the following remarks.
1. The previous approaches can be extended to more general models including vector parameters and more complicated dependence structures including irregular graphs.
2. Different kinds of conditioning in the MLE (conditional maximum likelihood) and edge corrections can be also adopted after modifications of the Gibbs sampler.

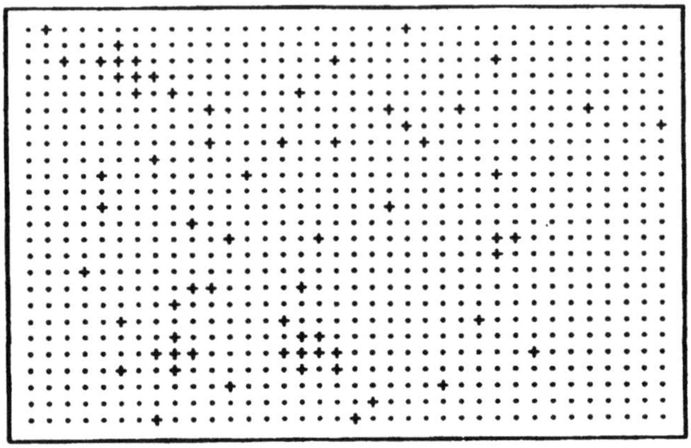

Fig. 1. *Sirex*-infested trees (+ -marked) in a 25×36 square lattice peantation of *Pinus radiata* from (Verhagen, 1971).

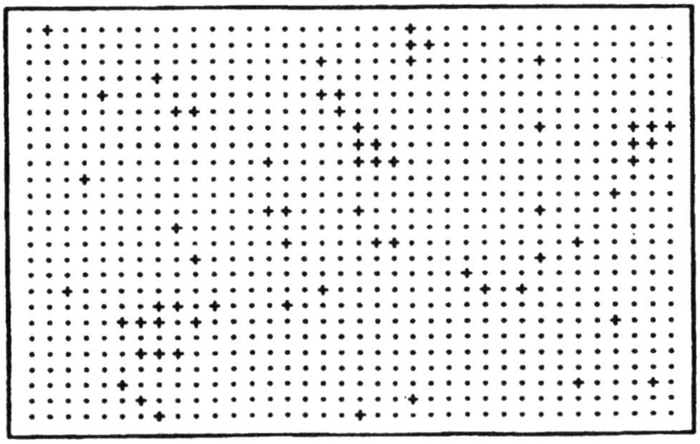

Fig. 2. A realization of the Markov field model ($\theta = 0.58$) conditioned by the marginal rows and columns and by the number of infested trees as in Fig. 1.

3. Applications in data analysis with dependent observations are under study. These include logit models in cluster sampling, item analysis in educational studies, and logit models in field experiments.

4. AN APPLICATION

We illustrate these techniques by the example due to Verhagen (Verhagen, 1971). The data consists of *Sirex*-infested trees in a 25×36 square lattice plantation of

Pinus radiata (Fig. 1). We apply the model of §2 with the log-likelihood function

$$l(\theta; x) = -\log Z(\theta) + \theta Y,$$

where Y is the number of neighbouring pairs with one tree infested and the other one healthy. We also condition by the marginal rows and columns and by the number of infested trees. As a simulation method, the Gibbs sampler has been used.

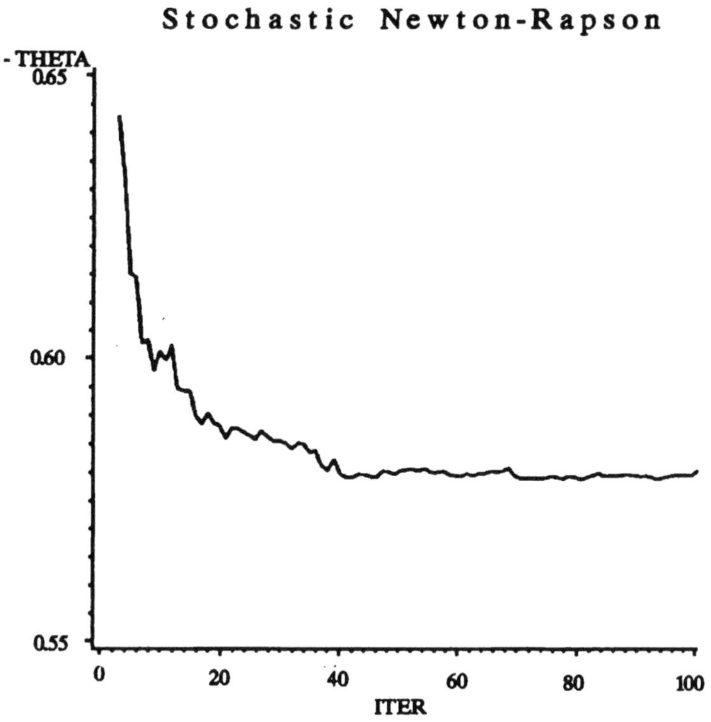

Fig. 3. Results of iterations in the stochastic Newton–Raphson algorithm. The graph describes the cumulative means of sequential parameter values which results from iteration. The initial value is $\hat{\theta}^{(0)} = 0$.

The first method of Geyer and Thompson (Geyer and Thompson, 1992) has given the following sequence of estimator: no solution was obtained when starting with $\theta^{(0)} = 0$ corresponding to independent site variables. Instead, the algorithm gives $\hat{\theta}^{(1)} = -0.585$ and $\hat{\theta}^{(2)} = -0.581$ when the starting value $\theta^{(0)} = -0.2$. This method seems to be efficient with respect to the number of simulations but some difficulties arise in the numerical solution of the Monte Carlo likelihood equation. We accept the estimator $\hat{\theta} = -0.581$ as the maximum likelihood solution.
The stochastic Newton–Raphson algorithm gives estimates which are presented in Fig. 3, where the estimates are cumulative means of the $\hat{\theta}^{(k)}$'s. Starting from $\theta = 0$, the maximum likelihood estimate $\hat{\theta} = -0.581$ was obtained.

The results of the Robbins–Monro algorithm are presented in Fig. 4. Starting from $\theta = 0$, the method yields the maximum likelihood estimator $\hat{\theta} = -0.580$.

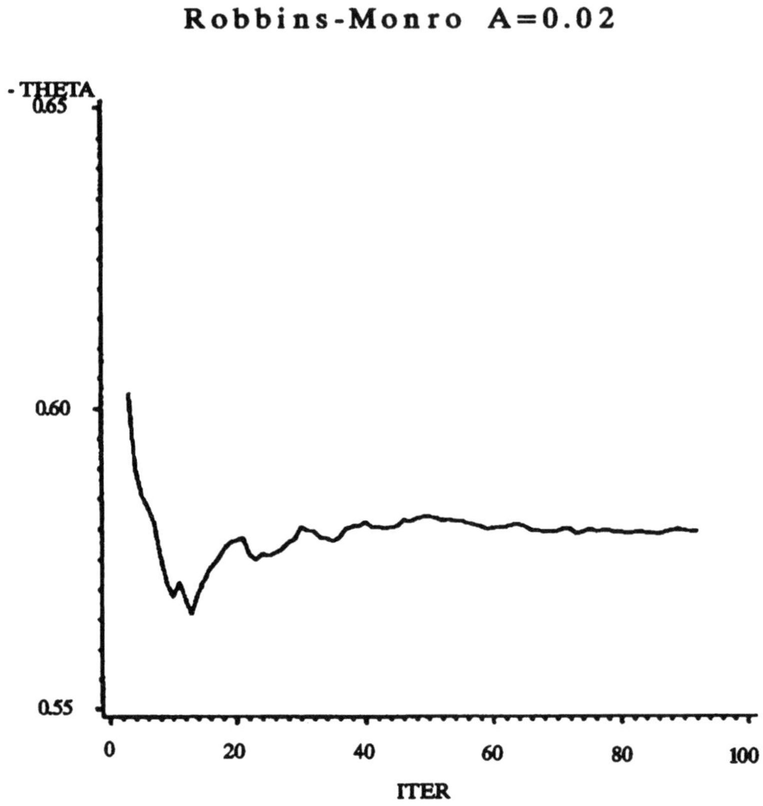

Fig. 4. Results of iterations in the Robbins–Monro algorithm. The initial value is $\hat{\theta}^{(0)} = 0$.

The comparison of the methods is not straightforward. In particular, we have not adopted any stopping rule and, hence, exact measurements of the CPU time have not been done. Something can be said about the performance of the methods discussed above. Firstly, all the methods give almost the same estimator, $\hat{\theta}_{ML} = -0.58$. Fig. 2 gives a realization of the MRF with the parameter value $\theta = -0.581$ and with the same boundary rows and columns as the original data. In a comparison, the maximum pseudo-likelihood estimator is $\hat{\theta}_{MPL} = -0.76$. The last two stochastic approximations seem to require the same amount of work in simulation whereas the first method is not so laborious.

Acknowledgements
I want to thank Dr. Charles Geyer for providing me with the manuscript (Geyer and Thompson, 1992) which has led to rewriting of §3.2. My thanks are also due

to Mr. Juha Heikkinen whose valuable assistance has made the presentation of the example possible.

REFERENCES

Besag, J. (1974). Spatial interaction and the statistical analysis of lattice systems (with Discussion). *J. Roy. Statist. Soc. Ser. B* **36**, 192–236.

Besag, J. (1975). Statistical analysis of nonlattice data. *The Statistician* **24**, 179–195.

Besag, J.E. (1986). On the statistical analysis of dirty pictures (with Discussion). *J. Roy. Statist. Soc. Ser B* **48**, 259–302.

Besag, J., York, J., and Mollie, A. (1991). Bayesian image restoration with two applications in spatial statistics (with Discussion). *Ann. Inst. Statist. Math.* **43**, 1–59.

Geman, S. and Geman, D. (1984). Stochastic relaxation, Gibbs distributions and the Bayesian restoration of images. *IEEE Trans. PAM* **6**, 721–741.

Geyer, C. and Thompson, E.A. (1992). Constrained Monte Carlo maximum likelihood for dependent data. *J. Roy. Statist. Soc. Ser. B* (to appear).

Hall, P. and Heyde, P. (1984). *Martingale Limit Theory and Its Applicatios.* Academy Press, New York, London.

Metropolis, N., Rosenbluth, A.W., Rosenbluth, M.N., Teller, A.H., and Teller, E. (1953). Equations of state calculations by fast computing machines. *Chem. Phys.* **21**, 1087–1092.

Moyeed, R.A. and Baddeley, A.J. (1991). Stochastic approximation of the MLE for a spatial point pattern. *Scand. J. Statist.* **18**, 39–50.

Ogata, Y. and Tanemura, M. (1981). Estimation of interaction potentials of spatial point patterns through the maximum likelihood procedure. *Ann. Inst. Statist. Math.* **33**, 315–338.

Ogata, Y. and Tanemura, M. (1984). Likelihood analysis of spatial point patterns. *J. Roy. Statist. Soc. Ser. B* **46**, 496–518.

Penttinen, A. (1984). Modelling interactions in spatial point patterns: parameter estimation by the maximum likelihood method. *Jyväskylä Studies in Computer Science, Economics and Statistics* **7.**

Ripley, B.D. (1988). *Statistical Inference for Spatial Processes.* Cambridge Univ. Press, Cambridge.

Sampson, A.R. (1988). Stochastic approximation. In: *Encyclopedia of Statistical Sciences,* Vol. 8, John Wiley, New York, pp. 784–789.

Swendsen, R.H. and Wang, J. (1987). Nonuniversal critical dynamics in Monte Carlo similations *Physical Rev. Letters* **58**, 86–88.

Särkkä, A. (1990). *Applications of Gibbs Point Processes: Pseudo-Likelihood Estimation Method with Comparisons.* Reports from the Department of Statistics, University of Jyväskylä, 10/1990.

Wasan, M.T. (1969). *Stochastic Approximation.* Cambridge Univ. Press, Cambridge.

Verhagen, A.M. (1971). Clustering of attributes on regular point lattices (with an Appendix illustrating the aplication of the procedure to Sirex infestations in plantations of Pinus radiata). *Appl. Ecol.* **8**, 665–682.

Frontiers in Pure and Applied Probab. 1, pp. 206 – 208
H. Niemi et al. (Eds.)
1993 TVP/VSP

THE EDGEWORTH EXPANSION IN $\mathbf{R}^k$

Yu. V. PROKHOROV
Steklov Mathematical Institute, Vavilov Street 42, 117966 Moscow GSP-1, Russia

ABSTRACT

Bounds for the total variation of signed measures which appear in the multivariate Edgeworth expansion are obtained. Some of these estimates cannot be improved (in accodance of dependence of space dimension).

Let $X = (X^{(1)}, \dots, X^{(k)})$ be a random vector taking values in the k-dimensional Euclidian space $\mathbf{R}^k$. We assume that this vector has zero expectation, covariance matrix with positive diagonal elements $\sigma_1^2, \sigma_2^2, \dots, \sigma_k^2$ and that its moments of any order are finite.
We consider a sequence

$$X, X_1, X_2, \dots, X_n, \dots \tag{1}$$

of independent identically distributed random vectors, i.e. $X_n \stackrel{d}{=} X$. It is a well-known fact that the characretistic function $f_n(t)$ of the normalized sum

$$S_n = \frac{X_1 + \ldots + X_n}{\sqrt{n}}$$

admits an asymptotic expansion in power series of $1/\sqrt{n}$

$$f_n(t) = e^{-s/2}\Big\{1 - \frac{1}{\sqrt{n}}\frac{i}{6}\mathbf{E}(t,X)^3 + \frac{1}{n}\Big(-\frac{1}{72}(\mathbf{E}(t,X)^3)^2 + \frac{1}{24}(\mathbf{E}(t,X)^4 - s^2)\Big) + \dots\Big\}, \tag{2}$$

where $s = s(t) = \mathbf{E}(t,X)^2$, $t = (t_1, t_2, \dots, t_k) \in \mathbf{R}^k$, and (t,X) means the scalar product of t and X. The coefficient of $n^{-\nu/2}$, $\nu = 1, 2, 3, \dots$, is a linear combination with constant coefficients of ν homogeneous polynomials in $t_1, t_2, \dots, k_k$ of $\nu + 2$, $\nu + 4, \dots, 3\nu$.
The polynomial of power $\nu + 2h$, $h = 1, \dots, \nu$, depends on $s, \mathbf{E}(t,X)^3, \dots, \mathbf{E}(t,X)^{\nu-h+3}$. Applying the inverse Fourier transform to every term of series (2) we get a formal expansion of probability $\mathbf{P}\{S_n \in A\}$

$$\mathbf{P}\{S_n \in A\} = \Phi(A) + \frac{1}{\sqrt{n}}\Phi_1(A) + \frac{1}{n} + \Phi_2(A) + \dots,$$

where Φ is the normal distribution with characteristic function $e^{-s/2}$ and $\Phi_1, \Phi_2, \ldots$ are signed measures. These measures have the densities with respect to Lebesque measure in $\mathbf{R}^k$

$$\Phi_\nu(A) = \int_A q_\nu(x)dx,$$

and their corresponding total variations are

$$V_\nu = \frac{1}{2}\int_{\mathbf{R}^k} |q_\nu(x)|dx, \qquad \nu = 1, 2, 3, \ldots .$$

Let $Z_1, Z_2, \ldots, Z_k$ be independent $(0,1)$ normal random variables. Then it is easy to see that $2V_\nu$ can be expressed as the expectation of absolute values of a polynomial in $Z_1, Z_2, \ldots, Z_k$ (e.g., (Prokhorov, 1990, 1991)).

Example 1. If X has a spherically symmetric distribution and $s(t) = |t|^2$, then $\Phi_1 = \Phi_3 = \ldots = 0$ and (e.g., (Prokhorov, 1990))

$$V_2 = \frac{1}{6}|A_1| \cdot \mathbf{E}\Big|L_2^{k/2-1}\Big(\frac{1}{2}(Z_1^2 + \ldots + Z_k^2)\Big)\Big|,$$

where $A_1 = \mathbf{E}(X^{(1)})^4 - 3$, and

$$V^4 = \frac{1}{2}\mathbf{E}\Big|\frac{1}{3}A_1^2 L_4^{k/2-1}\Big(\frac{1}{2}(Z_1^2 + \ldots + Z_k^2)\Big) - \frac{1}{15}A_2 L_3^{k/2-1}\Big(\frac{1}{2}(Z_1^2 + \ldots + Z_k^2)\Big)\Big|,$$

where $A_2 = \mathbf{E}((X^{(1)})^6 - 15(X^{(1)})^4 + 30)$ and so on.
Here $L_m^\alpha(z)$ is a generalized Laguerre polynomial, $m = 0, 1, 2, \ldots$,

$$L_m^\alpha(z) = \frac{1}{m!}z^{-\alpha}e^z\frac{d^m}{dz^m}(z^{m+\alpha}e^{-z}).$$

Example 2. If X has independent components, then (see (Prokhorov, 1991))

$$V_1 = \frac{1}{2}\mathbf{E}\Big|\sum_{j=1}^k \gamma_j H_3(Z_j)\Big|,$$

$$V_2 = \frac{1}{2}\mathbf{E}\Big|\sum_{j=1}^k (\alpha_j H_4(Z_j)\gamma_j H_6(Z_j)) + \sum_{1\le j<l\le k} \gamma_j\gamma_l H_3(Z_j)H_3(Z_l)\Big|,$$

and so on. Here $H_m(z)$ is a Hermite polynomial,

$$H_m(z) = (-1)^m e^{z^2/2}\frac{d^m}{dz^m}[e^{-z^2/2}], \qquad m = 0, 1, 2, \ldots,$$

and

$$\alpha_j = \frac{1}{24}\Big(\frac{\mathbf{E}(X^{(j)})^4}{\sigma_j^4} - 3\Big), \qquad \beta_j = \frac{1}{2}\gamma_j^2, \quad \gamma_j = \frac{1}{6}\frac{\mathbf{E}(X^{(j)})^3}{\sigma_j^3}, \quad \sigma_j^2 = \mathbf{E}(X^{(j)})^2.$$

The upper bounds for the total variation V_ν are simply obtained using the Cauchy–Bunyakovsky inequality $\mathbf{E}|\xi| \leq \sqrt{\mathbf{E}\xi^2}$. The summands under the modulo sign, encountered in the both examples are pairwise uncorrelated. Then direct calculations give us (in Example 2)

$$V_1 \leq \frac{1}{2}\sqrt{\sum \gamma_j^2 \cdot 3!}, \tag{3}$$

$$V_2 \leq \frac{1}{2}\Big(\sum_j (\alpha_j^2 \cdot 4! + \beta_j^2 \cdot 6!) + \sum_{j<l} \gamma_j^2 \gamma_l^2 \cdot 3!3!\Big)^{1/2},$$

and (for Example 1)

$$V_2 \leq \frac{1}{6}|A_1|\sqrt{\frac{1}{2!}\frac{k}{2}\Big(\frac{k}{2}+1\Big)}, \tag{5}$$

$$V_4 \leq \frac{1}{2}\Big(\frac{1}{9}A_1^4\frac{1}{4!}\frac{k}{2}\Big(\frac{k}{2}+1\Big)\Big(\frac{k}{2}+2\Big)\Big(\frac{k}{2}+2\Big)+\frac{1}{225}A_2^2\frac{1}{3!}\frac{k}{2}\Big(\frac{k}{2}+1\Big)\Big(\frac{k}{2}+2\Big)\Big)^{1/2} \tag{6}$$

We have in addition to (3) (see (Prokhorov, 1991))

$$V_1 \geq 0.008\sqrt{\sum \gamma_j^2 \cdot 3!} \tag{3'}$$

and in addition to (5)

$$V_2 \geq \frac{2}{e\sqrt{\pi}} \cdot \frac{1}{6}|A_1|\sqrt{\frac{1}{2!}\frac{k}{2}\Big(\frac{k}{2}+1\Big)}. \tag{5'}$$

The author's hypothesis is that in general the upper bound based on Cauchy–Bunyakovsky inequality multiplied by appropriate constants give the lower bounds for the corresponding total variation.

The proof of inequalities like (3′) and (5′) is rather complicated even in the particular cases. Here the following inequalities

$$\mathbf{P}\Big(\Big|\frac{Q-\mathbf{E}Q}{\sqrt{\mathbf{D}Q}}\Big| > x\Big) \geq k_1(1-\Phi(k_2 x))$$

can be used, where $Q = \sum_{j=1}^{k} \lambda_j P(Z_j)$, $P(z)$ is some polynomial, λ_j are any real numbers, $k_1 > 0$ and $k_2 > 0$ are constants depending on $P(z)$ only (and sometimes on the degree n of the polynomial $P(z)$ only, for example, when $n = 2$ and 3).

REFERENCES

Prokhorov, Yu.V. (1990). Sums of random vectors with increasing dimension. *Theory Probab. Appl.* **35**, 755–757.

Prokhorov, Yu.V. (1991). An estimate of the total variation of signed measures arizing in connection with central limit theorem. *Theory Probab. Appl.* **36**, 388–390.

Frontiers in Pure and Applied Probab. 1, pp. 209 – 222
H. Niemi et al. (Eds.)
1993 TVP/VSP

AN IMPROVED NONUNIFORM CONVERGENCE RATE ESTIMATE IN THE CENTRAL LIMIT THEOREM IN $\mathbf{R}^k$

V. V. SAZONOV*
Steklov Mathematical Institute, Vavilov Street 42, 117966 Moscow GSP-1, Russia

ABSTRACT

It has been known for some time (see, e.g., (Sazonov, 1981)) that the coefficient $c(k)$ in the nonuniform estimate of the rate of convergence on convex sets in the Central Limit Theorem in $\mathbf{R}^k$ increases as $k \to \infty$ not faster than $k^{5/2}$. In this paper we construct a new estimate which implies, in particular, that $c(k)$ increases not faster than k^2.

1. INTRODUCTION AND MAIN NOTATION

Let $X_1, X_2, \ldots$ be independent identically distributed (i.i.d.) random variables with values in $\mathbf{R}^k$. Assume that $\mathbf{E}|X_1|^2 < \infty$ and that the covariance matrix V of X_1 is nondegenerate. Denote the probability measure corresponding to the normalized sum $n^{-1/2}\sum_1^n V^{-1/2}(X_i - EX_i)$ by P_n, and let N be the standard Gaussian probability measure in $\mathbf{R}^k$.
Define $B_r(x) = \{y \in \mathbf{R}^k : |y - x| < r\}$ and put

$$E^\varepsilon = \bigcup \{B_\varepsilon(x) \colon x \in E\}, \qquad E^{-\varepsilon} = \bigcup \{x \colon B_\varepsilon(x) \subset E\}.$$

for any $E \subset \mathbf{R}^k$, $\varepsilon > 0$.
Let $\mathcal{A}$ be a class of convex Borel subsets of $\mathbf{R}^k$ such that

$$A + x \in \mathcal{A}, \qquad \varepsilon A \in \mathcal{A}, \qquad A^\varepsilon \in \mathcal{A}, \qquad A^{-\varepsilon} \in \mathcal{A} \tag{1}$$

for any $A \in \mathcal{A}$, $x \in \mathbf{R}^k$, and $\varepsilon > 0$ and

$$\int\limits_{(\partial A)^\varepsilon} N(dx) \le c'_k \varepsilon, \qquad \int\limits_{(\partial A)^\varepsilon} |x|^3 N(dx) \le c''_k \varepsilon, \tag{2}$$

where ∂A is the boundary of A. It is known that the class of all convex Borel sets satisfies these conditions with $c'_k = ck^{1/2}, c''_k = ck^2$ (see (von Bahr, 1967;

*This work was done while the author was visiting the Statistical Research Section, SMS, the Australian National University, Canberra, February–March 1991.

Bhattacharya and Ranga Rao, 1967)). Of course, the class of all balls satisfies (1), and it was observed in (Sazonov, 1976) that it also satisfies the first condition in (2) with c'_k equal to an absolute constant. In the same way as in (Sazonov, 1976), it is easy to check that the second condition in (2) is satisfied with $c''_k = ck^{3/2}$ for balls.

It was shown in (Bentkus, 1986) that, if $\mathcal{A}$ is a class of convex sets satisfying (1) and the first condition in (2), then

$$\sup\left\{|P_n(A) - N(A)|\colon A \in \mathcal{A}\right\} \le K(1 + c'_k)\beta n^{-1/2}, \tag{3}$$

where K is an absolute constant and $\beta = \mathbf{E}\left|V^{-1/2}(X_1 - EX_1)\right|^3$. In obtaining this estimate, Bentkus used the smoothness properties of the superposition of a smooth function and the distance function to a convex set pointed out by Senatov (Senatov, 1984). Otherwise, his approach is basically the convolution method (see, e.g., (Sazonov, 1981)). Bentkus also indicated the possibility of extending the convolution method modified by him to obtain nonuniform estimates and estimates in terms of truncated moments and pseudomoments (consideration of pseudomoments see in (Sazonov, 1981)).

In the present paper we give the proof of the nonuniform estimate based on this approach which turns out to be technically a nontrivial extension of Bentkus' proof. Our estimate is valid for sets A belonging to a class $\mathcal{A}$ satisfying (1) and (2).

The results of the present paper can be further extended to the case when the third moment does not exist, and the moments can be replaced by pseudomoments. Such extensions will be published elsewhere.

In what follows N_T, $T > 0$, denote the probability measure $N_T(\cdot) = N(T\cdot)$ and φ_T the corresponding density function; we write simply φ instead of φ_1. We introduce the following destinations for a non-negative integral vector $a \in \mathbf{R}^k$ (i.e., with non-negative integer components a_i): $\|a\| = \sum_1^k a_j$, $a! = a_1!, \ldots, a_k!$, D_a denotes the differential operator $\partial^{\|a\|}/\partial x_1^{a_1} \cdots \partial x_k^{a_k}$, and if $x \in \mathbf{R}^k$, then $x^a = x_1^{a_1} \cdots x_k^{a_k}$. Symbols c and K, with or without indices, stand for absolute constants, and I denotes the $k \times k$-identity matrix. Finally, $r(E) = \inf\left\{|x|\colon x \in E\right\}$ denotes the distance from a set $E \subset \mathbf{R}^k$ to the origin.

2. THE MAIN RESULT

The aim of the present paper is to prove the following

THEOREM. *In the above notation, for any A belonging to a class $\mathcal{A}$ of convex Borel subsets of $\mathbf{R}^k$ satisfying* (1) *and* (2) *the following inequality*

$$\left|P_n(A) - N(A)\right| \le \frac{c(k)\beta n^{-1/2}}{c_k + r^3(A)},$$

where $c(k) = K_1\left(k^{3/2}(1 + c'_k) + c''_k\right)$, $c_k = k^{3/2} + c''_k(1 + c'_k)^{-1}$, holds.

Remark 1. It follows from what was said in the Introduction concerning (2) that, if $\mathcal{A}$ is the class of all convex Borel subsets of $\mathbf{R}^k$, then $c(k) \le ck^2$, $k^{3/2} \le c_k \le ck^{3/2}(1 + k^{1/2}(1 + c'_k)^{-1})$ and, if $\mathcal{A}$ is the class of all balls, then $c(k) \le ck^{3/2}$, $c_k \ge k^{3/2}$.

Remark 2. Since $r(aA) = ar(A)$, $a > 0$, the relation

$$r^3(A)\big|P_n(A) - N(A)\big| \le c'(k)\beta n^{-1/2}, \quad A \in \mathcal{A},$$

is equivalent to

$$r^3(A)\big|P_n(k^{1/2}A) - N(k^{1/2}A)\big| \le c'(k)k^{-3/2}\beta n^{-1/2}, \quad A \in \mathcal{A}. \tag{4}$$

Thus, by (3), it is enough to prove (4) with $c'(k) = K'\big(k^{3/2}(1+c'_k)+c''_k\big)$ or, since $V^{-1/2}(X_1 - \mathbf{E}X_1)$ has the zero mean and the unit covariance matrix, it is enough to prove that, if $X_1, X_2, \ldots$ is a sequence of i.i.d. $\mathbf{R}^k$-valued random variables with the zero mean and the covariance matrix $k^{-1}I$, $P'_n(\cdot) = \mathbf{P}(n^{-1/2}\sum_1^n X_i \in \cdot)$, $N'(\cdot) = N_{k^{1/2}}(\cdot)$, then

$$r^3(A)\big|P'_n(A) - N'(A)\big| \le c'(k)\beta' n^{-1/2}$$

with the same $c'(k)$ and $\beta' = \mathbf{E}|X_1|^3$. In the sequel, we assume that $X_1, X_2, \ldots$ is such a sequence.

3. SOME AUXILIARY RESULTS

According to the assumptions made above, $X_1, X_2, \ldots$ are i.i.d. $\mathbf{R}^k$-valued random variables with the zero mean and the covariance matrix $k^{-1}I$. For $u \ge 0$, $t \ge 0$, $n \ge 1$ denote

$$\overline{S}_{n,u} = n^{-1/2}\sum_1^n X_j 1_{\{|X_j|<u\}}, \qquad \overline{S}_n = \overline{S}_{n,n^{1/2}},$$

$$\widehat{S}_{n,t} = n^{-1/2}\sum_1^n X_j 1_{\{|X_j|<(1+t)n^{1/2}\}}.$$

Denote also $\overline{P}_n(\cdot) = \mathbf{P}(\overline{S}_n \in \cdot)$.

LEMMA 1. 1. *For any $u \ge 0$, integer $n \ge 1$, and any Borel set E*

$$\big|P'_n(E) - \mathbf{P}(\overline{S}_{n,u} \in E)\big| \le n\mathbf{P}(|X_1| \ge u).$$

2. *For any $t \ge 0$, $p \ge 0$, integer $n \ge 1$, and any Borel set E*

$$\big|\overline{P}_n(E) - \mathbf{P}(\widehat{S}_{n,t} \in E)\big| \le c(p)\big(1 + r(E)\big)^{-p} n^{1-p/2}\mathbf{E}\big(|X_1|^p 1_{\{n^{1/2}\le|X_1|<(1+t)n^{1/2}\}}\big).$$

This lemma was proved in (Sazonov et al., 1987).

COROLLARY (cf. Lemma 5 in (Rotar, 1970)). *If* $u = (1 + r(E)) \times n^{1/2}$, $t = r(E)$, $p = 3$ *in Lemma 1, then*

$$|P_n'(E) - \overline{P}_n(E)| \le \frac{c\beta' n^{-1/2}}{1 + r^3(E)}$$

for any Borel set $E \subset \mathbf{R}^k$.

This lemma also implies that for any $u \ge n^{1/2}$

$$|P_n'(E) - \mathbf{P}(\overline{S}_{n,u} \in E)| \le \frac{c\beta' n^{-1/2}}{1 + r^3(E)}.$$

LEMMA 2. *For any* $p \ge 0$ *and any* i, $1 \le i \le n$,

$$\int |x|^p \overline{P}^{\,i}_{(n)}(dx) \le c(p) \quad \text{and} \quad \int |x|^p N'(dx) \le c(p)$$

(here and below a power of a measure is understood in the sense of convolution), where $\overline{P}_{(n)}(\cdot) = \mathbf{P}\Big(n^{-1/2} X_1 1_{\{|X_1| \le n^{1/2}\}} \in \cdot\Big)$.

The first of these inequalities follows easily from Pinelis' inequality (Pinelis, 1980), the second one is proved, e.g., in (Sazonov, 1981).

LEMMA 3. *For any* $E \subset \mathbf{R}^k$, $x \in \mathbf{R}^k$ $\gamma \in \mathbf{R}$, *and* $\delta > 0$,

$$\gamma E^{\pm\delta} = (\gamma E)^{\pm\gamma\delta}; \qquad E^{\pm\delta} - x = (E - x)^{\pm\delta}; \qquad \gamma(\partial E) = \partial(\gamma E);$$
$$(\partial E)^\delta = E^\delta \backslash E^{-\delta}; \qquad |r(E) - r(E + x)| \le |x|; \qquad r(\gamma E) = |\gamma| r(E);$$
$$r(E) \le r(E^\delta) + \delta; \qquad r(E) \le r(E^{-\delta}) + \delta.$$

The proof of Lemma 3 is elementary and, hence, is omitted.

LEMMA 4. (cf. (Kuelbs and Kurtz, 1974)). *Let* $E \subset \mathbf{R}^k$ *be a Borel set and* g_i, $i = 1, 2$, *be real Borel measurable functions defined on* $\mathbf{R}^k$ *such that* $0 \le g_i \le 1$, $i = 1, 2$, *and for some* $\varepsilon > 0$,

$$g_1(x) = \begin{cases} 1, & \text{if } x \in E, \\ 0, & \text{if } x \in (E^\varepsilon)^c, \end{cases} \qquad g_2(x) = \begin{cases} 1, & \text{if } x \in E^{-\varepsilon}, \\ 0, & \text{if } x \in E^c. \end{cases}$$

Then for any two Borel measures P, Q *on* $\mathbf{R}^k$

$$|P(E) - Q(E)| \le \max\Big\{\Big| \int g_i(x)(P - Q)(dx)\Big| \colon i = 1, 2\Big\} + Q((\partial E)^\varepsilon).$$

Proof. If $(P - Q)(E) \ge 0$, then

$$(P - Q)(E) \le \int g_1(x)(P - Q)(dx) + \int (g_1(x) - 1_E(x))\, Q(dx)$$
$$\le \Big| \int g_1(x)(P - Q)(dx)\Big| - Q(E^\varepsilon \backslash E).$$

If $(P-Q)(E) \le 0$, then, similarly,

$$(P-Q)(E) \ge \int g_2(x)\,(P-Q)(dx) + \int \big(g_2(x) - 1_E(x)\big)\,Q(dx)$$
$$\ge -\Big|\int g_2(x)\,(P-Q)(dx)\Big| - Q(E\backslash E^{-\varepsilon}),$$

and the lemma is thereby proved.

LEMMA 5. *Let A be a convex subset of $\mathbf{R}^k$ and $d(x) = d(x, A) = \inf\big\{|x - y|\colon y \in A\big\}$, $x \in \mathbf{R}^k$. Let $f(y)$ be a real differentiable function on $\mathbf{R}$ such that $f'(0) = 0$, $|f'(y_1) - f'(y_2)| \le \bar{c}|y_1 - y_2|$. Then $g(x) = f\big(d(x)\big)$ is differentiable and*

$$\big|\nabla g(x_1) - \nabla g(x_2)\big| \le 3\bar{c}|x_1 - x_2|.$$

This lemma was proved essentially in (Senatov, 1984).

4. PROOF OF THE THEOREM

By Remark 2, it is enough to prove that

$$\Delta_n = r^3(A)\big|P_n'(A) - N'(A)\big| \le c'(k)\beta' n^{-1/2}, \quad A \in \mathcal{A}, \tag{5}$$

with $c'(k) = K'\big(k^{3/2}(1+c_k') + c_k''\big)$. Observe also that (see Lemma 2) for any Borel $E \subset \mathbf{R}^k$

$$r^3(E)\big|P_1'(E) - N'(E)\big| \le r^3(E')(P_1'(E') + N'(E'))$$
$$\le \beta' + c$$
$$\le c\beta',$$

where $E' = E$, if $0 \notin E$, and $= E^c$, if $0 \in E$. This proves the theorem when $n = 1$, and, in what follows, we assume $n \ge 2$.

Let $f(y)$ be a real twice continuously differentiable nonincreasing function defined on $\mathbf{R}$ such that $f(y) = 1$, if $y \le 0$, $= 0$, if $y \ge 1$. For example, we can take $f(y) = (2\pi)^{-1}\sin 2\pi y - y + 1$, if $0 \le y \le 1$. Choose and fix a set $A \in \mathcal{A}$ and define $d(x)$ as in Lemma 5. For an $\varepsilon > 0$ let

$$g_{1,\varepsilon}(x) = f\big(d(x, A)\varepsilon^{-1}\big), \qquad g_{2,\varepsilon}(x) = f\big(d(x, A^{-\varepsilon})\varepsilon^{-1}\big).$$

Note that, by Lemma 5, for $j = 1, 2$

$$\big|\nabla g_{j,\varepsilon}(x)\big| \le c\varepsilon^{-1} 1_{(\partial A)^\varepsilon}(x),$$
$$\big|\nabla g_{j,\varepsilon}(x_1) - \nabla g_{j,\varepsilon}(x_2)\big| \le c\varepsilon^{-2}|x_1 - x_2|\big\{1_{(\partial A)^\varepsilon}(x_1) + 1_{(\partial A)^\varepsilon}(x_2)\big\}. \tag{6}$$

Applying Lemma 4, we have

$$r^3(A)\big|\overline{P}_n(A) - N'(A)\big| \le r^3(A)\Big[\max_{j=1,2}\Big\{\Big|\int g_{j,\varepsilon}\,(\overline{P}_n - N')(dx)\Big|\Big\} + N'((\partial A)^\varepsilon)\Big]. \tag{7}$$

Now, if $x \in (\partial A)^\varepsilon$, then there exists $y \in \partial A$ such that $|x-y| < \varepsilon$, and we have $r(A) \le |y| \le |x| + |y-x| < |x| + \varepsilon$. Hence, by Lemma 3 and (2),

$$\begin{aligned} r^3(A)N'((\partial A)^\varepsilon) &= k^{-3/2} r^3(k^{1/2}A) N\Big((\partial(k^{1/2}A))^{\varepsilon k^{1/2}} \Big) \\ &\le k^{-3/2} \int\limits_{(\partial k^{1/2}A)^{\varepsilon k^{1/2}}} \big(|x| + k^{1/2}\varepsilon\big)^3 N(dx) \\ &\le 4k^{-1}(c_k' k^{3/2}\varepsilon^3 + c_k'')\varepsilon. \end{aligned} \tag{8}$$

Furthermore, with $\overline{P}_{(n)}$ denoting the same probability measure as in Lemma 2, we can write

$$\overline{P}_n - N' = \overline{P}^n_{(n)} - N^n_{(kn)^{1/2}} = \sum_{i=0}^{n-1} \overline{W}_i * \overline{H},$$

where

$$\overline{W}_i = \overline{P}^i_{(n)} * N^{n-i-1}_{(kn)^{1/2}} = \overline{P}^i_{(n)} * N_{k^{1/2}t_i}, \qquad t_i = \big(n/(n-i-1)\big)^{1/2},$$

$$\overline{H} = \overline{P}_{(n)} - N_{(kn)^{1/2}}, \qquad i = 0, 1, \ldots, n-1,$$

and, hence, if g is any of the functions $g_{i,\varepsilon}$, $j = 1, 2$, we have

$$r^3(A) \int g(x)(\overline{P}_n - N')(dx) = \sum_{i=0}^{n-1} I_i, \tag{9}$$

where $I_i = r^3(A) \int g(x)\, \overline{W}_i * \overline{H}(dx)$.
To estimate I_i, we consider several cases separately.

1. *Estimate of* I_{n-1}
Using the Taylor expansion

$$g(x+y) = g(x) + \int_0^1 (\nabla g(x + \theta y), y)\, d\theta, \tag{10}$$

we can write

$$\begin{aligned} I_{n-1} &= r^3(A) \int g(x+y)\, \overline{P}^{n-1}_{(n)}(dx)\, \overline{H}(dy) \\ &= r^3(A) \int_0^1 \int (\nabla g(x+\theta y), y)\, \overline{P}^{n-1}_{(n)}(dx)\, \overline{H}(dy)\, d\theta. \end{aligned}$$

Hence, by (6),

$$|I_{n-1}| \le c r^3(A) \varepsilon^{-1} \int_0^1 \int \overline{P}^{n-1}_{(n)}\big((\partial A)^\varepsilon - \theta y\big) |y| (\overline{P}_{(n)} + N')(dy)\, d\theta.$$

Now, using Lemma 3, we can write for any $u \in \mathbf{R}^k$ and $i = 1, \ldots, n-1$

$$\overline{P}^i_{(n)}((\partial A)^\varepsilon - u) = (P''_i - N')(s_i(A^\varepsilon - u)) - (P''_i - N')(s_i(A^{-\varepsilon} - u)) + N'\Big(\big(\partial(s_i(A-u))\big)^{s_i\varepsilon}\Big),$$

where $P''_i(E) = \mathbf{P}(S''_i \in E)$, $S''_i = i^{-1/2}\sum_1^i X_i 1_{\{|X_j|<n^{1/2}\}}$, $s_i = (n/i)^{1/2}$. Furthermore, by Lemma 3, $r(A) \le |u| + \varepsilon + r(A^{\pm\varepsilon} - u)$, and, hence,

$$r^3(A) \le c\Big[|u|^3 + \varepsilon^3 + s_i^{-3} r^3\big(s_i(A^{\pm\varepsilon} - u)\big)\Big].$$

However for any Borel set $E \subset \mathbf{R}^k$, by the corollary to Lemma 1,

$$\big(1 + r^3(E)\big)\big|P''_i(E) - P'_i(E)\big| \le c\beta' i^{-1/2}.$$

Furthemore, by Lemma 3,

$$r^3(A) \le c\Big(|u|^3 + s_i^{-3} r^3\big(s_i(A-u)\big)\Big),$$

and we find, using (3) and (8), that for any $u \in \mathbf{R}^k$

$$\begin{aligned} r^3(A)\overline{P}^i_{(n)}((\partial A)^\varepsilon - u) &\le c\Big[\big(|u|^3 + \varepsilon^3\big)\bar{c}(k)\beta' i^{-1/2} + s_i^{-3}(\beta' i^{-1/2} + \Delta_i) \\ &\quad + |u|^3 c'_k k^{1/2}\varepsilon + s_i^{-3}k^{-1}(c'_k k^{3/2}\varepsilon^3 + c''_k)\varepsilon\Big], \\ r^3(A)\big|(P''_i - N')(s_i(A^{\pm\varepsilon} - u))\big| &\le c\Big[\big(|u|^3 + \varepsilon^3\big)\bar{c}(k)\beta' i^{-1/2} \\ &\quad + s_i^{-3}(\beta' i^{-1/2} + \Delta_i)\Big], \end{aligned} \tag{11}$$

where $\bar{c}(k) = K(1 + c'_k)k^{3/2}$, $\Delta_i = \sup_{A\in\mathcal{A}}\Big\{r^3(A)\big|P'_i(A) - N'(A)\big|\Big\}$.
Note also for future reference that, as follows from Lemma 3, using the inequality $r(A) \le r\big((A^{-\varepsilon})^{t\varepsilon}\big) + (1+t)\varepsilon$, ε, $t > 0$, one can show in the same way that

$$\begin{aligned} r^3(A)&\Big|(P''_i - N')\Big(s_i\big((A^{-\varepsilon})^{t\varepsilon} - u\big)\Big)\Big| \\ &\le c\Big[\big(|u|^3 + (1+t)^3\varepsilon^3\big)\bar{c}(k)\beta' i^{-1/2} + s_i^{-3}(\beta' i^{-1/2} + \Delta_i)\Big]. \end{aligned} \tag{12}$$

However (see Lemma 2)

$$\begin{aligned} \int |y|^\ell\, \overline{P}_{(n)}(dy) &\le \begin{cases} \beta' n^{-\ell/2}, & 0 \le \ell \le 3, \\ \beta' n^{-3/2}, & \ell \ge 3, \end{cases} \\ \int |y|^\ell\, N_{(kn)^{1/2}}(dy) &\le c(\ell) n^{-\ell/2}. \end{aligned} \tag{13}$$

Consequently, since $1 \le t_0 \le 2^{1/2}$,

$$|I_{n-1}| \le c\varepsilon^{-1}\Big[\big(\bar{c}(k)\beta' n^{-1/2} + c'_k k^{1/2}\varepsilon\big)\beta' n^{-3/2} + \big(\varepsilon^3\bar{c}(k)\beta' n^{-1/2} + \beta' n^{-1/2} + \Delta_{n-1} + k^{-1}(c'_k k^{3/2}\varepsilon^3 + c''_k)\varepsilon\big)\beta' n^{-1/2}\Big].$$

2. *Estimate of* I_i, $[n/2] < n-m \le i \le n-2$

Denoting $I'_i = r^{-3}(A)I_i$ we can write

$$I'_i = \int g(x+y+z)\, N_{k^{1/2}t_i}(dx)\, \overline{H}(dy)\, \overline{P}^i_{(n)}(dz) = \int g(u+z)\varphi_{k^{1/2}t_i}(u-y)\,du\, \overline{H}(dy)\, \overline{P}^i_{(n)}(dz). \tag{14}$$

Applying (10) to $\varphi_{k^{1/2}t_i}(u-y)$ and noting that for $T>0$

$$\frac{\partial\varphi_T}{\partial x_j}(x) = -T^2 x_j \varphi_T(x),$$

so that $(\nabla\varphi_{k^{1/2}t_i}(u-\theta y), y) = -kt_i^2(u-\theta y, y)\varphi_{k^{1/2}t_i}(u-\theta y)$, we obtain

$$I'_i = -k^{1/2}t_i \int_0^1 \int g(k^{-1/2}t_i^{-1}x + \theta y + z)\varphi(x)(x,y)\,dx\, \overline{H}(dy)\, \overline{P}^i_{(n)}(dz).$$

Applying (10) again with $k^{-1/2}t_i^{-1}x + z$ instead of x and θy instead of y, we represent I'_i in the following form:

$$\begin{aligned} I'_i = {} & -k^{1/2}t_i \int g(k^{-1/2}t_i^{-1}x + z)\varphi(x)(x,y)\,dx\, \overline{H}(dy)\, \overline{P}^i_{(n)}(dz) \\ & -k^{1/2}t_i \int_0^1\int_0^1\int \Big[(\nabla g(k^{-1/2}t_i^{-1}x + z + \theta_1\theta y), y) - (\nabla g(k^{-1/2}t_i^{-1}x + z), y)\Big] \\ & \qquad \times\theta(x,y)\varphi(x)\,dx\, \overline{H}(dy)\, \overline{P}^i_{(n)}(dz)\,d\theta\,d\theta_1 \\ & -\frac{1}{2}k^{1/2}t_i \int (\nabla g(k^{-1/2}t_i^{-1}x + z), y)(x,y)\varphi(x)\,dx\, \overline{H}(dy)\, \overline{P}^i_{(n)}(dz) \\ = {} & J_1 + J_2 + J_3. \end{aligned} \tag{15}$$

Observe now that, if $0 \notin A^\varepsilon$, then (see Lemma 3)

$$r^3(A)g(k^{-1/2}t_i^{-1}x+z) \le 4\big(r^3(A)+\varepsilon^3\big)g(k^{-1/2}t_i^{-1}x+z) \le 4\big(|k^{-1/2}t_i^{-1}x+z|^3+\varepsilon^3\big),$$

and, if $0 \in A^\varepsilon\backslash A^{-\varepsilon}$, then $r(A) \le \varepsilon$ and, hence, $r^3(A)g \le \varepsilon^3$. This fact together with the obvious relations (P'_1 and N' have the same first and second moments)

$$\int (x,y)^{\ell_1}(z,y)^{\ell_2}\,\overline{H}(dy) = n^{-(\ell_1+\ell_2)/2} \int\limits_{|y|\ge n^{1/2}} (x,y)^{\ell_1}(z,y)^{\ell_2}\,P_1'(dy),$$

$$\int |y|^2\,\overline{H}(dy) = n^{-1} \int\limits_{|y|\ge n^{1/2}} |y|^2\,P_1'(dy), \tag{16}$$

for $\ell_1,\ \ell_2 = 0,1,2$ and $1 \le \ell_1+\ell_2 \le 2$ implies, if $0 \notin A^{-\varepsilon}$,

$$r^3(A)|J_1| \le ck^{1/2}t_i n^{-1/2} \int\limits_{|y|\ge n^{1/2}} \left(|k^{-1/2}t_i^{-1}x+z|^3+\varepsilon^3\right)$$
$$\times\,|(x,y)|\varphi(x)\,dx\,P_1'(dy)\,\overline{P}^{\,i}_{(n)}(dz). \tag{17}$$

If $0 \in A^{-\varepsilon}$, then (see Lemma 3)

$$\begin{aligned} r^3(A)\big(1-g(k^{-1/2}t_i^{-1}x+z)\big) &\le 4\big(r^3(A^{-\varepsilon})+\varepsilon^3\big)\big(1-g(k^{-1/2}t_i^{-1}x+z)\big)\\ &\le 4\big(|k^{-1/2}t_i^{-1}x+z|^3+\varepsilon^3\big), \end{aligned}$$

and, since $g(k^{-1/2}t_i^{-1}x+z)$ in the definition of J_1 can be replaced by $g(k^{-1/2}t_i^{-1}x+z)-1$, (17) is valid in all the cases.
To estimate the integral in (17), we observe that

$$\begin{aligned} \int |x|^{\ell_1}|(x,y)|^{\ell_2}\varphi(x)\,dx &\le \Big(\int |x|^{2\ell_1}\varphi(x)\,dx\Big)^{1/2}\Big(\int (x,y)^{2\ell_2}\varphi(x)\,dx\Big)^{1/2}\\ &\le c(\ell_1,\ell_2)k^{\ell_1/2}|y|^{\ell_2}, \quad \ell_1,\ell_2\ge 0, \end{aligned}$$

$$\int\limits_{|y|\ge n^{1/2}} |y|^{\ell}\,P_1'(dy) \le \beta' n^{(\ell-3)/2}, \quad \ell = 0,1,2, \tag{18}$$

and, by Lemma 2, $\int |z|^3\,\overline{P}^{\,i}_{(n)}(dz) \le c$. Being combined, these relations imply

$$r^3(A)|J_1| \le ck^{1/2}t_i\beta' n^{-1/2}(1+\varepsilon^3). \tag{19}$$

Furthermore, by (6),

$$\begin{aligned} |J_2| &\le ck^{1/2}t_i\varepsilon^{-2}\int |y|^2|(x,y)|\big[1_{(\partial A)^\varepsilon}(k^{-1/2}t_i^{-1}x+z+\theta\theta_1 y)\\ &\quad + 1_{(\partial A)^\varepsilon}(k^{-1/2}t_i^{-1}x+z)\big]\varphi(x)\,dx\,(\overline{P}_{(n)}+N_{(kn)^{1/2}})(dy)\,\overline{P}^{\,i}_{(n)}(dz)\\ &= ck^{1/2}t_i\varepsilon^{-2}\int \Big[\bar{P}^{i}_{(n)}\big((\partial A)^\varepsilon - k^{-1/2}t_i^{-1}x-\theta\theta_1 y\big)+\overline{P}^{\,i}_{(n)}\big((\partial A)^\varepsilon - k^{-1/2}t_i^{-1}x\big)\Big]\\ &\quad \times |y|^2|(x,y)|\varphi(x)\,dx\,\big(\bar{P}_{(n)}(dy)+N_{(kn)^{1/2}}(dy)\big). \end{aligned}$$

Using (11), (13), and (18), we obtain

$$r^3(A)|J_2| \le ck^{1/2}t_i\varepsilon^{-2}\beta' n^{-3/2}\big[\bar{c}(k)\beta' n^{-1/2}(1+\varepsilon^3)+c_k'k^{1/2}(\varepsilon+\varepsilon^4)+c_k''k^{-1}\varepsilon+\Delta_i\big].$$

To estimate $r^3(A)|J_3|$, we observe that, by (16) and (6),

$$
\begin{aligned}
&r^3(A)|J_3| \\
&= ck^{1/2}t_i n^{-1} r^3(A) \Big| \int\limits_{|y|\geq n^{1/2}} (\nabla g(k^{-1/2}t_i^{-1}x + z), y)(x,y)\varphi(x)\,dx\,P_1'(dy)\,\overline{P}^{i}_{(n)}(dz) \Big| \\
&\leq ck^{1/2}t_i n^{-1} r^3(A)\varepsilon^{-1} \int\limits_{|y|\geq n^{1/2}} 1_{(\partial A)^\varepsilon}(k^{-1/2}t_i^{-1}x + z)|y| \\
&\qquad \times |(x,y)|\varphi(x)\,dx\,P_1'(dy)\,\overline{P}^{i}_{(n)}(dz) \\
&\leq ck^{1/2}t_i n^{-1} r^3(A)\varepsilon^{-1} \int_{|y|\geq n^{1/2}} \overline{P}^{i}_{(n)}((\partial A)^\varepsilon - k^{-1/2}t_i^{-1}x)|y|\,|(x,y)|\varphi(x)\,dx\,P_1'(dy).
\end{aligned}
$$

Using (11), (13), and (18) again, we deduce that

$$
r^3(A)|J_3| \leq ck^{1/2}t_i\varepsilon^{-1}\beta' n^{-3/2}\big(\bar{c}(k)\beta' n^{-1/2}(1+\varepsilon^3) + c_k' k^{1/2}(\varepsilon+\varepsilon^4) + c_k'' k^{-1}\varepsilon + \Delta_i\big).
$$

Summing up the above estimates, we have

$$
\begin{aligned}
|I_i| \leq ck^{1/2}t_i\Big[1 + \varepsilon^3 + (\varepsilon^{-1} + \varepsilon^{-2})\big(\bar{c}(k)\beta' n^{-1/2}(1+\varepsilon^3) \\
+ c_k' k^{1/2}(\varepsilon + \varepsilon^4) + c_k'' k^{-1}\varepsilon + \Delta_i\big)\Big]\beta' n^{-3/2}.
\end{aligned}
$$

3. *Estimate of* $I_i, [n/2] \leq i < n - m$

Once again, we use representation (14). Nothing that for any $T > 0$

$$
\begin{aligned}
&\sum_{\|a\|=1} y^a (a!)^{-1} D_a \varphi_T(u) = T(Tu, y)\varphi_T(u), \\
&\sum_{\|a\|=2} y^a (a!)^{-1} D_a \varphi_T(u) = \frac{1}{2}T^2\big((Tu,y)^2 - |y|^2\big)\varphi_T(u), \\
&\sum_{\|a\|=3} y^a (a!)^{-1} D_a \varphi_T(u) = \frac{1}{6}T^3\big((Tu,y)^3 - 3|y|^2(Tu,y)\big)\varphi_T(u) \qquad (20)
\end{aligned}
$$

and expanding $\varphi_{k^{1/2}t_i}(u - y)$ by Taylor's formula up to the terms of the third order, we obtain

$$
\begin{aligned}
I_i' &= -k^{1/2}t_i \int g(k^{-1/2}t_i^{-1}x + z)(x,y)\varphi(x)\,dx\,\overline{H}(dy)\,\overline{P}^{i}_{(n)}(dz) \\
&+ \frac{1}{2}kt_i^2 \int g(k^{-1/2}t_i^{-1}x + z)\big((x,y)^2 - |y|^2\big)\varphi(x)\,dx\,\overline{H}(dy)\,\overline{P}^{i}_{(n)}(dz) \\
&- \frac{1}{2}k^{3/2}t_i^3 \int_0^1 \int g(k^{-1/2}t_i^{-1}x + y\theta + z)(1-\theta)^2\big((x,y)^3 - 3|y|^3(x,y)\big) \\
&\times \varphi(x)\,dx\,\overline{H}(dy)\,\overline{P}^{i}_{(n)}(dz)\,d\theta = L_1 + L_2 + L_3. \qquad (21)
\end{aligned}
$$

The term L_1 coincides with J_1 in (15), and, thus, we have proved (19) for $r^3(A)|J_1|$. To estimate L_2, we represent it in the form

$$L_2 = M_1 + M_2,$$

where M_1 is determined with $\overline{P}^i_{(n)} - N^i_{(kn)^{1/2}}$ instead of $\overline{P}^i_{(n)}$ in the definition of L_2. By (16), we have

$$|M_1| \le ckt_i^2 n^{-1}\Big| \int\limits_{|y|>n^{1/2}} g(k^{-1/2}t_i^{-1}x + z)((x,y)^2 - |y|^2)$$
$$\times \varphi(x)\,dx\, P_1'(dy)\,(\overline{P}^i_{(n)} - N^i_{(kn)^{1/2}})(dz)\Big|.$$

Moreover, since $\mathbf{E}X = \int_0^\infty \big(1-\mathbf{P}(X \le v)\big)\,dv$ for any non-negative random variable X (see e.g. (Feller, 1971)), for any $u \in \mathbf{R}^k$

$$\Big|\int g(u+z)\,(\overline{P}^i_{(n)} - N^i_{(kn)^{1/2}})(dz)\Big| = \Big|\int g(u+s_i^{-1}z)\,(P_i'' - N')(dz)\Big|$$
$$= \Big|\int\limits_0^1 h(u,v)\,dv\Big|, \tag{22}$$

where $h(u,v) = \mathbf{P}\big(g(u+s_i^{-1}s_i'') < v\big) - \mathbf{P}\big(g(u+s_i^{-1}Y') < v\big)$ and Y' is a random variable with probability measure N'. But, if $t = \inf\{t'\colon\ f(t') = v,\ t' \ge 0\}$, it follows easily from the definition of g that $\{z\colon\ g(u+s_i^{-1}z) \le v\} = \Big\{s_i\big((A_1^{t\varepsilon})^c - u\big)\Big\}$, where $A_1 = A$, if $g = g_{1,\varepsilon}$, and $= A^{-\varepsilon}$, if $g = g_{2,\varepsilon}$. Thus, $h(u,v) = (P_i''-N')\big(s_i(A_1^{t\varepsilon} - u)\big)$, and (11), (12) imply

$$r^3(A)\big|h(k^{-1/2}t_i^{-1}x, v)\big| \le c\Big[(k^{-3/2}|x|^3 + \varepsilon^3)\bar{c}(k)\beta' i^{-1/2} + \beta' i^{-1/2} + \Delta_i\Big]. \tag{23}$$

This inequality together with (18) yields

$$r^3(A)|M_1| \le ckt_i^2\big[\bar{c}(k)\beta' i^{-1/2}(1+\varepsilon^3) + \Delta_i\big]\beta' n^{-3/2}.$$

Furthermore, using (20) and integrating by parts, we have

$$M_2 = \frac{1}{2}kt_i^2 \int g\big(k^{-1/2}(t_i^{-1}x + s_i^{-1}z)\big)\big((x,y)^2 - |y|^2\big)\varphi(x)\varphi(z)\,dx\,\overline{H}(dy)\,dz$$
$$= kt_i^2 \int g\big(k^{-1/2}(t_i^{-1}x + s_i^{-1}z)\big)\Big(\sum_{\|a\|=2} y^a (a!)^{-1} D_a\varphi(x)\Big)\varphi(z)\,dx\,\overline{H}(dy)\,dz$$
$$= kt_i^2 \int g(k^{1/2}w)\Big(\sum_{\|a\|=2} y^a (a!)^{-1} D_a\varphi(x)\Big)\varphi(s_i w - s_i t_i^{-1}x)s_i^k\,dx\,\overline{H}(dy)\,dw$$
$$= ks_i^2 \int g(k^{1/2}w)\Big(\sum_{\|a\|=2} y^a (a!)^{-1} D_a\varphi(s_i w - s_i t_i^{-1}x)\Big)\varphi(x)s_i^k\,dx\,\overline{H}(dy)\,dw$$
$$= ks_i^2 \int g\big(k^{-1/2}(t_i^{-1}x + s_i^{-1}z)\big)\big((z,y)^2 - |y|^2\big)\varphi(x)\varphi(z)\,dx\,\overline{H}(dy)\,dz. \tag{24}$$

Now, exactly in the same way as we have deduced (17), and noting that $\int ((z,y)^2 - |y|^2)\varphi(z)\,dz = 1$, we obtain

$$r^3(A)|M_2| \le cks_i^2 n^{-1} \int\limits_{|y|\ge n^{1/2}} \left(k^{-3/2}(|x|^3+|z|^3)+\varepsilon^3\right)$$
$$\times \left((z,y)^2+|y|^2\right)\varphi(x)\varphi(z)\,dx\,P_1'(dy)\,dz;$$

together with (18), this estimate gives $r^3(A)|M_2| \le ck\beta' n^{-3/2}(1+\varepsilon^3)$, and, thus, $r^3(A)|L_2| \le ck\beta' n^{-\frac{3}{2}}(1+\varepsilon^3) + ckt_i^2\left[\bar{c}(k)\beta' n^{-1/2}(1+\varepsilon^3)+\Delta_i\right]\beta' n^{-3/2}$.
The estimation of L_3 goes along the same lines as the estimation of L_2. We represent L_3 as $L_3 = Q_1 + Q_2$, where Q_1 is determined with $\overline{P}^i_{(n)} - N^i_{(kn)^{1/2}}$ instead of $\overline{P}^i_{(n)}$ in the definition of L_3. Using (22), (23), (13), and (18), we have

$$r^3(A)|Q_1| \le ck^{3/2}t_i^3 \int_0^1\!\!\int \left|\int_0^1 r^3(A)h(k^{-1/2}t_i^{-1}x+\theta y, v)\,dv\right|(1-\theta)^2$$
$$\times \left(\left|(x,y)\right|^3 + 3|y|^2\left|(x,y)\right|\right)\varphi(x)\,dx\,\left(\overline{P}_{(n)}+N_{(kn)^{1/2}}\right)(dy)\,d\theta$$
$$\le ck^{3/2}t_i^3\left[\bar{c}(k)\beta' n^{-1/2}(1+\varepsilon^3)+\Delta_i\right]\beta' n^{-3/2}.$$

Furthermore, similarly to (24), we have (see (20))

$$Q_2 = \frac{1}{2}k^{3/2}t_i^3 \int_0^1\!\!\int g(k^{-1/2}(t_i^{-1}x+s_i^{-1}z)+\theta y)(1-\theta)^2$$
$$\times \left(\sum_{\|a\|=3} y^a (a!)^{-1} D_a\varphi(x)\right)\varphi(z)\,dx\,\overline{H}(dy)\,dz\,d\theta$$
$$= \frac{1}{2}k^{3/2}t_i^3 \int_0^1\!\!\int g(k^{-1/2}(t_i^{-1}x+s_i^{-1}z)+\theta y)(1-\theta)^2$$
$$\times \left((z,y)^3 - 3|y|^2(z,y)\right)\varphi(x)\varphi(z)\,dx\,\overline{H}(dy)\,dz.$$

Now, since $\int ((z,y)^3 - 3|y|^2(z,y))\varphi(z)\,dz = 0$, using the same argument that led to (17), we obtain

$$r^3(A)|Q_2| \le ck^{3/2}s_i^3 \int \left(k^{-3/2}(|x|^3+|z|^3)+|y|^3+\varepsilon^3\right)$$
$$\times \left(\left|(z,y)\right|^3 + 3|y|^2\left|(z,y)\right|\right)\varphi(x)\varphi(z)\left(\overline{P}_{(n)}+N_{(kn)^{1/2}}\right)(dy).$$

Together with (13) and (18), this estimate gives

$$r^3(A)|Q_2| \le ck^{3/2}\beta' n^{-3/2}(1+\varepsilon^3).$$

Hence,

$$r^3(A)|L_3| \le ck^{3/2}\beta' n^{-3/2}(1+\varepsilon^3) + ck^{3/2}t_i^3\left[\bar{c}(k)\beta' n^{-1/2}(1+\varepsilon^3) + \Delta_i\right]\beta' n^{-3/2}.$$

It follows from the above estimates it follows that

$$|I_i| \le c(k^{1/2}t_i + k^{3/2})\beta' n^{-3/2}(1+\varepsilon^3) + ck^{3/2}t_i^3\left[\bar{c}(k)\beta' n^{-1/2}(1+\varepsilon^3) + \Delta_i\right]\beta' n^{-3/2}.$$

4. *Estimate of* I_i, $0 \le i < [n/2]$

As in the previous case, we have representation (21). It has been pointed out above that L_1 coincides with J_1 in (15), and we have proved (19) for $r^3(A)|J_1|$. In the same way as we have obtained (17), we deduce

$$r^3(A)|L_2| \le ckt_i^2 n^{-1} \int\limits_{|y|\ge n^{1/2}} \left(|k^{-1/2}t_i^{-1}x + z|^3 + \varepsilon^3\right) \times \left((x,y) + |y|\right)\varphi(x)\,dx\, P_1'(dy)\, \overline{P}^{\,i}_{(n)}(dz).$$

This estimate together with (18) implies $r^3(A)|L_2| \le ckt_i^2\beta' n^{-3/2}(1+\varepsilon^3)$. Similarly to (17), we also have

$$r^3(A)|L_3| \le ck^{3/2}t_i^3 \int \left(k^{-3/2}t_i^{-3}|x|^3 + |y|^3 + |z|^3 + \varepsilon^3\right) \times \left(\left|(x,y)\right|^3 + 3|y|^2\left|(x,y)\right|\right)\varphi(x)\,dx\,\left(\overline{P}_{(n)} + N_{(kn)^{1/2}}\right)(dy)\,\overline{P}^{\,i}_{(n)}(dz),$$

and, using (13) and (18), we obtain $r^3(A)|L_3| \le ck^{3/2}t_i^3\beta' n^{-3/2}(1+\varepsilon^3)$. Thus, since $t_i \le 2^{1/2}$ in the case we are considering, $|I_i| \le ck^{3/2}\beta' n^{-3/2}(1+\varepsilon^3)$. From the corollary to Lemma 1, (7), (8), and the above estimates of I_i for $0 \le i \le n-1$, it follows, since

$$\sum_{n-m\le i\le n-2} t_i \le 2m^{1/2}n^{1/2}, \qquad \sum_{[n/2]\le i<n-m} t_i < 2n,$$

$$\sum_{[n/2]\le i<n-m} t_i^3 \le cm^{-1/2}n^{3/2},$$

that for any $\varepsilon \le 1$

$$\begin{aligned} r^3(A)|(P_n' - N')(A)| \le c\Big\{&\varepsilon^{-1}\left[\bar{c}(k)(\beta' n^{-1/2})^2 + (\Delta_{n-1} + \tilde{c}(k)\varepsilon)\beta' n^{-1/2}\right] \\ &+ k^{1/2}m^{1/2}\Big[1 + \varepsilon^{-2}\big(\bar{c}(k)\beta' n^{-1/2} + \tilde{c}(k)\varepsilon + \max_{n-m\le i\le n-2}\Delta_i\big)\Big]\beta' n^{-1} \\ &+ k^{3/2}m^{-1/2}\left[\bar{c}(k)\beta' n^{-1/2} + \max_{[n/2]\le i<n-m}\Delta_i\right]\beta' \\ &+ k^{3/2}\beta' n^{-1/2}\Big\} + 4\tilde{c}(k)\varepsilon, \end{aligned}$$

where $\tilde{c}(k) = c_k' k^{1/2} + c_k'' k^{-1}$. Now we use induction on n and prove (5) for a given n assuming that it is true for all n less than this given value.
It follows from the inductive hypothesis that

$$\Delta_i \leq 3^{1/2} c'(k) \beta' n^{-1/2}, \quad [n/2] \leq i \leq n-1.$$

Now, take b so large that $c3^{1/2}b^{-1} \leq 1/11$. Then, take a so large that $c3^{1/2}a^{-2}b \leq 1/11$ and $c3^{1/2}a^{-1} \leq 1/11$. Put $\varepsilon = ak\beta' n^{-1/2}$, $m = b^2 k^3 (\beta')^2$, and suppose that $\varepsilon \leq 1$, $m < n - [n/2]$. Substituting these values for ε, m in the right-hand side of (25) and assuming that $c'(k)$ is so large that each summand not containing Δ_i is less than $(1/11)c'(k)\beta' n^{-1/2}$, we obtain (5). This is achieved, if $c'(k)$ is not less than $K'(k^{3/2}(1 + c_k') + c_k'')$, where K' is an appropriate absolute constant.
If $\varepsilon = ak\beta' n^{-1/2} \geq 1$ or $m = b^2 k^3 (\beta')^2 \geq n - [n/2]$ (i.e., $2^{1/2} b k^{3/2} \beta' n^{-1/2} \geq 1$), then, since for any Borel set $E \subset \mathbf{R}^k$

$$r^3(E) \left| P_n'(E) - N'(E) \right| \leq \tilde{c}(1 + k^{1/2} \beta' n^{-1/2})$$

(see, for example, (Sazonov, 1981, p. 33)), we have

$$r^3(A) \left| P_n'(A) - N'(A) \right| \leq \tilde{c} k^{3/2} \beta' n^{-1/2}.$$

The proof of the theorem is complete.

REFERENCES

Bahr, B. von (1967). Multidimensional integral limit theorems. *Ark. Mat.* **7**, 71–88.

Bentkus, V. (1986). On the dependence on the dimension of the estimate of Berry–Esseen. *Lithuanian Math. J.* **26**, 205–210 (in Russian).

Bhattacharya, R. N. and Ranga Rao, R. (1976). *Normal Approximation and Asymptotic Expansions.* Jonh Wiley, New York.

Feller, W. (1971). *An Introduction to Probability Theory and its Applications.* Vol. **2**, Jonh Wiley, New York.

Kuelbs, J. and Kurtz, T. (1974). Berry–Esseen estimates in Hilbert space and application to the law of the iterated logarithm. *Ann. Probab.* **2**, 387–407.

Pinelis, I. F. (1980). Estimates for moments of infinite-dimensional martingales. *Math. Notes* **27**, 953–958 (in Russian).

Rotar, V. I. (1970). A nonuniform estimate for the convergence speed in the multidimensional central theorem. *Theory Probab. Appl.* **15**, 630–648.

Sazonov, V. V. (1972). On a bound for the rate of convergence in the multidimensional Central Limit Theorem. In: *Proceedings of the Sixth Berkeley Symposium Math. Statist. Probab.*, Vol. **2**, California Univ. Press, Berkeley, pp. 563–581.

Sazonov, V. V. (1981). *Normal Approximation — Some Recent Advances.* Springer-Verlag, Berlin.

Sazonov, V.V., Ulyanov, V.V., and Zalesskii, B.A. (1987). On normal approximation in Hilbert space. In: *Probab. Theory and Math. Statist. Proceedings of the Fourth Vilnius Conference*, Vol. **2**, VNU Sci. Press, Utrecht, pp. 561–580.

Senatov, V. V. (1984). An estimate of the Lévy–Prokhorov metric. *Theory Probab. Appl.* **29**, 109–114.

Frontiers in Pure and Applied Probab. 1, pp. 223 - 227
H. Niemi et al. (Eds.)
1993 TVP/VSP

ASYMPTOTIC PROPERTIES OF THE MAXIMUM LIKELIHOOD ESTIMATORS UNDER RANDOM NORMALIZATION FOR A FIRST ORDER AUTOREGVESSIVE MODEL

A. N. SHIRYAEV and V. G. SPOKOINY
Steklov Mathematical Institute, Vavilov Street 42, 117966 Moscow GSP-1, Russia
Institute for Problems of Information Transmission, Ermolov Street 19, 101447 Moscow, Russia

ABSTRACT

Theorem 3 of this paper states that a seguential maximum likelihood estimator under normalization by stopped stochastic Fisher information is asympotically normal for all values of the unknown parameter θ, $-\infty < \theta < \infty$.

1. Let $\mathcal{E} = (\mathbf{R}^\infty, \mathcal{B}(\mathbf{R}^\infty); \mathbf{P}_\theta, \theta \in \mathbf{R})$ be a statistical experiment generated by an autoregressive model of the first order

$$x_k = \theta\, x_{k-1} + \varepsilon_k, \qquad k \geq 1, \tag{1}$$

where $x_0 = 0$; ε_k, $k \geq 1$ is a sequence of independent identically distributed (i.i.d.) random variables with standard normal distribution $\mathcal{N}(0,1)$. The maximum likelihood estimator $\widehat{\theta}_k$ of the unknown parameter θ based on observations $x = (x_1, x_2, \ldots, x_n)$ has the form

$$\widehat{\theta}_n = \left(\sum_{k=1}^{n} x_{k-1}x_k\right)\left(\sum_{k=1}^{n} x_{k-1}^2\right)^{-1}. \tag{2}$$

For every parameter value θ from (1) and (2), we have ($\mathbf{P}_\theta$-a.s.)

$$\widehat{\theta}_n = \theta + \frac{M_n}{\langle M \rangle_n},$$

where

$$M_n = \sum_{k=1}^{n} x_{k-1}\varepsilon_k,$$

$$\langle M\rangle_n = \sum_{k=1}^{n} x_{k-1}^2.$$

It is clear that $M = (M_n, \mathcal{F}_n, \mathbf{P}_\theta)$ with $F_n = \sigma(x_1, \dots, x_n)$ is a square integrable martingale with a quadratic characteristic $\langle M\rangle = (\langle M\rangle_n, \mathcal{F}_{n-1}, \mathbf{P}_\theta), \mathcal{F}_0 = \{\varnothing, \Omega\}$.
Let

$$I_n(\theta) = \mathbf{E}_\theta\Big(- \frac{\partial^2 p_\theta(x_1, \dots, x_n)}{\partial\theta^2}\Big)$$

be the Fisher information in $(x_1, \dots, x_n)$ with respect to θ, where $p_\theta(x_1, \dots, x_n)$ is a normal density of $(x_1, \dots, x_n)$ for a given parameter value θ.
It is easy to see that

$$I_n(\theta) = \mathbf{E}_\theta \langle M\rangle_n.$$

By a straightforward calculation we find

$$I_n(\theta) = \begin{cases} \dfrac{1}{1-\theta^2}\Big(n - \dfrac{1-\theta^{2n}}{1-\theta^2}\Big), & |\theta| \neq 1, \\ \dfrac{n(n-1)}{2}, & |\theta| = 1. \end{cases}$$

Hence, for large n,

$$I_n(\theta) \sim \begin{cases} \dfrac{n}{1-\theta^2}, & |\theta| < 1, \\ \dfrac{n^2}{2}, & |\theta| = 1, \\ \dfrac{\theta^2 n}{(\theta^2 - 1^2}), & |\theta| > 1. \end{cases}$$

By the strong law of large numbers for martingales (Shiryaev, 1984),

$$\frac{M_n}{\langle M\rangle_n} \to 0 \qquad (\mathbf{P}_\theta\text{-a.s. }),$$

because

$$\langle M\rangle n \longrightarrow \infty \qquad (\mathbf{P}_\theta\text{-a.s }), \quad n \to \infty.$$

Therefore, the maximum likelihood estimators $\widehat{\theta}_n$, $n \geq 1$, are strongly consistent:

$$\mathbf{P}_\theta(\lim_n \widehat{\theta}_n = \theta) = 1, \quad -\infty < \theta < \infty.$$

2. The problem of the limit distribution of $\widehat{\theta}_n - \theta$ was considered in many papers (Mann and Wald, 1943; White, 1958; Rao, 1978; Evans and Savin, 1981). The following theorem summarizes "the classical" results.

THEOREM 1. *Let $\varepsilon = (\varepsilon_k)_{k\geq 1}$ be a sequence of i.i.d., $\mathcal{N}(0,1)$-random variables, and let $x_0 = 0$. Then*

$$\lim_{n\to\infty} \mathbf{P}_\theta\Big(\sqrt{I_n(\theta)}(\widehat{\theta}_n - \theta) \leq x\Big) = \begin{cases} \Phi(x), & |\theta| < 1, \\ H_\theta(x), & |\theta| = 1, \\ C(x), & |\theta| > 1, \end{cases}$$

where

$$\Phi(x) = \frac{1}{\sqrt{2\pi}} \int_{-\infty}^{x} e^{-u^2/2} du$$

is the standard normal distribution,

$$C(x) = \frac{1}{\pi} \int_{-\infty}^{x} \frac{du}{1+u^2}$$

is the standard normal distribution, and

$$H_\theta(x) = \mathbf{P}\left(\theta \frac{W^2(1)-1}{2\sqrt{2}\int_0^1 W^2(s)\,ds} \le x\right)$$

for a standard Wiener process $W = (W(s))_{s \le 1}$ *given on a probability space* $(\Omega, \mathcal{F}, \mathbf{P})$.

3. Three different limit distributions lead to search of another kind of normalization of $\widehat{\theta}_n - \theta$ to obtain less number of the limit distribution. In (Mann and Wald, 1943; White, 1958; Rao, 1978; Evans and Savin, 1981) the following result has been proved.

THEOREM 2. *Under the assumption of Theorem* 1

$$\lim_n \mathbf{P}_\theta\left(\sqrt{\langle M\rangle_n}(\widehat{\theta}_n - \theta) \le x\right) = \begin{cases} \Phi(x), & |\theta| \ne 1, \\ H_\theta(\sqrt{2}x), & |\theta| = 1. \end{cases}$$

Of course, it is interesting to find a normalization that limit distribution is the same for all $\theta \in \mathbf{R}$. The main result of the present paper (Theorem 3) asserts that sequential maximum likelihood estimators permit to make the number of limit distributions as small as one. Let $h > 0$ and

$$\tau(h) = \inf\{n \ge 1\colon \langle M\rangle_n \ge h\}$$

is a stopping time. It is well-known that $\tau(h) \to \infty$ ($\mathbf{P}_\theta$-a.s.), $h \to \infty$, for all $\theta \in \mathbf{R}$. Moreover, (Lai and Siegmund, 1983)

$$\lim_{n\to\infty} \frac{\tau(h)}{h} = 1 - \theta^2 \ (\mathbf{P}_\theta\text{-a.s. }), \quad |\theta| < 1;$$

$$\frac{\tau(h)}{\sqrt{h}} \xrightarrow{\mathrm{d}} \inf\left\{t \ge 0 : \int_0^t W^2(s)ds \ge 1\right\}, \ |\theta| = 1$$

(d - distribution) and (Konev and Pergamenshchikov, 1988)

$$\lim \frac{\tau(h)}{log_\theta^2(h)} = 1 \quad (\mathbf{P}_\theta\text{-a.s. }), \quad |\theta| > 1.$$

Denote by

$$\widehat{\theta}_{\tau(h)} = \left(\sum_{k=1}^{\tau(h)} x_{k-1}x_k\right)\left(\sum_{k=1}^{\tau(h)} x_{k-1}^2\right)^{-1}$$

a sequential maximum likelihood estimator based on observations $(x_1, x_2, \ldots, x_{\tau(h)})$.

THEOREM 3. *Under assumptions of Theorem 1, for all $\theta \in \mathbf{R}$,*

$$\sup_x \left| \mathbf{P}_\theta\Big(\sqrt{\langle M\rangle_{\tau(h)}}(\widehat{\theta}_{\tau(h)} - \theta) \le x\Big) - \Phi(x)\right| \longrightarrow 0, \qquad h \to \infty. \tag{3}$$

Furthermore,

$$\sup_{|\theta|\le 1} \sup_x \left| \mathbf{P}_\theta\Big(\sqrt{\langle M\rangle_{\tau(h)}}(\widehat{\theta}_{\tau(h)} - \theta) \le x\Big) - \Phi(x)\right| \longrightarrow 0, \qquad h \to \infty, \tag{4}$$

and, for arbitrary numbers b and B such that $1 < b \le B < \infty$,

$$\sup_{b\le|\theta|\le B} \sup_x \left| \mathbf{P}_\theta\Big(\sqrt{\langle M\rangle_{\tau(h)}}(\widehat{\theta}_{\tau(h)} - \theta) \le x\Big) - \Phi(x)\right| \longrightarrow 0.$$

The proof of (4) can be found in (Lai and Siegmund, 1983), see also (Greenwood and Shiryaev, 1992). For (3) when $|\theta| > 1$ we give here only a sketch of the proof. For a complete proof, see our monograph (Spokoiny and Shiryaev, 1993). We have (for $|\theta| > 1$)

$$\langle M\rangle_n^{1/2}(\widehat{\theta}_n - \theta) = \frac{M_n}{\langle M\rangle_n^{1/2}} = \frac{\xi_n(\theta)}{\eta_n(\theta)},$$

where

$$\xi_n(\theta) = \frac{M_n}{|\theta|^n}, \qquad \eta_n(\theta) = \frac{\langle M\rangle_n^{1/2}}{|\theta|^n}.$$

Define

$$Z(\theta) = \sum_{k-1}^{\infty} \theta^{-k}\varepsilon_k, \quad \xi_n(\theta) = \operatorname{sign} Z(\theta)\sqrt{1-\theta^{-2}} \sum_{k-1}^{n} \theta^{-(n-k)}\varepsilon_k.$$

From (1), we obtain $x_n = \sum_{k=0}^{n-1} \theta^k \varepsilon_{n-k}$. Therefore, we have $\theta^{-n}x_n \sim Z(\theta)$, where $Z(\theta)$ has a normal distribution $\mathcal{N}(0, (1-\theta^{-2})^{-1})$, and

$$\frac{\xi_n(\theta)}{\eta_n(\theta)} \sim \xi_n(\theta).$$

It gives a reason to conclude that the limit distribution for $\langle M\rangle_\tau(h)^{1/2}(\widehat{\theta}_{\tau(h)} - \theta)$ can be reduced to a study of the limit distribution of $\xi_{\tau(h)}(\theta)$, $h \to \infty$. Up to sign, the random variables $\zeta_{\tau(h)}(\theta)$ coincide with

$$S_{\tau(h)}(\theta) = \sqrt{1-\theta^{-2}} \sum_{k=1}^{\tau(h)} \theta^{(\tau(h)-k}\varepsilon_k.$$

Our approach to the analysis of the asymptotic distribution of $S_{\tau(h)}(\theta)$ is as follows. If $\tau(h)$ and $\varepsilon = (\varepsilon_k)_{k\le 1}$ were independent, then the conditional law

$\mathcal{L}(S_{\tau(h)}(\theta)|\tau(h))$ would be normal $\mathcal{N}(0, 1 - \theta^{-2\tau(h)})$ which is "close" to $\mathcal{N}(0, 1)$ for large h. Of course, $\tau(h)$ and $\varepsilon = (\varepsilon_k)_{k\geq 1}$ are dependent. But, we see that the main contribution to $S_{\tau(h)}$ is given by ε_k with numbers k close to $\tau(h)$, and $\tau(h)$ depends mainly on ε_k with small k. So, for large h, the random variables $\tau(h)$ and ε_k, which give "essential" contribution to $S_{\tau(h)}(\theta)$ are "almost" independent. The above consideration show that the limit distribution for $S_{\tau(h)}(\theta)$ and, hence for $\zeta_{\tau(h)}(\theta)$, is a standard normal, $\mathcal{N}(0, 1)$.

REFERENCES

Evans, G.B. and Savin, N.E. (1981). The calculation of the limiting distribution of the least square estimator of the parameter in a random walk model. *Ann. Statist.* **9**, 1114–1118.

Greenwood, P.E. and Shiryaev, A.N (1992). Asymptotic minimaxity of a sequential estimator for a first order autoregressive model. *Stochastics and Stichastics Reports* **38**, 49–65.

Konev, V.V. and Pergamenshchikov, S.M. (1988). On guaranteed parameter estimation for unstable dynamic systems. *Autimation Remote Control* **11**, 130–141 (in Russian).

Lai, T.L. and Siegmund, D. (1983). Fixed accuracy estimation of an autoregressive parameter. *Ann. Statist.* **11**, 478–485.

Mann, H.B. and Wald, A. (1943). On the statistical treatment of linear stochastic difference equations. *Econometrica* **11**, 173–200.

Rao, M.M (1978). Asymptotic distribution of an estimator of the boundary parameter of an unstable process. *Ann. Statist.* **6**, 185–190.

Shiryaev, A.N. (1984). *Probability*. Springer-Verlag, New York.

Spokoiny, V.G. and Shiryaev, A.N. (1993). *Statistics of Stochastic Processes.* I. *Statistical Experiments and Decisions (Asymptotic Theory).* VINITI, Moscow (in Russian).

White, J.S. (1958). The limit distribution of the serial correlation coefficient in the explosive case. *Ann. Math. Statist.* **29**, 1188–1197.

Frontiers in Pure and Applied Probab. 1, pp. 228 – 237
H. Niemi et al. (Eds.)
1993 TVP/VSP

INFINITE SYSTEMS OF DISSIMILAR PARTICLES WITH RANDOM INTERACTIONS

A. V. SKOROKHOD
Institute of Mathematics, Repin Street 4, 252601 Kiev GSP, Ukraine

ABSTRACT

We have considered recently (see (Skorokhod, 1992)) interacting particles which are situated at points of the integer-valued lattice $\mathbf{Z}^d \subset \mathbf{R}^d$ and have constructed, under some restrictions, a Markov process which describes the evolution of the whole system by means of characteristics of the system in finite subsets, where single interactions take place. Now we generalize these results to systems of particles which have another "inner" phase space and change their positions in it at the moment of interactions. The characteristics of interactions depend on the positions of the interacting particles in the inner phase space.

1. We have considered recently in the article (Skorokhod, 1992) interacting randomly particles which are situated at the points of the integer-valued latice $\mathbf{Z}^d \subset \mathbf{R}^d$. It was supposed that all $z \in \mathbf{Z}^d$ contained a finite number of the particles, particles of some finite subset $A \subset \mathbf{Z}^d$ could interact at random moments, and after these interactions the particles changed in a random way for $z \in A$. The main result of (Skorokhod, 1992) is the construction of a Markov process which discribes the evolution of the whole system by means of characteristics of the system in finite subsets, where single interactions take place.

The aim of the present article is the generalizations of the results of (Skorokhod, 1992) to systems of particles which have another "inner" phase space and change their positions in it at the moment of interacting. The characteristic of the interactions depends on the positions of the interacting particles in the inner phase space. Thus, we come to the necessity to consider dissimilar particles which have different positions in the inner phase space.

Since further investigations are connected with the results of (Skorokhod, 1992), we describe them here. At first, we introduce the characteristics of the processes. The state of a system is defined by the set $\{k_z, z \in \mathbf{Z}^d\}$, $k_z \in \mathbf{Z}_+$. Therefore, the phase space of the Markov process which describes the evolution of a system is $X = \mathbf{Z}_+^{\mathbf{Z}^d}$. We denote points of X by $x = \{k_z, z \in \mathbf{Z}\}$ and put $k_z = k_z(x)$ which is the number of particles at the point z in the state x. We suppose that an instantaneous interaction can take place in the balls $B_a(z) = \{u \in \mathbf{Z}^d\colon |u-z| \le a\}$, $z \in \mathbf{Z}^d$, where $a > 0$ is fixed. The interaction in the ball $B_a(z)$ changes the values $k_u(x)$, $u \in B_a(z)$, to some different $\bar{k}_u(x)$, and the whole system changes its state. Probabilistic characteristics of these changes are determined as the intensities of the corresponding events. We recall that the value λ for which the equation

$$\mathbf{P}\{B \text{ appears at } [t, t+h]\} = \lambda h + o(h) \tag{1}$$

holds is the intensity of a random event B which occurs at some random time. We remark that such intensities exist for Markov processes with a denumerable phase space.

Let $X_A = \mathbf{Z}_+^A$, $A \subset \mathbf{Z}^d$, $x' \in X_A$, if $x' = \{k_u(x), u \in A\}$. We define the function $a(z, x', y')\colon \mathbf{Z}^d \times X_{B_a(z)} \times X_{B_a(z)} \to \mathbf{R}_+$ which has the property

$$a(z, x') = \sum_{y' \in B_a(z)} a(z, x', y') < \infty. \tag{2}$$

We the projection x on X_A: $\pi(A)x = \{k_u(x), u \in A\}$, $x \in X$ by denote $\pi(A)x$. The meaning of the function a is the following: $a(z, x', y')$ is the intensity of the interaction in the ball $B_a(z)$, under which the state $x' = \pi(A)x$ transfers into the state y'; $a\Big(z, \pi\big(B_a(z)\big)x\Big)$ is the intensity of the interaction in $B_a(z)$, if the system is at the state x.

A function $f\colon X \to \mathbf{R}$ is called cylindrical, if there are a finite set A_f and a function $g\colon X_A \to \mathbf{R}$ such that $f(x) = g\big(\pi(A)x\big)$. The fact that $a(z, x', y')$ is the intensity of the event mentioned above means that the relation

$$\lim_{t \to 0} t^{-1}\big[\mathbf{E}_x f(x_t) - f(x)\big] = \mathfrak{A} f(x), \tag{3}$$

where

$$\mathfrak{A} f(x) = \sum_{B_a(z) \cap A_f \ne \varnothing} \sum_{y' \in B_a(z)} \Big[f\Big(T\big(x, B_a(z), y'\big)\Big) - f(x)\Big] \times a\Big(z, \pi\big(B_a(z)\big)x, y'\Big), \tag{4}$$

is true for all bounded cylindrical functions f, if x_t is the state at the moment t, $\mathbf{E}_X$ is the expectation under the condition that the initial state of the system is x,

$$T(x, A, y') = \{k_u(x), u \in \mathbf{Z}^d \backslash A(z), k_u(y'), u \in A\}, \quad y' \in X_A.$$

The intensities of transitions from a state to a state do not always define a Markov process even in the denumerable phase spaces. In our case the phase space is more complicated, and we need some special conditions for the existence of a Markov process.

THEOREM 1. *Suppose there exists $b > 0$ such that*

$$a(z, x', 0) + \sum_{y' \in B_a(z)} a(z, x', y')|y'| \le b|x'| \tag{5}$$

for all $z \in \mathbf{Z}^d$, $x' \in B_a(z)$, and put

$$X^\lambda = \Big\{x \in X\colon \sum_{u \in \mathbf{Z}^d} \lambda^{|u|} k_x(u) = |x|_\lambda < \infty\Big\};$$

X^λ is a complete separable space with the metric $|x-y|_\lambda = \sum_{u\in Z^d} \lambda^{|u|} \big|k_u(x) - k_u(y)\big|$. Then there exists a Markov process $x(t)$ with the phase space X^λ for which (3) and (4) are valid.

2. We consider some constructions which were used in (Skorokhod, 1992). We connect a Poisson measure $\mu_z(dt \times d\theta)$ on $\mathbf{R}_+ \times \mathbf{R}_+$ (μ_z is a random measure with independent values, $\mu_z(B)$ is a Poisson random variable with the mean $m(B)$ which is the Lebesgue measure on $\mathbf{R}_+^2$) with each point $z \in \mathbf{Z}^d$. We assume the measures $\{\mu_z, z \in \mathbf{Z}^d\}$ to be independent. The measure μ_z is used for constructions of random events which occur in the ball $B_a(z)$. We denote the probability space generated by the measures $\{\mu_z, z \in \mathbf{Z}^d\}$ by $\{\Omega, \mathcal{F}, \mathbf{P}\}$, $\mathcal{F}_t$ is the σ-algebra generated by the values of μ_z, $z \in \mathbf{Z}^d$, on $[0,t] \times \mathbf{R}_+$. We show how a process with the intensities $a(z,x',y')$ can be constructed, if there is a finite number of particles at the initial time. We put $\Delta_{x,z} = \big[0, a\big(z, \pi(B_a(z))x\big)\big]$, $\mu_z\big([0,t] \times \Delta_{x,z}\big)$ is the Poisson process (in t); here x is the initial state of the system. The set of balls containing particles at the initial time is finite. Let τ_1 be the first moment for which $\sum_z \mu_z\big([0,\tau_1] \times \Delta_{x,z}\big) = 1$ and $z_1 \in \mathbf{Z}^d$ be such that $\mu_{z_1}\big([0,\tau_1] \times \Delta_{x,z_1}\big) = 1$. We say that the system has the first interaction at the time τ_1 in the ball $B_a(z_1)$. We divide the interval $\Delta_{x,z}$ into nonintersecting intervals

$$\Delta_{x,z} = \bigcup_{y' \in B_a(z)} \Delta_{x,z}(y'), \qquad |\Delta_{x,z}(y')| = a\Big(z, \pi\big(B_a(z)\big)x, y'\Big),$$

$$\mu_z\big([0,t] \times \Delta_{x,z}\big) = \sum_{y'} \mu_z\big([0,t] \times \Delta_{x,z}(y')\big).$$

There exists $y' \in B_a(z_1)$ such that $\mu_z\big([0,\tau_1] \times \Delta_{x,z_1}(y')\big) = 1$. We say that after the interaction at the moment τ_1 the state x transfers into the state $T\big(x, B_a(z), y'\big)$. We note that this state has a finite number of particles, and we can continue our construction. There is a possibility that at some finite time the number of particles in the system will grow to infinity. This construction can be used for the determination of the processes $x_R(t)$, for which interactions take place within the ball $B_R(0)$. These processes can be constructed in such a way that $\pi\big(B_R(0)\big)x_{R_1}(t) = \pi\big(B_R(0)\big)x_{R_2}(t)$, $R < R_1 < R_2$, until "perturbations" in the process $x_{R_2}(t)$ begining in the region $\{u\colon |u| > R_1\}$ reach the ball $B_R(0)$. The last fact means that there is a chain of balls $\{B_a(z_i), i = 1, \ldots, n\}$ which connects the sets $\{u\colon |u| > R_1\}$ and $B_R(0)$, and interactions move along this chain towards $B_R(0)$ and reach it before the time t. It is possible that perturbations begining everywhere reach $B_R(0)$ instantaneuosly. These difficulties are the most important ones in the construction of the Markov process.

3. Now we consider a system of different particles. We suppose that there is a complementary phase space (Y, C) which determines the inner state of the particle. Different particles are located at different points of the space Y. The state of a system is determined by the numbers $\{k_u, u \in \mathbf{Z}^d\}$ and the positions of each of k_u particles in the space Y for all $u \in \mathbf{Z}^d$. We denote the kth Cartesian symmetric degree of the space Y (we set $(y_1, \ldots, y_k) = (y_{i_1}, \ldots, y_{i_k})$ if $i_1, i_2, \ldots, i_k$ is a permutation of $1, 2, \ldots, k$) by $\widetilde{Y}^k$. Let $\widetilde{Y} = \bigcup_{k=0}^{\infty} \widetilde{Y}^k$, where $\widetilde{Y}^0$ is a fixed one-point set. We put $k(y) = l$, if $y \in \widetilde{Y}^l$. The interaction changes just as the numbers of particles at points $z \in \mathbf{Z}^d$, so the positions of these particles in the space Y. Markov processes will be used again for a probabilitistic description of these systems. The space $\widetilde{X} = \widetilde{Y}^{\mathbf{Z}^d}$ with the points $x = \{\tilde{y}_u, u \in \mathbf{Z}^d\}$, where $\tilde{y}_u \in \widetilde{Y}$, must be the phase space of the process. Let $\widetilde{C}$ be the σ-algebra in $\widetilde{Y}$ generated by C, $\pi(A)\tilde{x} = \{\tilde{y}_u, u \in A\} \in \widetilde{Y}^A$, $A \subset \mathbf{Z}^d$, $\tilde{x} \in \widetilde{X}$, C^A be the σ-algebra in $\widetilde{Y}^A$ generated by $\widetilde{C}$. We define the function

$$a\Big(z, \pi\big(B_a(z)\big)\tilde{x}, V\Big) \colon \mathbf{Z}^d \times \widetilde{Y}^{B_a(z)} \times \widetilde{C}^{B_a(z)} \to \mathbf{R}_+ \tag{6}$$

which is a measure as a function of $V \in \widetilde{C}^{B_a(z)}$, where $\widetilde{C}^{B_a(z)}$ is measurable in $\big(B_a(z)\big)x$. The function $a(z, x', V)$ is the intensity of events: the interaction takes place in the ball $B_a(z)$, and the state of the system in this ball changes from x' to some state of the set V.
We call a $\widetilde{C}^{\mathbf{Z}^d}$-measurable function $f(\tilde{x})$ cylindrical, if a finite set $A_f \subset \mathbf{Z}^d$ such that $f(\tilde{x}) = f_1\big(\pi(A_f)\tilde{x}\big)$, where f_1 is $\widetilde{C}^A$-measurable, exists. We put

$$\mathfrak{A} f(\tilde{x}) \doteq \sum_{B_a(z) \cap A_f \neq \varnothing} \int_{\widetilde{Y}^{B_a(z)}} \Big[f_1(\tilde{v}) - f_1\big(\pi(A)\tilde{x}\big)\Big] a\Big(z, \pi\big(B_a(z)\big)\tilde{x}, d\tilde{v}\Big), \tag{7}$$

for bounded cylindrical functions $f(\tilde{x}) = f_1\big(\pi(A_f)\tilde{x}\big)$. The aim of this article is to construct a Markov process in the phase space $\widetilde{X}$ with the characteristic operator $\mathfrak{A}$.
First we consider a process with uniformly bounded intensities. We suppose that the condition

$$\sup_{z,\tilde{x}} a\Big(z, \pi\big(B_a(z)\big)\tilde{x}, \widetilde{Y}^{B_a(z)}\Big) \leq \alpha < \infty \tag{8}$$

holds. Let $R > a$. By $\tilde{x}^R(t)$ we denote the Markov process which has the characteristic operator $\widetilde{\mathfrak{A}}$ defined by (7) but we suppose that $a(z, \cdot, \cdot) = 0$, if $z \notin B_{R-a}(0)$. This means that interactions take place in the balls $B_a(z) \subset B_R(0)$ only. It is easy to verify that $\widetilde{\mathfrak{A}}$ is the characteristic operator of a jump Markov process in the phase space $\widetilde{Y}^{B_R(0)}$. We want to extend $\tilde{x}^R(t)$ as $R \to \infty$. For this purpose we consider the process $x^*(t)$ in the phase space X which is the process of transfering the perturbations. Its intensities are determined by the equalities

$$a^*(z,x',y')=\alpha,\quad x'\neq 0,\quad k_u(y')=1,\quad u\in B_a(z);\quad a^*(z,x',y')=0,\quad x'=0.$$

This process was introduced in (Skorokhod, 1992). It was proved there that

$$\mathbf{E}_x \sum_u k_u\big(x^*(t)\big) \leq \Big(\sum_u k_u(x)\Big)e^{ct}, \tag{9}$$

where c depends on α and a only. Thus, the process $x^*(t)$ has a finite number of particles at all moments t, if it has a finite number of particles at the initial time. Let

$$\mathbf{P}^*_{R_1,R_2}(t)=\mathbf{P}_x\Big\{\sum_{u\in B_{R_1}(0)} k_u\big(x^*(t)\big)\geq 1\Big\},\quad k_u(x)=1-I_{B_{R_2}(0)}(u),$$

$\mathbf{P}^*_{R_1,R_2}(t)$ being the probability that perturbations in the process $x^*(t)$ reach the ball $B_{R_1}(0)$ from the region $\{|u|>R_2\}$. Then $\lim_{R_2\to\infty}\mathbf{P}^*_{R_1,R_2}(t)=0$. Now we can construct both the processes $\tilde{x}^R(t)$ and $x^*(t)$ with the same initial condition on the probability space $\{\Omega,\mathcal{F},\mathbf{P}\}$ in such a way that the inequality $k_u\big(x^*(t)\big)>0$, if $k_u\big(\tilde{x}^R(t)\big)>0$, holds for all $u\in\mathbf{Z}^d$. Thus, we have a uniform in R bound for the probability $\mathbf{P}^R_{R_1,R_2}(t)$ of the event that perturbations in the process $x^R(t)$ reach the ball $B_{R_1}(0)$ from the region $\{u\colon |u|>R_2\}$: $\mathbf{P}^R_{R_1,R_2}(t)\leq\mathbf{P}^*_{R_1,R_2}(t)$, if $R_1<R_2<R$.
We can prove the following result using this inequality and repeating, practically, the proof of Theorem 1 in (Skorokhod, 1992).

THEOREM 2. *Let $a(\cdot,\cdot,\cdot)$ satisfy inequality (8), $f(x)$ be a bounded cylindrical function, $\widetilde{X}^1=\{\tilde{x}\colon \sum_u k_u(\tilde{x})<\infty\}$. Then*

$$\lim_{R\to\infty}\mathbf{E}_{\tilde{x}}f\big(\tilde{x}^R(t)\big)=\int f(\tilde{v})\mathbf{P}^1(\tilde{x},t,d\tilde{v}) \tag{10}$$

exists for all $\tilde{x}\in\widetilde{X}^1$, and $\mathbf{P}^1(\cdot,\cdot,\cdot)$ is the transition probability of a homogeneous Markov process with the phase space $\widetilde{X}^1$.

4. Now we consider processes with unbounded intensities. Instead of condition (5), we have now the following one: there exists a positive b such that

$$\int_{\widetilde{Y}^{B_a(z)}} a\Big(z,\pi\big(B_a(z)\big)\tilde{x},d\tilde{v}\Big)\sum_{u\in B_a(z)}k_u(\tilde{v})$$

$$+a\Big(z,\pi\big(B_a(z)\big)\tilde{x},\widetilde{Y}^0\Big)\leq b\sum_{u\in B_a(z)}k_u(\tilde{x}) \tag{11}$$

for all $z \in \mathbf{Z}^d$, $\tilde{x} \in \widetilde{X}$. We put

$$\widetilde{X}^\lambda = \left\{\tilde{x} \in \widetilde{X}\colon \sum_u \lambda^{|u|} k_u(\tilde{x}) < \infty\right\} \quad \text{for } 0 < \lambda \le 1. \tag{12}$$

THEOREM 3. *Let condition (11) hold. Then for all $\lambda \in (0,1]$ there exists a homogeneous Markov process $\tilde{x}^\lambda(t)$ with the phase space $\widetilde{X}^\lambda$ and the characteristic operator $\mathfrak{A}$ which is determined by formula (7). This process is regular in the following sense: for all $u \in \mathbf{Z}^d$ the process $k_u(\tilde{x}^\lambda(t))$ is a jump process.*

Proof. Let

$$a^N\Big(z, \pi(B_a(z))\tilde{x}, V\Big) = I_{\left\{\sup_{u\in B_a(z)} k_u(\tilde{x}) \le N\right\}}$$
$$\times \int_V I_{\left\{\sup_{u\in B_a(z)} k_u(\tilde{v}) \le N\right\}} a\Big(z, \pi(B_a(z))\tilde{x}, d\tilde{v}\Big), \tag{13}$$

where $N \in \mathbf{Z}_+$, $V \in \widetilde{C}^{B_a(z)}$. It is easy to see that $a^N(\cdot)$ satisfies condition (8). Theorem 2 implies the existence of a Markov process $\tilde{x}_N^1(t)$ in $\widetilde{X}^1$ with the characteristic operator $\mathfrak{A}^N$ determined by (7), if we substitute a^N for a there. We set

$$\|\tilde{x}\|_\lambda = \sum_{u\in\mathbf{Z}^d} \lambda^{|u|} k_u(\tilde{x}). \tag{14}$$

There exists an independent on N constant $c(\lambda)$ such that

$$\mathfrak{A}^N \|\tilde{x}\|_\lambda \le c(\lambda)\|\tilde{x}\|_\lambda. \tag{15}$$

Therefore,

$$\mathbf{E}_{\tilde{x}} \|\tilde{x}_N^1(t)\|_\lambda \le \|\tilde{x}\|_\lambda \exp\{c(\lambda)t\}. \tag{16}$$

Now we evaluate the probability that perturbations in the process $\tilde{x}_N^1(t)$ reach the ball $B_R(0)$ from the region $\{u\colon |u| > R_1\}$ during the time t. We denote the last event by $U(R_1, R, t, \tilde{x}_N^1(\cdot))$. At first, we consider the number of interactions in the process $\tilde{x}_N^1(t)$ in the ball $B_{R_1}(0)$ during the time t. Denote it by $\nu_N(R_1, t)$. It is easy to see that

$$\nu_N(R_1, t) = \lim_{h\to 0} \sum_{ih<t} I_{\bigcup_{|u|\le R_1} \left\{k_u\left(\tilde{x}_N^1(ih+h)\right) \ne k_u\left(\tilde{x}_N^1(ih)\right)\right\}}.$$

Therefore,

$$\mathbf{E}_{\tilde{x}}\, \nu_N(R_1, t) \le \lim_{h\to 0} \mathbf{E}_{\tilde{x}} \sum_{ih<t} \sum_{|u|\le R_1} I_{\left\{k_u\left(\tilde{x}_N^1(ih+h)\right) \ne k_u\left(\tilde{x}_N^1(ih)\right)\right\}}$$
$$= \lim_{h\to 0} \sum_{ih<t} \sum_{|u|\le R_1} \mathbf{E}_{\tilde{x}} \Big(a(u, \tilde{x}_N^1(ih))h + o(h)\Big)$$

$$\le b \sum_{|z|\le R_1} \sum_{u\in B_a(z)} \mathbf{E}_{\tilde{x}} \int_0^t k_u(\tilde{x}_N^1(s))\,ds \le c_2\,\mathbf{E}_{\tilde{x}} \sum_{|u|\le R_1+a} \int_0^t k_u(\tilde{x}_N^1(s))\,ds$$

$$\le \frac{c_2}{\lambda^{R_1+|a|}} \int_0^t \mathbf{E}_{\tilde{x}} \sum_u \lambda^{|u|}\,k_u(\tilde{x}_N^1(s))\,ds \le \frac{c_2}{\lambda^{R_1+|a|}} \|\tilde{x}\|_\lambda \frac{\exp\{c(\lambda)t\}-1}{c(\lambda)}.$$

Thus,

$$\mathbf{E}_{\tilde{x}}\,\nu_N(R_1,t) \le \frac{c_1(\lambda)}{\lambda^{R_1}} \exp\{c(\lambda)t\}, \tag{17}$$

where $c_1(\lambda)$ is independent of N and R_1. Now we consider the maximal number of particles which appear at the points of the ball $B_R(0)$ during the time t. We have

$$\mathbf{P}_{\tilde{x}}\Big\{\sup_{|u|\le R_1}\sup_{s\le t} k_u(\tilde{x}_N^1(s)) > l\Big\} \le \sum_{|u|\le R_1} \mathbf{P}_{\tilde{x}}\Big\{\sup_{s\le t} k_u(\tilde{x}_N^1(s)) > l\Big\}. \tag{18}$$

We evaluate one summand in the right-hand side of (18). We suppose that the function $a^*(z,x,V)$ satisfies the condition: $k_u(t)$ does not change after any moment s, if $k_u(s) > l$. Besides, $a^* = a$. We can choose the function a^* in such a way that it satisfies condition (11) with the same b. It is clear that

$$\mathbf{P}_{\tilde{x}}\Big\{\sup_{s\le t} k_u(\tilde{x}_N^1(s)) > l\Big\} \le \mathbf{P}_{\tilde{x}}\Big\{k_u(\tilde{x}_N^{1*}(s)) > l\Big\}$$

$$\le l^{-1}\lambda^{-|u|}\,\mathbf{E}_{\tilde{x}}\,\|\tilde{x}_N^{1*}(t)\| \le l^{-1}\lambda^{-|u|}\|\tilde{x}\|_\lambda e^{c(\lambda)t},$$

where $\tilde{x}_N^{1*}(t)$ is a process with the intensities of interactions a^*. Using (18), we can write

$$\mathbf{P}_{\tilde{x}}\Big\{\sup_{|u|\le R_1}\sup_{s\le t} k_u(\tilde{x}_N^1(s)) > l\Big\} \le c_2 R_1^d l^{-1}\lambda^{-R_1}\|\tilde{x}\|_\lambda e^{c(\lambda)t}. \tag{19}$$

We note that

$$\mathbf{P}_{\tilde{x}}\Big(U(R_1,R,t,\tilde{x}_N^1(\cdot))\bigcap\Big\{\sup_{|u|\le R_1}\sup_{s\le t} k_u(\tilde{x}_N^1(s)) \le l\Big\}\Big)$$

$$\le \mathbf{P}_{\tilde{x}}\Big(U(R_1,R,t,\tilde{x}_l^1(\cdot))\Big) \tag{20}$$

for $N > l$ and $\sup_{\tilde{x},z} a^l\Big(z,\pi\big(B_a(z)\big)\tilde{x}\Big) \le \alpha_l < \infty$. Let $x^*(t)$ be the process defined in §3 under the condition $\alpha = \alpha_l$. Then

$$\mathbf{E}_x \sum_u \lambda^{|u|} k_u(x^*(t)) \le \Big(\sum_u \lambda^{|u|} k_u(\tilde{x})\Big) e^{c_l^*(\lambda)t} \quad \text{for } 0 < \lambda \le 1.$$

We put $k_u(x) = I_{\{|u|>R_1\}}$. Then

$$\mathbf{E}_x \sum_{u\in B_R(0)} \lambda^{|u|} k_u(x^*(t)) \leq \sum_{|u|>R_1} \lambda^{|u|} e^{c_l^*(\lambda)t},$$

$$\mathbf{E}_x \sum_{u\in B_R(0)} k_u(x^*(t)) \leq c_2 \lambda^{R_1-R} e^{c_l^*(\lambda)t}. \tag{21}$$

The left-hand side of (21) is greater than $\mathbf{P}_{\tilde{x}}\Big\{U(R_1, R, t, \tilde{x}_l^1(\cdot))\Big\}$. Thus,

$$\mathbf{P}_{\tilde{x}}\Big(U(R_1, R, t, \tilde{x}_l^1(\cdot))\Big) \leq c_2 \lambda^{R_1-R} \exp\{c_l^*(\lambda)t\}. \tag{22}$$

Let $N_1 > N \geq l$. We can construct a process $\tilde{x}^1_{N_1}(t)$ such that $k_u(\tilde{x}^1_N(s)) = k_u(\tilde{x}^1_{N_1}(s))$ for $u \in B_R(0)$ and $s \leq t$, if $\sup_{u\in B_{R_1}(0)} \sup_{s\leq t} k_u(\tilde{x}^1_{N_1}(s)) \leq N$ and the event $U(R_1, R, t, \tilde{x}^1_{N_1}(\cdot))$ does not occur. Let $f(x)$ be a cylindrical bounded function and $A_f \subset B_R(0)$, $R_1 > R + a$. Then

$$\Big|\mathbf{E}_{\tilde{x}}\, f(\tilde{x}^1_{N_1}(t)) - \mathbf{E}_{\tilde{x}}\, f(\tilde{x}^1_N(t))\Big| \leq \|f\| \Big(\mathbf{P}_{\tilde{x}}\Big\{U(R_1, R, t, \tilde{x}_l^1(\cdot))\Big\}$$

$$+\mathbf{P}_{\tilde{x}}\Big\{\sup_{|u|\leq R_1} \sup_{s\leq t} k_u(\tilde{x}^1_N(s)) > l\Big\}\Big)$$

$$\leq \|f\| \Big(c_2 \lambda_1^{R_1-R} \exp\{c_l^*(\lambda_1)t\} + c_2 l^{-1} \lambda^{-R_1} R_1^d \|\tilde{x}\|_\lambda \exp\{c_l^*(\lambda)t\}\Big), \tag{23}$$

where λ_1, $\lambda \in (0,1]$. We can prove using (23) that

$$\lim_{N\to\infty} f(\tilde{x}^1_N(t)) = T_t^1 f(\tilde{x}) \tag{24}$$

exists for all cylindrical bounded functions and

$$T_t^1 f(\tilde{x}) = \int f(\tilde{v}) \mathbf{P}^1(\tilde{x}, t, d\tilde{v}), \tag{25}$$

where $\mathbf{P}^1(t, \tilde{x}, V)$ is the transition probability for some Markov process in the phase space $\widetilde{X}^1$. We denote this process by $\tilde{x}^1(t)$. We can pass to the limit in inequalities (17), (19), and (20) and conclude that they are true, if $\tilde{x}^1_N(t)$ is replaced by $\tilde{x}^1(t)$. Let f be a cylindrical bounded function, $A_f \subset B_R(0)$, $R_1 > R + a$, $\tau_{R_1,R}$ be the moment when perturbations in the process $\tilde{x}^1(t)$ reach $B_R(0)$ from the region $\{u\colon |u| > R_1\}$: $U(R_1, R, t, x^1(\cdot)) = \{\tau_{R_1,R} < t\}$.
Then $k_u(x^1(t)) = k_u(x^{R_1}(t))$, $t < \tau_{R_1,R}$, $|u| \leq R$, where $\tilde{x}^{R_1}$ is the process defined in §3. Since

$$\lim_{t\to 0} t^{-1}\Big(\mathbf{E}_{\tilde{x}}\, f(\tilde{x}^R(t)) - f(\tilde{x})\Big) = \mathfrak{A} f(\tilde{x})$$

and $\mathbf{E}_{\tilde{x}}\, f(\tilde{x}^R(t)) - \mathbf{E}_{\tilde{x}}\, f(\tilde{x}^1(t)) = o(t)$, $\mathfrak{A}$ is the characteristic operator of the process $x^1(t)$.

Let $\tilde{x} \in \widetilde{X}^{\lambda}$, $0 < \lambda \le 1$, $\tilde{x} = \{\tilde{y}_u, u \in \mathbf{Z}^d\}$. We define $\tilde{x}^R = \{\tilde{y}'_u, u \in \mathbf{Z}^d\}$, $\tilde{y}'_u = y_u$, $|u| \le R$; $k_u(\tilde{x}^R) = 0$, $|u| > R$. Then $\tilde{x}^R \in \widetilde{X}^1$. We can construct a process $\tilde{x}(t,\omega,R)$ on the probability space $\{\Omega, \mathcal{F}, \mathbf{P}\}$ with the initial value $\tilde{x}^R$ and the transition probability $\mathbf{P}^1(\tilde{x}, t, V)$ such that

$$k_u(\tilde{x}(t,\omega,R_1)) = k_u(\tilde{x}(t,\omega,R_2)), \quad |u| \le R, \quad \text{if} \quad t < \tau^{R_2}_{R_1,R},$$

for $R < R_1 < R_2$, the last expression being the moment when perturbations in the process $x(t,\omega,R_2)$ reach $B_R(0)$ from the region $\{u\colon |u| > R_1\}$. We can establish using inequality (20) for the process $\tilde{x}^{R_1}$ that

$$\lim_{R\to\infty} \mathbf{E}_{\tilde{x}^R} f(\tilde{x}^R(t)) = T_t^{\lambda} f(\tilde{x}) = \int \mathbf{P}^{\lambda}(\tilde{x}, t, d\tilde{v}) f(\tilde{v})$$

exists for all cylindrical bounded functions and $\mathbf{P}^{\lambda}$ is the transition probability for some Markov process $\tilde{x}^{\lambda}(t)$ with the phase space $\widetilde{X}^{\lambda}$. Let $\lambda < \lambda_1$, then $\widetilde{X}^{\lambda_1} \subset \widetilde{X}^{\lambda}$, $\widetilde{X}^{\lambda_1}$ is an invariant set for the process $\tilde{x}^{\lambda}(t)$, and its transition probability on this set coinsides with $\mathbf{P}^{\lambda_1}(x,t,V)$. Let $R_1 > R$, $\tau_{R_1,R}(\tilde{x}^{\lambda}(\cdot))$ be the time when perturbations in the process $\tilde{x}^{\lambda}(t)$ reach the ball $B_R(0)$ from the region $\{u\colon |u| > R_1\}$. We can construct the process $x(t,\omega,R_1)$ so that

$$\mathbf{P}_{\tilde{x}}\Big\{k_u(\tilde{x}^{\lambda}(s)) = k_u(\tilde{x}(s,\omega,R_1)), s \le \tau_{R_1,R}(\tilde{x}^{\lambda}(\cdot)), |u| \le R\Big\} = 1.$$

Now we can prove that $\mathfrak{A}$ is the characteristic operator of the process $\tilde{x}^{\lambda}(t)$.

5. Now we consider a system of particles with random interactions in the space $\mathbf{R}^d$. We suppose that all the balls contain finite numbers of particles. The state S of a system is determined by an integer-valued measure $m_S(A)$, $A \in B(\mathbf{R}^d)$, where $B(\mathbf{R}^d)$ is the Borel σ-algebra in $\mathbf{R}^d$. Let $B \in B(\mathbf{R}^d)$ be a bounded set. We denote the subsystem of particles situated in B by S_B and the corresponding measure by m_S^B. Further, let $\widetilde{B} = \bigcup_{k=0}^{\infty} \widetilde{B}^k$. The measure m_S^B is determined by the point $\tilde{x} \in \widetilde{B}$: $m_S^B(B) = k$, if $\tilde{x} \in B^k$,

$$m_S^B(A) = \sum_{i=1}^{k} I_A(x_i), \quad \tilde{x} = (x_1, \dots x_k), \quad k > 0. \tag{26}$$

We use the norm $|\cdot|_1$ in $\mathbf{R}^d$: $\left|(x^1,\dots,x^d)\right|_1 = \sum_{i=1}^d |x^i|$. Put

$$Q_a(z) = \prod_{i=1}^{d} \Big(z^i - \frac{a}{2}, z^i + \frac{a}{2}\Big), \quad z \in \mathbf{Z}^d \subset \mathbf{R}^d, \quad a > 0, \tag{27}$$

where $\prod$ means the Decart product. We suppose that interactions take place in the sets $Q_a(z)$, $a > 0$ being fixed, $a \in \mathbf{Z}$, $z \in \mathbf{Z}^d$. The intensities of interactions are determined by function

$$a(z, \tilde{x}, V)\colon \mathbf{Z}^d \times \widetilde{Q_a(z)} \times B(\widetilde{Q_a(z)}) \to \mathbf{R}_+ \tag{28}$$

which is a finite measure in V and $B(\widetilde{Q_a(z)})$-measurable in $\tilde{x}$.

Let M be the set of locally finite integer-valued measures on $B(\mathbf{R}^d)$, $B(M)$ be the σ-algebra generated by sets $\{m \in M\colon m(A) = k\}$, $A \in B(\mathbf{R}^d)$, $k \in \mathbf{Z}_+$.

A function $f(m) = g(m(A_1), \ldots, m(A_k))$, where $g\colon \mathbf{Z}_+^k \to \mathbf{R}$, $A_I \in B(\mathbf{R}^d)$, is called cylindrical, $A_f = \bigcup_{i=1}^k A_i$.

Our aim is to construct a Markov process in $(M, B(M))$ with the characteristic operator defined on all bounded cylindrical functions $f(m)$ by the equality

$$\mathfrak{A}f(m) = \sum_{Q_a(z)\cap A_f \neq \varnothing} \int\limits_{\widetilde{Q_a(z)}} [f(m^{\mathbf{R}^d - Q_a(z)} + \delta_{\tilde{v}}) - f(m)]\, a(z, \tilde{x}_{Q_a(z)}(m), d\tilde{v}), \quad (29)$$

where $\delta_{\tilde{v}}(A) = \sum_{i=1}^{k} I_A(v_i)$, if $\tilde{v} = (v_1, \ldots, v_k) \in \widetilde{Q}_a^k(z)$, $\delta_{\tilde{v}}(Q_a(z)) = 0$, if $\tilde{v} \in \widetilde{Q}_a^0(z)$, and for $m \in M$ $\tilde{x}_{Q_a(z)}(m) \in \widetilde{Q}_a^0(z)$, if $m(Q_a(z)) = 0$, $\tilde{x}_{Q_a(z)}(m) = (x_1, \ldots, x_k) \in \widetilde{Q}_a^k(z)$, if $m(A \cap Q_a(z)) = \sum_{i=1}^k I_A(x_i)$.

THEOREM 4. *Let $a(z, \tilde{x}, V)$ satisfy the following condition: the inequality*

$$a(\tilde{z}, \tilde{x}, \widetilde{Q}_a^0(z)) + \int\limits_{\widetilde{Q}_a(z)} \delta_{\tilde{v}}(Q_a(z))\, a(z, \tilde{x}, d\tilde{v}) \le b\delta_{\tilde{x}}(\widetilde{Q}_a(z)) \quad (30)$$

holds for some $b > 0$. By M^λ, $0 < \lambda \le 1$, we denote the set of the measures $m \in M$ such that

$$\int\limits_{\mathbf{R}^d} \lambda^u m(du) < \infty. \quad (31)$$

A Markov process with the phase space M and the characteristic operator $\mathfrak{A}$ given by (29) exists for all $0 < \lambda \le 1$.

Proof. Let $Y = Q_{1/2}(0)$. We connect the state $\tilde{x} \in \widetilde{X} = Y^{\mathbf{Z}^d}$ with the state $m_u \in M$ in the following way: $\tilde{x} = \{y_u, u \in \mathbf{Z}^d\}$, where $\tilde{y}_S = (y_1, \ldots, y_k) \in \widetilde{Y}^k$, if

$$m_S^{Q_{1/2}(u)}(A) = \sum_{i=1}^{k} I_{A\cap Q_{1/2}(u)}(u + y_i) \quad (32)$$

and the right-hand side in (32) equals zero for $k = 0$. It is easy to see that we have constructed a one-to-one correspondence between M and $\widetilde{X}$. We can verify that the function a transitioned to the phase space $\widetilde{X}$ satisfies condition (11), and formula (29) is transformed into (7). Thus, the proof of Theorem 4 follows from Theorem 3.

REFERENCES

Skorokhod, A.V. (1992). Infinite systems of randomly interacting particles. *Random Oper. Stoch. Equat.* 1, 127–138.

Frontiers in Pure and Applied Probab. 1, pp. 238 – 255
H. Niemi et al. (Eds.)
1993 TVP/VSP

ON SEQUENTIAL DETECTION OF A SMALL DISORDER

V. G. SPOKOINY
Institute for Problems of Information Transmission, Ermolov Street 19, 101447 Moscow, Russia

ABSTRACT

The problem of quickest (sequential) detection of a change of probability properties of time series is studied under the assumption that the value of change is small in some sense. We establish a lower bound of asymptotically minimax risk of arbitrary decisions (which are stopping times) and describe the structure of asymptotically optimal decisions. These decisions can be treated as modifications of the well-known cumulative sums method and the Girshich–Rubin–Shiryaev method.

INTRODUCTION

Assume we observe a process $Y = (Y_n, n \geq 1)$ with independent variables Y_n and let m be a "disorder" time. This means that before the moment m the observations Y_n have a common distribution F and they are i.i.d. with an another distribution F' after the moment m.

Our goal is to detect the moment of disorder as soon as possible, i.e., we deal with the problem of sequential detection of the moment of a change in the probability characteristics of the observed process.

This problem has been very well studied and there are a lot of papers devoted to it. For the most significant results and ideas we refer to (Page, 1954; Shiryaev, 1961, 1963a,b, 1965; Lorden, 1970; Pollak and Siegmund, 1975, 1985). More detailed historical remarks and bibliography can be found, for example, in (Nikiforov, 1983) and (Kolmogorov, Prokhorov, and Shiryaev, 1990).

The decision of this problem is a stopping time ν with respect to the filtration generated by the observations $(Y_n, n \geq 1)$.

The main characteristics of this decision are the values $\mathbf{E}_\infty \nu$ and $\mathbf{E}_m[\nu - m | \nu \geq m]$, $m \geq 1$, where $\mathbf{E}_\infty \nu$ is the average of ν under the condition the disorder does not occur, and $\mathbf{E}_m$ means expectation if the disorder occurs at a moment $m \geq 1$. The value $\mathbf{E}_\infty \nu$ is treated as the mean time before a false alarm, and the values $\mathbf{E}_m[\nu - m | \nu \geq m]$ are called the conditional mean delay time, $m \geq 1$. Namely, $\nu - m$ is a delay time and $\mathbf{E}_m[\nu - m | \nu \geq m]$ is the mean delay time under the condition that there was no a false alarm before the moment m. The reasons for such a choice of characteristics of a decision ν can be found, e.g., in (Kolmogorov, Prokhorov, and Shiryaev, 1990).

There are various approaches to formulate the optimization problem for this model (see,e.g., (Nikiforov, 1983)). More useful are the Bayessian approach investigated by (Girshich and Rubin, 1952) and, especially, by (Shiryaev, 1961, 1963a,b, 1964), and the minimax one, see (Lorden, 1970; Lai, 1975; Pollak and Siegmund, 1975, 1985), etc. In the sequel we follow the minimax approach. One of the possible variants to formulate such a problem is as follows. For a prescribed number $T > 0$ find a decision ν^* such that $\mathbf{E}_\infty \nu^* \geq T$ and

$$R(\nu^*) = \inf_{\nu:\mathbf{E}_\infty \nu \geq T} R(\nu) \tag{1}$$

with $R(\nu) = \sup_{m\geq 1} \mathbf{E}_m[\nu - m | \nu \geq m]$. The complete solution of this problem is unknown but the approximately optimal solution can be described for large T. The main result in this direction is

$$\inf_{\nu:\mathbf{E}_\infty \nu \geq T} R(\nu) = I^{-1}(F, F') \log T(1 + o_T(1)) \tag{2}$$

where $o_T(1) \to 0$ as $T \to \infty$ and $I(F, F')$ is the Kullback–Leibler information for the distributions F, F',

$$I(F, F') = \int \log \frac{dF'}{dF} dF.$$

There are two most popular algorithms for the problem considered, both are optimal in the sense (1)–(2). The first one was proposed in (Page, 1954) and is called the cumulative sum method (CUSUM). Let the process (S_n) be defined by

$$S_n = \max\left\{0,\ S_{n-1} + \log \frac{dF'}{dF}(Y_n)\right\}.$$

CUSUM corresponds to the decision rule $\tilde{\nu}_T$ of the form $\tilde{\nu}_T = \inf\{n\colon S_n \geq B\}$, where the threshold B is the root of the equation $e^B - B - 1 = T$.
The second method was proposed independently by (Girshich and Rubin, 1952) and (Shiryaev, 1961, 1963a,b) and corresponds to the generalized Bayessian approach with uniform a priori distribution. The related decision ν_T^* is defined as $\nu_T^* = \inf\{n\colon \Phi_n \geq T\}$ with

$$\Phi_n = \frac{1}{n} \sum_{k=1}^{n} \prod_{i=k}^{n} \frac{dF'}{dF}(Y_i).$$

The corresponding theory has been also developed for the analogous problem in continuous time. In this case we observe a step signal in a white noise and it is necessary to detect the unknown jump moment as soon as possible. In this situation even more precise result (than (2)) can be stated, (Shiryaev, 1961, 1963b; Pollak and Siegmund, 1985),

$$\inf_{\nu:\mathbf{E}_\infty \nu \geq T} R(\nu) = 2\sigma^2 \left[\log(2\sigma^{-2}T) - 1 - C + o_T(1)\right], \tag{3}$$

where $C = 0.577...$ is the Euler constant and σ^2 is the signal to noise ratio (see §4 below for more details).

Below we consider the original problem under the assumption that the value of the disorder is small. The first question here is what we treat as the value of a disorder. It follows from the preceeding discussion that a suitable characteristic is the Kulback–Leibler information $I(F, F')$ between F and F'. If F and F' both are normal with parameters (θ, σ^2) and (θ', σ^2) respectively, then $I(F, F') = \sigma^{-2}|\theta - \theta'|^2$. If both F and F' belong to some regular parametric family of distributions, $F = F_\theta$, $F' = F_{\theta'}$, then for θ' closed to θ one has $I(F, F') \approx I(F_\theta)|\theta - \theta'|^2$, where $I(F_\theta)$ is the Fisher information for the prescribed parametric family at the point θ. Thus for such situations the value $|\theta - \theta'|$ can be naturally understood as the value of a disorder. Therefore, the case of a small disorder corresponds to a small value of $|\theta - \theta'|$.

It is natural to try to use the model in continuous time as an approximation (in the sense of statistical experiment theory) of the original model with a small disorder value. This idea was suggested, for example, in (Kolmogorov, Prokhorov, and Shiryaev, 1990) but its accurate realization requires a very difficult technique.

The present paper is devoted to a systematic exposition of the corresponding theory. The main result can be writen in the form

$$\lim_{\varepsilon\to 0} \inf_{\nu:\, \varepsilon^{-2}\mathbf{E}_\infty \nu \geq T} \varepsilon^{-2} R(\nu) = 2\sigma^2 \Big[\log(2\sigma^{-2}T) - 1 - C + o_T(1)\Big], \tag{4}$$

where $\varepsilon = |\theta - \theta'|$ and $\sigma^2 = I^{-1}(F_\theta)$.

1. MODEL

Now we introduce a detailed description of our model. To give an accurate definition we introduce a small parameter ε which has the meaning of the disorder value. As before the observable process $Y = (Y_n,\ n \geq 1)$ is assumed to have a one-dimensional distribution F before the disorder and F^ε after it.

First we explain what is a decision for our model. Our problem is a particular case of the sequential analysis problem. At each moment n we choose a decision (based on the made observations) whether or not the disorder has already occurred. Therefore, the alarm signal ν is a stopping time with respect to the filtration $\mathbf{G} = (\mathcal{G}_n, n \geq 1)$ generated by the process Y, i.e., $\mathcal{G}_n = \sigma(Y_1 ... Y_n)$.

It is clear that in the problem of quickest detection of a small disorder our decision ν can depend on the value of the disorder (which is assumed to be known), $\nu = \nu_\varepsilon$. In other words, we must choose for each $\varepsilon > 0$ a stopping time ν_ε and then we are interested in the asymptotic properties of these decisions as $\varepsilon \to 0$.

To formulate the optimization problem we need a loss function. There are a lot of possibilities to choose a loss function (see, e.g., (Shiryaev, 1990)). But we use a somewhat different approach based on the ideas of the general Le Cam asymptotic theory of statistical experiments.

First we introduce for each $\varepsilon > 0$ the statistical experiment $E^\varepsilon = (\mathbf{R}^\infty, \mathcal{B}(\mathbf{R}^\infty); \mathbf{P}^\varepsilon_m, m \in \bar{\mathbf{N}})$ corresponding to the studied model with

$$X_n \backsim \begin{cases} F, & n < m, \\ F^\varepsilon, & n \geq m. \end{cases}$$

Here the sample space $(\mathbf{R}^\infty, \mathcal{B}(\mathbf{R}^\infty))$ is the space of sequences of observations, $\bar{\mathbf{N}} = \mathbf{N} \cup \{\infty\}$, $\mathbf{N}$ is the set of all integers, $\mathbf{P}_m^\varepsilon$ is the measure on the sample space which describes the distribution of the process X under the assumption (1.1), $m \in \mathbf{N}$, and $\mathbf{P}_\infty^\varepsilon = \mathbf{P}_\infty$ corresponds to the case without disorder (all the observations Y_n are i.i.d. with the distribution F).

Next we study asymptotic properties of the experiments E^ε as $\varepsilon \to 0$. We shall show that after a reparametrization the experiments E^ε converge (in some sense) to a limit experiment $\mathcal{E}$ which corresponds to the observation of a step signal in the white noise. After this we solve an optimization problem for the limit experiment $\mathcal{E}$. If we can find the value of the minimax risk for the limit experiment, then this value is the lower bound of the asymptotically minimax risk for the original model (for the same optimization problem). Moreover, this lower bound is usually attainable and the structure of the asymptotically optimal decisions is similar to the structure of the optimal decision for the limit experiment.

This approach is very useful and fruitful for the classical statistical theory. But now we deal with a nonclassical situation because our decisions are stopping times. Such kind of decisions is a particular case of filtered decisions. Other examples of filtered decisions arise in problems of filtration, sequential estimation, sequential analysis and sequential control, and so on.

Unfortunately classical minimax results (in particular, the famous Le Cam–Hajek theorem) are not applicable to filtered decisions. Therefore, we need some extension of Le Cam's theory to a model with filtration and with filtered decisions. In the next section we give the main definitions and the corresponding results. For the sake of brevity and simplicity, we consider only the sequential analysis problems for which our decisions include stopping times.

In §3 we consider the problem of quickest detection of a disorder in the asymptotic framework and prove the convergence of the corresponding filtered experiments.

In §4 we study the sequential optimization problem for the limit filtered experiment.

At last, §5 contains the main result giving the value of the asymptotical minimax risk and the structure of asymptotically optimal decisions for the original problem.

2. CONVERGENCE OF FILTERED EXPERIMENTS

2.1. In this section we give definitions of filtered experiments, filtered decision problem and define convergence of filtered experiments. Then we state a version of the general Le Cam–Hajek theorem for the case of filtered experiments and filtered decisions. To simplify the exposition we confine ourself to the sequential analysis problem.

Definition 2.1. A *filtered space* is a triple $(\Omega, \mathcal{F}, \mathbf{F})$ including a measurable space $(\Omega, \mathcal{F})$ and a filtration $\mathbf{F} = (\mathcal{F}_t,\ t \in \mathbf{R}_+)$ where $\mathcal{F}_t$ is a σ-subfield in $\mathcal{F}$, $t \in \mathbf{R}_+$, and $\mathcal{F}_s \subseteq \mathcal{F}_t, s \leq t$. The filtration $\mathbf{F}$ is assumed to be right-continues, i.e., $\bigcap_{s>t} \mathcal{F}_s = \mathcal{F}_t$.

Definition 2.2. A *filtered experiment* $\mathcal{E} = (\Omega, \mathcal{F}, \mathbf{F}; \mathcal{P})$ consists of a filtered space $(\Omega, \mathcal{F}, \mathbf{F})$ and a family of probability measures $\mathcal{P} = (\mathbf{P}_\theta, \theta \in \Theta)$, Θ being a *parameter space.*

Below we suppose Θ to be a topological space. Denote by $\mathbf{P}_{\theta,t}$ the restriction of the measure $\mathbf{P}_\theta$ to the σ-field $\mathcal{F}_t$, $\mathbf{P}_{\theta,t} = \mathbf{P}_\theta | \mathcal{F}_t$, $\theta \in \Theta$, $t \in \mathbf{R}_+$. A filtered experiment $\mathcal{E}$ is *locally dominated* if the family $\mathcal{P}$ is locally dominated by some measure $\mathbf{P}$ on $(\Omega, \mathcal{F})$, i.e., $\mathbf{P}_{\theta,t} \ll \mathbf{P}_t$ with $\mathbf{P}_t = \mathbf{P}|\mathcal{F}_t$, $\theta \in \Theta$, $t \in \mathbf{R}_+$. In this situation we can define for each $\theta \in \Theta$ an important characteristic of the filtered experiment $\mathcal{E}$ called a likelihood process. Denote

$$Z(\theta, t) = \frac{d\mathbf{P}_{\theta,t}}{d\mathbf{P}_t}, \ \mathcal{Z}(\theta) = (Z(\theta, t), \ t \in \mathbf{R}_+), \quad \theta \in \Theta.$$

$\mathcal{Z}(\theta)$ is the *likelihood process* for the filtered experiment $\mathcal{E}$ at the point θ with respect to the locally dominating measure $\mathbf{P}$. This process can be chosen to be F-adapted and right-continues.

Now we introduce the concept of filtered decisions and filtered decision problem (for the case of sequential analysis). First we recall general notions of a Markov time and a Markov decision.

A *Markov time* τ for a filtered space $(\Omega, \mathcal{F}, \mathbf{F})$ is a map $\Omega \longmapsto \bar{\mathbf{R}}_+$ ($\bar{\mathbf{R}}_+ = [0, +\infty]$) such that the set $\{\tau \le t\}$ belongs to $\mathcal{F}_t$ for any $t \in \bar{\mathbf{R}}_+$ (with $\mathcal{F}_\infty = \mathcal{F}$). If τ is a Markov time, then $\mathcal{F}_\tau$ is the σ-subfield of $\mathcal{F}$ such that

$$A \in \mathcal{F}_\tau \Leftrightarrow A \cap \{\tau \le t\} \in \mathcal{F}_t, \quad t \in \bar{\mathbf{R}}_+.$$

Let $(D, \mathcal{D})$ be a *decision space* which is assumed to be Polish (i.e., D is a complete separable metric space with Borel σ-field $\mathcal{D}$). A nonrandomized Markov decision ρ from $(\Omega, \mathcal{F})$ to $(D, \mathcal{D})$ is a measurable map $\rho : (\Omega, \mathcal{F}) \longmapsto (D, \mathcal{D})$. To construct a complete decision theory it is necessary to consider randomized and even generalized decisions. But we do not give more details about these constructions and refer to (Le Cam, 1986) and (Strasser, 1985). Below we suppose all the decisions ρ to be nonrandomized. The generalization to the case of randomized decisions is straightforward.

Definition 2.3. A *sequential decision* for a filtered experiment $\mathcal{E} = (\Omega, \mathcal{F}, \mathbf{F}; \mathcal{P})$ is a pair (ρ, τ) including a Markov time τ with respect to the filtration $\mathbf{F}$ and a Markov decision ρ with values in a decision space D which is $\mathcal{F}_\tau$-measurable.

The latter means that the final decision ρ depends only on the observations made before the moment τ.

The set of all sequential decisions for a filtered experiment $\mathcal{E}$ and a decision space D is denoted by $\mathcal{S}(\mathcal{E}, D)$ or simply by $\mathcal{S}$.

Finally we define a sequential decision problem.

Definition 2.4. A *sequential decision problem* (Θ, D, W, C) consists of a parameter space Θ, a decision space D, a loss function $W = W(\theta, r, t)$ and a constraint function $C = C(\theta, r, t)$.

The loss function $W = W(\theta, r, t)$ and the constraint function $C = C(\theta, r, t)$ are assumed to be measurable on the product $\Theta \times D \times \bar{\mathbf{R}}_+$ and, moreover, the loss

function W is assumed to be non-negative. The value $W(\theta, r, t)$ is treated as the loss if we choose the decision $(\rho, \tau) = (r, t)$ and the parameter value is θ.
The constraint function C determines some restrictions on the chosen sequential decisions. More precisely, we must choose our decisions from the subclass $\mathcal{S}(C)$ (or $\mathcal{S}(\mathcal{E}, D; C)$) which is defined as follows

$$\mathcal{S}(C) = \Big\{(\rho, \tau)\colon \mathbf{E}_\theta C(\theta, \rho, \tau) \leq 0, \quad \theta \in \Theta\Big\},$$

where $\mathbf{E}_\theta$ means integration with respect to $\mathbf{P}_\theta$.
The mean value of $W(\theta, \rho, \tau)$ with respect to $\mathbf{P}_\theta$ is called the *risk of the sequential decision* (ρ, τ) at the point θ, $R_\theta(\rho, \tau) = \mathbf{E}_\theta W(\rho, \tau, \theta)$, $\theta \in \Theta$.
A sequential decision $(\rho^*, \tau^*) \in \mathcal{S}(C)$ is optimal for the experiment $\mathcal{E}$ and sequential decision problem (Θ, D, W, C) if it minimizes the corresponding risk function in the minimax sense, i.e.,

$$\sup_{\theta \in \Theta} \mathbf{E}_\theta W(\theta, \rho^*, \tau^*) = \inf_{(\rho, \tau) \in \mathcal{S}(C)} \sup_{\theta \in \Theta} \mathbf{E}_\theta W(\theta, \rho, \tau).$$

Example 2.5. *Sequential estimation.* If the decision space D coincides with the parameter space Θ, then the corresponding decision problem is an estimation problem. In such situations the space Θ is usually a subspace of a Banach space B, the loss function W does not depend on t and it is of the form $W(\theta, r, t) = w(r - \theta)$, where w is a non-negative function on B. The most typical restriction on sequential estimates is $\mathbf{E}_\theta \tau \leq T$, $\theta \in \Theta$, i.e., the mean observation time is limited by some positive number T. Thus, the constraint function is of the form $C(\theta, r, t) = t - T$. The problem is to minimize the risk $\sup_{\theta \in \Theta} \mathbf{E}_\theta w(\rho - \theta)$ under the constraints $\sup_{\theta \in \Theta} \mathbf{E}_\theta(\tau - T) \leq 0$.

Example 2.6. *Sequential testing.* Let the decision space D consist of two points and let Θ_1, Θ_2 be a partition of the parameter space Θ, i.e., $\Theta_1 \cap \Theta_2 = \varnothing$, $\Theta_1 \cup \Theta_2 = \Theta$. For the problem of sequential testing the error probabilities are limited and the mean observation time is minimized. Thus we must consider decisions (ρ, τ) such that

$$\mathbf{P}_\theta(\rho = 2) \leq \alpha, \quad \theta \in \Theta_1, \qquad \mathbf{P}_\theta(\rho = 1) \leq \beta, \quad \theta \in \Theta_2,$$

with prescribed positive numbers α, β (such that $\alpha + \beta < 1$), and the risk function is $\sup_{\theta \in \Theta} \mathbf{E}_\theta \tau$. In other words the loss function is $W(\theta, r, t) = t$ and the constraint function is

$$C(\theta, r, t) = \Big[\mathbf{1}(\rho = 2) - \alpha\Big] \mathbf{1}(\theta \in \Theta_1) + \Big[\mathbf{1}(\rho = 1) - \beta\Big] \mathbf{1}(\theta \in \Theta_2),$$

$\mathbf{1}(A)$ being the indicator function of an event A.

Example 2.7. *Optimal stopping.* Let D consist of one point, i.e., a decision is simply a stopping time. In such situations a loss function and a constraint function depend only on t and θ, $W = W(\theta,t)$, $C = C(\theta,t)$. The considered in this paper problem of quickest detection of a change point belongs to this class. Precise definitions of loss and constraint functions for this case see in §5.

2.2. Later we will need a more general notion of sequential decision problem including a conditional risk.

A *conditional loss function* is a pair (W, S) of measurable functions $W = W(\theta,r,t)$, $S = S(\theta,r,t)$, W being non-negative. The risk of a decision (ρ,τ) for a conditional loss function (W,S) is defined as

$$R(\rho,\tau;W,S) = \sup_{\theta\in\Theta} \mathbf{E}_\theta\Big[W(\theta,\rho,\tau)|S(\theta,r,t) \le 0\Big].$$

As before the goal is to find $(\rho^*,\tau^*) \in \mathcal{S}(C)$ with

$$R(\rho^*,\tau^*;W,S) = \inf_{(\rho,\tau)\in\mathcal{S}(C)} R(\rho,\tau;W,S).$$

2.3. *Asymptotic theory.* Now we turn to the asymptotical aspects of the theory of filtered experiments. We give the definition of λ-convergence of filtered experiments and state a version of Le Cam–Hajek theorem for filtered decisions. Below we consider a set (also called a sequence) of filtered experiments $\mathcal{E}^\varepsilon = (\Omega^\varepsilon, \mathcal{F}^\varepsilon, \mathbf{F}^\varepsilon; \mathbf{P}^\varepsilon_\theta, \theta \in \Theta)$ with a common parameter space Θ, $\varepsilon > 0$. The filtered space $\mathcal{E}^\varepsilon = (\Omega^\varepsilon, \mathcal{F}^\varepsilon, \mathbf{F}^\varepsilon)$ with a filtration $\mathbf{F}^\varepsilon = (\mathcal{F}^\varepsilon_t, t \in \mathbf{R}_+)$ can depend on ε. The experiments $\mathcal{E}^\varepsilon$ are assumed to be locally dominated by measures $\mathbf{P}^\varepsilon, \varepsilon > 0$. Denote

$$Z^\varepsilon(\theta,t) = \frac{d\mathbf{P}^\varepsilon_{\theta,t}}{d\mathbf{P}^\varepsilon_t}, \qquad \theta\in\Theta, \quad t\in\mathbf{R}_+$$

with

$$\mathbf{P}^\varepsilon_{\theta,t} := \mathbf{P}^\varepsilon_\theta|\mathcal{F}^\varepsilon_t, \quad \mathbf{P}^\varepsilon_t = \mathbf{P}^\varepsilon_t|\mathcal{F}^\varepsilon_t, \qquad \theta\in\Theta, \quad t\in\mathbf{R}_+.$$

Let D be a decision space. An *asymptotical sequential decision* for the filtered experiments $\mathcal{E}^\varepsilon, \varepsilon > 0$, is a set (a sequence) of sequential decisions $(\rho_\varepsilon, \tau_\varepsilon) \in \mathcal{S}(\mathcal{E}^\varepsilon, D)$, $\varepsilon > 0$. The class of all asymptotical sequential decisions is denoted by $\mathcal{AS}$.

Let now (Θ, D, W, C) be a sequential decision problem (see Definition 2.4). Denote by $\mathcal{AS}(C)$ the class of such asymptotical decisions that

$$\limsup_{\varepsilon} \sup_{\theta\in\Theta} \mathbf{E}^\varepsilon_\theta C(\theta,\rho_\varepsilon,\tau_\varepsilon) \le 0. \tag{2.1}$$

Note that it does not require for the decisions $(\rho_\varepsilon, \tau_\varepsilon; \varepsilon > 0)$ to belong to the class $\mathcal{S}(\mathcal{E}^\varepsilon, D; C)$ but (2.1) implies that $(\rho_\varepsilon,\tau_\varepsilon) \in \mathcal{S}(\mathcal{E}^\varepsilon, D; C-\delta)$ for any $\delta > 0$ and sufficiently small $\varepsilon > 0$.

The asymptotical risk of an asymptotical sequential decision $(\rho_\varepsilon, \tau_\varepsilon; \varepsilon > 0)$ for experiments $\mathcal{E}^\varepsilon, \varepsilon > 0$, and a sequential decision problem (Θ, D, W, C) is denoted by the formula

$$R\Big((\rho_\varepsilon,\tau_\varepsilon;\varepsilon>0), W\Big) = \limsup_{\varepsilon} \sup_{\theta\in\Theta} \mathbf{E}^\varepsilon_\theta W(\theta,\rho_\varepsilon,\tau_\varepsilon). \tag{2.2}$$

The value of expression $\inf_{\mathcal{AS}(C)} R\Big((\rho_\varepsilon, \tau_\varepsilon; \varepsilon > 0), W\Big)$ is called the *asymptotical minimax risk* for the filtered experiments $\mathcal{E}^\varepsilon, \varepsilon > 0$, and the sequential decision problem (Θ, D, W, C). An asymptotical decision $(\rho_\varepsilon^*, \tau_\varepsilon^*; \varepsilon > 0)$ is *asymptotically optimal* in $\mathcal{AS}(C)$ if it belongs to $\mathcal{AS}(C)$ and

$$R\Big((\rho_\varepsilon^*, \tau_\varepsilon^*; \varepsilon > 0), W\Big) = \inf_{\mathcal{AS}(C)} R\Big((\rho_\varepsilon, \tau_\varepsilon; \varepsilon > 0), W\Big).$$

The standard problems are to find the value of asymptotical minimax risk and to describe the structure of asymptotically optimal decisions.
For the first problem the approach of general Le Cam's theory of statistical experiments is widely used. Let experiments $\mathcal{E}^\varepsilon$ converge (in some sense) to an experiment $\mathcal{E}$. Then the corresponding asymptotical minimax risk for any decision problem (Θ, D, W, C) is not less than the minimax risk for the limit experiment $\mathcal{E}$ and the same decision problem. Only a few weak regularity conditions are needed to prove this fact for the classical nonfiltered theory. Below we claim a similar result for the filtered experiments theory.
Now we define the concept of convergence of the filtered experiments $\mathcal{E}^\varepsilon$ to some limit filtered experiment $\mathcal{E}$. One way is to extend the common and useful definitions of the weak and strong (or Δ-) convergence on the case of filtered experiments. But even in the nonfiltered case an accurate description of the asymptotical theory requires rather difficult methods of functional and convex analysis (see e.g. (Strasser, 1985) or (Le Cam, 1986)). For filtered experiments still more additional problems arise in this way. However the corresponding theory can be constructed and it has recently been done by D.Kramkov (Kramkov, 1992). These results are very significant and directly applicable to the described problem of disorder detection. But we shall not use these results below and our approach is based on another definition of convergence for filtered experiments called λ-convergence. A similar definition for nonfiltered experiments generalizes the well-known asymptotical normality condition from the case of a Gaussian limit experiment to the case of an arbitrary regular limit experiment. A detailed exposition of asymptotical properties of statistical experiments under the condition of λ-convergence (in nonfiltered case) can be found in (Shiryaev and Spokoiny, 1992). The definitions and the main result for the filtered case see below.
First we describe the structure of a limit experiment. Let E be a locally compact set (a state space) and let $\mathbf{X} = \mathbf{D}(E)$ be the space of right-continuous E-valued functions $x = (x_t, t \in \mathbf{R}_+)$ on $\mathbf{R}_+$ with left limits. Let $\mathcal{X}$ be the Borel σ-field on $\mathbf{X}$ for the Skorokhod topology and let $\mathbf{X} = (\mathcal{X}_t, t \in \mathbf{R}_+)$ be the natural filtration in $\mathcal{X}$, i.e., σ-subfield $\mathcal{X}_t$ is generated by the coordinate functions x_s with $s \le t$, $t \in \mathbf{R}_+$. Such a filtered space $(X, \mathcal{X}, \mathbf{X})$ is called a standard filtered space.

Definition 2.8. A filtered experiment $\mathcal{E} = (X, \mathcal{X}\mathbf{X}, X; \mathbf{P}_\theta, \theta \in \Theta)$ is regular if the filtered space $(X, \mathcal{X}, \mathbf{X})$ is standard, $\mathcal{E}$ is locally dominated and there exist a dominating measure $\mathbf{P}$ such that the corresponding likelihood function $Z(\theta, x, t) = d\mathbf{P}_{\theta,t}/d\mathbf{P}_t(x)$ is continuous in $x \in \mathbf{X}$, $\theta \in \Theta$, $t \in \mathbf{R}_+$.
To state the central definitions of λ-convergence we need an additional notion. Let $(X, \mathcal{X}, \mathbf{X})$ be a standard filtered space and let $(\Omega, \mathcal{F}, \mathbf{F})$ be an arbitrary filtered space. A map $\lambda : \Omega \longmapsto \mathbf{X}$ is $F/\mathbf{X}$-adapted if for any $t \in \mathbf{R}_+$ the set $\lambda^{-1}(A)$

belongs $\mathcal{F}_t$ for $A \in \mathcal{X}_t$. Such a map can be treated as a $\mathbf{F}$-adapted E-valued random process (because $\mathbf{X} = \mathbf{D}(E)$), $\lambda(\omega) = (\lambda_t(\omega), t \in \mathbf{R}_+), \omega \in \Omega$.

Definition 2.9. Let $\mathcal{E}^\varepsilon = (\Omega^\varepsilon, \mathcal{F}^\varepsilon, \mathbf{F}^\varepsilon; \mathbf{P}^\varepsilon_\theta, \theta \in \Theta)$ be a filtered experiment locally dominated by a measure $\mathbf{P}^\varepsilon, \varepsilon > 0$, and let $\mathcal{E} = (X, \mathcal{X}, \mathbf{X}; \mathbf{P}_\theta, \theta \in \Theta)$ be a regular experiment. We say that the λ-convergence of $\mathcal{E}^\varepsilon$ to $\mathcal{E}$ is fulfilled as $\varepsilon \to 0$ if there exists for each $\varepsilon > 0$ a $F^\varepsilon/\mathbf{X}$-adapted map $\lambda^\varepsilon : \Omega^\varepsilon \longmapsto \mathbf{X}$ such that the following conditions hold

(i) For any $\theta \in \Theta$, $\delta > 0$, and for each $T > 0$

$$\lim_\varepsilon \mathbf{P}^\varepsilon_T\Big(|Z^\varepsilon(\theta,t) - Z(\theta,\lambda^\varepsilon,t)| > \delta\Big) = 0,$$

where $\lim_\varepsilon$ means $\lim_{\varepsilon\to 0}$, $\mathbf{P}^\varepsilon_T = \mathbf{P}^\varepsilon|\mathcal{F}^\varepsilon_T$,

$$Z^\varepsilon(\theta,t) = d\mathbf{P}^\varepsilon_{\theta,t}/d\mathbf{P}^\varepsilon_t, \quad Z(\theta,x,t) = d\mathbf{P}_{\theta,t}/d\mathbf{P}_t(x), \quad \theta \in \Theta, \quad t \in \mathbf{R}_+, \quad x \in \mathbf{X}.$$

(ii) For each $T > 0$ the distributions of the random processes $\lambda^\varepsilon = (\lambda^\varepsilon_t,\ t \in [0,T])$ corresponding to the measure $\mathbf{P}^\varepsilon_T = \mathbf{P}^\varepsilon|\mathcal{F}^\varepsilon_T$ weakly converge to the measure $\mathbf{P}_T = \mathbf{P}|\mathcal{F}_T$.

(iii) The filtration $\mathbf{F}^\varepsilon$ for any $\varepsilon > 0$ is generated by the process λ^ε, i.e.,

$$\mathcal{F}^\varepsilon_t = \sigma\{\lambda^\varepsilon_s,\ s \le t\}.$$

Remark 2.10. The last condition (iii) is very significant for us and it is essentially used in the proof of the main result. This condition is not necessary and it can be weaken but we can not drop it.

THEOREM 2.11. *Let the λ-convergence of filtered experiments $\mathcal{E}^\varepsilon$ to a regular experiment $\mathcal{E}$ hold. Let (Θ, D, W, C) be such a sequential decision problem that the loss function W and the constraint function C satisfy the following condition (WC): For each $\theta \in \Theta$ the function $W(\theta,r,t)$ and $C(\theta,r,t)$ are continuous in $t \in \mathbf{R}_+$ and $r \in D$, and*

$$\lim_{t\to\infty} W(\theta,r,t) + \big|C(\theta,r,t)\big| = +\infty, \qquad r \in D, \qquad \theta \in \Theta.$$

Then for any asymptotical decision $(\rho^\varepsilon, \tau^\varepsilon; \varepsilon > 0) \in \mathcal{AS}(C)$

$$R\Big((\rho_\varepsilon, \tau_\varepsilon; \varepsilon > 0), W\Big) = \liminf_\varepsilon \sup_{\theta\in\Theta} \mathbf{E}^\varepsilon_\theta W(\theta, \rho_\varepsilon, \tau_\varepsilon) \ge \mathcal{R}(\mathcal{E}, D, W, C)$$

with

$$\mathcal{R}(\mathcal{E}, D, W, C) = \inf_{(\rho,\tau)\in\mathcal{S}(C)} \sup_{\theta\in\Theta} \mathbf{E}_\theta W(\rho, \tau, \theta).$$

The result of this theorem is immediately extended to the case of conditional loss functions (W, S) (see above).

THEOREM 2.12. *Let $\mathcal{E}^\varepsilon \xrightarrow{\lambda_f} \mathcal{E}$ for a regular experiment $\mathcal{E}$, let $(\Theta, D, (W, S), C)$ be a decision problem with a conditional loss function (W, S), all the functions W, S, C be lower semicontinuous in $r \in D$ and $t \in \mathbf{R}_+$ for each $\theta \in \Theta$ and*

$$\lim_{t\to\infty} \mathbf{1}(S(\theta,r,t) \le 0)\Big[W(\theta,r,t) + \big|C(\theta,r,t)\big|\Big] = +\infty, \quad r \in D, \quad \theta \in \Theta.$$

Then

$$\liminf_{\varepsilon} \sup_{\theta\in\Theta} \mathbf{E}^\varepsilon_\theta\Big[W(\theta,\rho_\varepsilon,\tau_\varepsilon)|S(\theta,\rho_\varepsilon,\tau_\varepsilon) \le 0\Big]$$

$$\ge \inf_{(\rho,\tau)\in\mathcal{S}(C)} \sup_{\theta\in\Theta} \mathbf{E}_\theta\Big[W(\theta,\rho,\tau)|S(\theta,\rho,\tau) \le 0\Big].$$

The proofs of Theorems 2.11, 2.12 are beyond the scope of this paper.

3. CONVERGENCE OF FILTERED EXPERIMENTS FOR THE DISORDER PROBLEM

Now we come back to the problem of sequential detection of a small disorder. First we change the parameter space and define a sequence of filtered experiments $\mathcal{E}^\varepsilon$ with the common parameter set $\Theta = \bar{\mathbf{R}}_+$. Then we show that these filtered experiments $\mathcal{E}^\varepsilon$ converge to the filtered experiment $\mathcal{E}$ corresponding to the usual disorder problem in continuous time.

Let $(\mathbf{R}^\infty, \mathcal{B}(\mathbf{R}^\infty))$ be a sample space and let $\mathbf{P}^\varepsilon_m$ be the joint distribution of observations $X_1, ..., X_n, ...$ under the assumption that the disorder value is ε and the disorder time is m (i.e., the observation $X_1, ..., X_{m-1}$ are i.i.d. with the density function $f(x)$ and $X_m, X_{m+1}, ...$ are i.i.d. with density function $f_\varepsilon(x)$.

Introduce a new parameter $\theta \in \mathbf{R}_+$ putting $\theta = \varepsilon^2 m$ and define measures $\mathbf{P}^\varepsilon_\theta = \mathbf{P}^\varepsilon_m$ with $m = [\varepsilon^{-2}\theta]$, $[a]$ being the integer part of a. Also let $\mathcal{F}^\varepsilon_t$ be the σ-subfield in $\mathcal{B}(\mathbf{R}^\infty)$ generated by $X_1, ..., X_{[\varepsilon^{-2}t]}$, and let $\mathbf{F}^\varepsilon = (\mathcal{F}^\varepsilon_t, t \in \mathbf{R}_+)$. At last, let $\mathcal{F}^\varepsilon_\infty = \mathcal{B}(\mathbf{R}^\infty)$ and $\mathbf{P}^\varepsilon_\infty$ correspond to the case without any disorder.

Below we establish λ-convergence of the filtered experiments $\mathcal{E}^\varepsilon = (\mathbf{R}^\infty, \mathcal{B}(\mathbf{R}^\infty), \mathbf{F}^\varepsilon; \mathbf{P}^\varepsilon_\theta, \theta \in \bar{\mathbf{R}}_+)$ to the filtered experiment $\mathcal{E}$ which is defined as follows.

Let $\mathbf{D}$ be the space of CADLAG (right-continuous and with left limits) functions on $\mathbf{R}_+$ with the Skorokhod topology (see (Billingsley, 1968)), let $\mathcal{B} = \mathcal{B}(\mathbf{D})$ be the Borel σ-field in $\mathbf{D}$, and let $B = (\mathcal{B}_t, t \in \mathbf{R}_+)$ be the natural filtration in $\mathcal{B}$. The filtered experiment $\mathcal{E} = (\mathbf{D}, \mathcal{B}(\mathbf{D}), \mathbf{B}; \mathbf{P}_\theta, \theta \in \bar{\mathbf{R}}_+)$ corresponds to the observable process $X = (X_t, t \in \mathbf{R}_+)$ which satisfies the stochastic differential equation

$$dX_t = \mathbf{1}(t \ge \theta)dt + \sigma dW_t, \tag{3.1}$$

where $W = (W_t,\ t \in \mathbf{R}_+)$ is a standard Wiener process, σ is a known positive constant, and θ is the estimated parameter. Thus the measure $\mathbf{P}_\theta$ coincides with the distribution in $(\mathbf{D}, \mathcal{B}(\mathbf{D}))$ of the process X satisfying (3.1), $\theta \in \mathbf{R}_+$, and $\mathbf{P}_\infty$ is the Wiener measure.

Now we state the main result of this section.

THEOREM 3.1. *Let the following condition hold*

(F) *The density function $f(y)$ is positive and there exists a function $g(y)$ such that*

$$I = \int_{-\infty}^{\infty} g^2(y) f(y) dy < \infty,$$

and

$$\lim_{\varepsilon} \int_{-\infty}^{\infty} \left[\sqrt{\frac{f_\varepsilon(y)}{f(y)}} - 1 - \frac{1}{2}\varepsilon g(y) \right]^2 f(y) dy = 0.$$

Then the filtered experiments $\mathcal{E}^\varepsilon$ λ-converge to the filtered experiment $\mathcal{E}$ with $\sigma^2 = I$. The corresponding maps $\lambda^\varepsilon : R^\infty \longmapsto \mathbf{D}$, which are the processes $X^\varepsilon = (X_t^\varepsilon,\ t \in \mathbf{R}_+)$, can be defined as follows

$$X_t^\varepsilon = \varepsilon \sum_{k=1}^{[\varepsilon^{-2} t]} \frac{g(Y_k)}{f(Y_k)}, \quad t \in \mathbf{R}_+.$$

Remark 3.2. Condition (F) is standard for the asymptotical statistical theory (see, e.g., (Ibragimov and Khasminskii, 1981)). In the typical situation one has

$$g(x) = \left. \frac{df_\varepsilon(x)}{d\varepsilon} \right|_{\varepsilon=0}.$$

In any case we obtain under condition (F)

$$\int_{-\infty}^{\infty} g(y) f(y) dy = 0. \tag{3.2}$$

Proof of Theorem 3.1. To begin with we consider the likelihood process for the experiments $\mathcal{E}$ and $\mathcal{E}^\varepsilon$, $\varepsilon > 0$. The experiment $\mathcal{E}$ is locally dominated by the measure $\mathbf{P}_\infty$ which corresponds to the case without a disorder. It follows from Girsanov's theorem that, for $t \geq \theta$,

$$Z(\theta, t) = \frac{d\mathbf{P}_{\theta,t}}{d\mathbf{P}_t} = \exp\left\{ \sigma^{-1} \int_\theta^t dX_t - \frac{1}{2}\sigma^{-2} \int_\theta^t dt \right\}$$

$$= \exp\left\{ \sigma^{-1}(X_t - X_\theta) - \frac{\sigma^{-2}(t-\theta)}{2} \right\}$$

and $Z_{\theta,t} = 1$ for $t < \theta$ (before the moment θ the process X coincides with σW). Similarly the experiment $\mathcal{E}^\varepsilon$, $\varepsilon > 0$, is locally dominated by the measure $\mathbf{P}_\infty^\varepsilon$,

$$Z(\theta, t) = \frac{d\mathbf{P}_{\theta,t}^\varepsilon}{d\mathbf{P}_t^\varepsilon} = \prod_{k=[\varepsilon^{-2}\theta]+1}^{[\varepsilon^{-2} t]} \frac{f_\varepsilon(Y_k)}{f(Y_k)}, \quad t \geq \theta,$$

and $Z_{\theta,t}^\varepsilon = 1$ for $t < \theta$.

Define the process $X^\varepsilon = (X_t^\varepsilon,\ t \in \mathbf{R}_+)$ with

$$X_t^\varepsilon = \varepsilon \prod_{k=1}^{[\varepsilon^{-2}t]} \frac{g(Y_k)}{f(Y_k)}.$$

To prove the λ-convergence of $\mathcal{E}^\varepsilon$ to $\mathcal{E}$ we should verify the following conditions:
(i) For each $\theta \in \mathbf{R}_+$ and $t > \theta$

$$Z_{\theta,t}^\varepsilon - \exp\left\{\sigma^{-1}(X_t - X_\theta) - \frac{\sigma^{-2}(t-\theta)}{2}\right\} \xrightarrow{\mathbf{P}_\infty^\varepsilon} 0, \quad \varepsilon \to 0.$$

(ii) $\mathcal{L}(X^\varepsilon | \mathbf{P}_{\infty,T}^\varepsilon) \xrightarrow{\bar{w}} \mathbf{P}_{\infty,T}$, where $T > 0$, $\mathbf{P}_{\infty,T} = \mathbf{P}_\infty | \mathcal{F}_T$ is the Wiener measure on $\mathbf{D}[0,T]$.
Asymptotical expansion (i) is well known for the case $t = 0$ (see, e.g., (Strasser, 1985) or (Ibragimov and Khasminskii, 1981)) and the case of an arbitrary $t > 0$ is proved in the same way. To explain the weak convergence (ii) we note that the process $X^\varepsilon = (X_t^\varepsilon,\ t \in \mathbf{R}_+)$ is a square-integrable martingale with respect to the measure $\mathbf{P}_\infty^\varepsilon$. In fact, the variables $g(Y_k)/f(Y_k)$ are i.i.d. under $\mathbf{P}_\infty^\varepsilon$ and from (F) and (3.2)

$$\mathbf{E}_\infty^\varepsilon \frac{g(Y_k)}{f(Y_k)} = \int_{-\infty}^{\infty} \frac{g(y)}{f(y)} f(y)dy = \int_{-\infty}^{\infty} g(y)dy = 0,$$

$$\mathbf{E}_\infty^\varepsilon \left[\frac{g(Y_k)}{f(Y_k)}\right]^2 = \int_{-\infty}^{\infty} \frac{g^2(y)}{f^2(y)} f(y)dy = I = \sigma^{-2} < \infty.$$

Thus (ii) follows from the usual invariance principle (see (Billingsley, 1968)).

4. OPTIMIZATION PROBLEM FOR THE LIMIT EXPERIMENT

In this section we explore the standard approach to the problem of detection of a disorder for model (3.1). Let τ be any decision which is a stopping time. The standard minimax optimization problem for this problem is formulated in the following way. Let $T > 0$ and let $\mathcal{S}_T$ be the class of all stopping times τ with $\mathbf{E}_\infty \tau \geq T$,

$$\mathcal{S}_T = \{\tau\colon\ \mathbf{E}_\infty \tau \geq T\}.$$

The goal is to find for a given T a stopping time τ^* from $\mathcal{S}_T$ minimizing the risk $R(\tau)$ over $\mathcal{S}_T$, $R(\tau^*) = \inf_{\tau \in \mathcal{S}_T} R(\tau)$ with $R(\tau) = \sup_{\theta \in \mathbf{R}_+} \mathbf{E}_\theta[\tau - \theta | \tau \geq \theta]$.
The structure of the optimal solutions τ^* for this problem is unknown but approximatly optimal decisions τ_T^* can be found for large T. To this purpose define the process Φ_t, $t \geq 0$, with $\Phi_0 = 0$ and

$$\Phi_t = \int_0^t \exp\left\{\sigma^{-1}(X_t - X_\theta) - \sigma^{-2}(t-\theta)/2\right\} d\theta. \tag{4.1}$$

THEOREM 4.1. *Let τ_T^* be the stopping time defined by*

$$\tau_T^* = \inf\{t:\ \Phi_t \geq T\}. \tag{4.2}$$

Then $\tau_T^* \in \mathcal{S}_T$, for each $T > 0$,

$$R(\tau_T^*) = 2\sigma^2 \log(\sigma^{-2}T/2) - 1 - C + \varepsilon(T), \tag{4.3}$$

C=0.577... being the Euler constant, $\varepsilon(T) \to 0$ as $T \to \infty$, and

$$\inf_{\tau \in \mathcal{S}_T} R(\tau) = R(\tau_T^*) + \varepsilon_1(T)$$

with $\varepsilon_1(T) \to 0$ as $T \to \infty$.

Instead of the decisions τ_T^* from (4.2) the following decisions $\tilde{\tau}_T$ are useful

$$\tilde{\tau}_T = \inf\{t : X_t - \frac{t}{2} - \min_{s \leq t}(X_s - \frac{s}{2}) \geq B\}, \tag{4.4}$$

where B is the root of the equation $e^B - B - 1 = T$. Pollak and Siegmund (Pollak and Siegmund, 1985) established that

$$R(\tilde{\tau}_T) = 2\sigma^2 \log(\sigma^{-2}T/2) - 3/2 + \varepsilon_2(T) \tag{4.5}$$

with $\varepsilon_2(T) \to 0$ as $T \to \infty$. Thus the decisions $\tilde{\tau}_T$ are not optimal even in the asymptotical sense but their risks $R(\tilde{\tau}_T)$ are closed to $R(\tau_T^*)$.

To follow §2 we redefine the considered sequential decision problem in terms of the conditional loss function (W, S) and a constraint function C. Put

$$W(\theta, t) = (t - \theta)^+ = \max\{t - \theta, 0\}, \quad t, \theta \in \overline{\mathbf{R}}_+, \tag{4.6}$$

$$S(\theta, t) = \theta - t, \quad t, \theta \in \overline{\mathbf{R}}_+, \tag{4.7}$$

$$C(\theta, t) = \begin{cases} T - t, & \theta = \infty \\ 0, & \theta \in \mathbf{R}_+ \end{cases}. \tag{4.8}$$

Now we can define the optimization problem in the form

$$\tau^* = \arg \inf_{\tau \in \mathcal{S}_T} R(\tau), \qquad \mathcal{S}_T = \{\tau : \mathbf{E}_\infty C(\theta, \tau) \leq 0,\ \theta \in \Theta\}$$

$$R(\tau) = \sup_{\theta \in \Theta} \mathbf{E}_\theta \Big[W(\theta, \tau) | S(\theta, \tau) \leq 0\Big]. \tag{4.9}$$

with $\Theta = \overline{\mathbf{R}}_+$. This representation is needed for us in the next section to apply Theorem 3.12 for the original problem.

5. ASYMPTOTICALLY OPTIMAL DECISIONS FOR THE PROBLEM OF DETECTION OF A SMALL DISORDER

5.1. In §3 we reformulated the original problem in terms of the new continuous parameter θ and showed that the corresponding filtered experiments $\mathcal{E}^\varepsilon$, $\varepsilon > 0$, converge to the filtered experiment $\mathcal{E}$ considered in §4. Now we can use the general results of §2. For this purpose we consider the same optimization problem for the experiments $\mathcal{E}$ as in §4 for $\mathcal{E}$. Theorem 2.12 allows us to establish lower bounds of the asymptotically minimax risk for the experiments $\mathcal{E}^\varepsilon$. Then we state the attainability of this lower bound end describe the structure of asymptotically optimal decisions.

Let the filtered experiments $\mathcal{E}^\varepsilon = (\mathbf{R}^\infty, \mathcal{B}(\mathbf{R}^\infty), \mathbf{F}^\varepsilon; \mathbf{P}^\varepsilon_\theta, \theta \in \Theta,), \varepsilon > 0$, be those introduced in §3 (recall that $\Theta = \overline{\mathbf{R}}_+$, the measure $\mathbf{P}^\varepsilon_\theta$ for $\theta \in \mathbf{R}_+$ corresponds to the case of a disorder at the moment $[\varepsilon^{-2}\theta]$ and the measure $\mathbf{P}^\varepsilon_\infty$ corresponds to the case without a disorder). Now we consider the asymptotical sequential decision problem $(\Theta, (W, S), C)$ for the experiments $\mathcal{E}^\varepsilon$, $\varepsilon > 0$, with the a conditional loss function (W, S) and the constraint function C defined in (4.6)–(4.8).

Let $T > 0$ and $\mathcal{AS}_T$ be the class of all asymptotical sequential decisions $(\tau_\varepsilon, \varepsilon > 0)$ such that

$$\lim_\varepsilon \mathbf{E}^\varepsilon_\infty \tau_\varepsilon \geq T. \tag{5.1}$$

An asymptotical decision $(\tau^*_\varepsilon, \varepsilon > 0)$ from $\mathcal{AI}_T$ is asymptotically optimal if

$$\lim_\varepsilon \sup_{\theta \in \mathbf{R}_+} \mathbf{E}^\varepsilon_\theta \Big[\tau^*_\varepsilon - \theta | \tau^*_\varepsilon \geq \theta\Big] = \inf_{(\tau_\varepsilon, \varepsilon > 0) \in \mathcal{AS}_T} \sup_{\theta \in \mathbf{R}_+} \mathbf{E}^\varepsilon_\theta \Big[\tau_\varepsilon - \theta | \tau_\varepsilon \geq \theta\Big]. \tag{5.2}$$

In terms of the decision problem $(\Theta, (W, S), C)$ expressions (5.1) and (5.2) can be rewritten in the form

$$\mathcal{AS}_T = \Big\{(\tau_\varepsilon, \varepsilon > 0) : \limsup_\varepsilon \sup_{\theta \in \Theta} \mathbf{E}^\varepsilon_\theta C(\theta, \tau_\varepsilon) \leq 0\Big\},$$

$$\limsup_\varepsilon \sup_{\theta \in \Theta} \mathbf{E}^\varepsilon_\theta \Big[W(\theta, \tau^*_\varepsilon) | S(\theta, \tau^*_\varepsilon) \leq 0\Big] \geq \inf_{\mathcal{AS}_T} \limsup_\varepsilon \sup_{\theta \in \Theta} \mathbf{E}^\varepsilon_\theta \Big[W(\theta, \tau_\varepsilon) | S(\theta, \tau_\varepsilon) \leq 0\Big].$$

Thus we can apply Theorems 4.1 and 3.12 which imply the following result.

THEOREM 5.1. *Let condition* (F) *of Theorem* 3.1 *hold. Then for any asymptotical decision* $(\tau_\varepsilon, \varepsilon > 0) \in \mathcal{AS}_T$

$$\lim_\varepsilon \sup_{\theta \in \mathbf{R}_+} \mathbf{E}^\varepsilon_\theta \Big[\tau_\varepsilon - \theta | \tau_\varepsilon \geq \theta\Big] \geq 2\sigma^2 \log(\sigma^{-2} T/2) - 1 - C + \varepsilon(T)$$

with $\varepsilon(T) \to 0$ *as* $T \to 0$, C *is the Euler constant.*

Proof. It is necessary to verify only that all the conditions of Theorem 3.12 are fullfilled, which is obvious.

5.2. Now we describe the structure of the asymptotically optimal decisions $(\tau^*_\varepsilon, \varepsilon > 0)$ which is similar to the structure of the decision τ^*_T from (4.2). More precisely, the decisions τ^*_ε are chosen as the compositions of the maps λ^ε (or some their modifications $\hat{\lambda}^\varepsilon$) from Theorem 3.1 with $\tau^*_T, \tau^*_\varepsilon = \tau^*_T \circ \hat{\lambda}^\varepsilon$.

To give more detailes consider the processes $X^\varepsilon = (X^\varepsilon_t, t \geq 0)$,

$$X^\varepsilon_t = \varepsilon \sum_{k=1}^{[\varepsilon^{-2} t]} \frac{g(Y_k)}{f(Y_k)}, \tag{5.3}$$

with $g(y)$ and $f(y)$ defined in condition (F) in §3. These processes weakly converge with respect to $\mathbf{P}_{\infty}^{\varepsilon}$ to the process $X = \sigma W$, W being a standard Wiener process (see §3). Below (in Lemma 5.2) we shall show how we can modify the processes X^{ε} in to the processes $\hat{X}^{\varepsilon} = (\hat{X}_t^{\varepsilon}, t \geq 0)$ so that $\hat{X}^{\varepsilon}$ are closed to X^{ε} but the distributions of $\hat{X}^{\varepsilon}$ under the measure $\mathbf{P}_{\infty}^{\varepsilon}$ coincide (nonasymptotically) with the distribution of σW. Then we define the processes $\hat{\Phi}^{\varepsilon}$ based on $\hat{X}^{\varepsilon}$ in the same way as the process Φ in (4.1) is constructed using X, i.e.,

$$\hat{\Phi}_t^{\varepsilon} = \int_0^t \exp\left\{\sigma^{-1}(\hat{X}_t^{\varepsilon} - \hat{X}_{\theta}^{\varepsilon}) - \sigma^{-2}(t-\theta)/2\right\} d\theta. \tag{5.4}$$

Finally, the stopping time τ_{ε}^{*}, for each $\varepsilon > 0$, is defined as the first moment when $\hat{\Phi}^{\varepsilon}$ attains the threshold T,

$$\tau_{\varepsilon}^{*} = \inf\{t : \hat{\Phi}_t^{\varepsilon} \geq T\}. \tag{5.5}$$

THEOREM 5.2. *Let condition* (F) *of Theorem* 3.1 *hold. Then the stopping times τ_{ε}^{*} from* (5.5) *are asymptotically optimal in the sense that $(\tau_{\varepsilon}^{*}, \varepsilon > 0) \in \mathcal{AS}_T$ and*

$$\lim_{\varepsilon} \sup_{\theta \in \mathbf{R}_+} \mathbf{E}_{\theta}^{\varepsilon}\left[\tau_{\varepsilon}^{*} - \theta | \tau_{\varepsilon}^{*} \geq \theta\right] = 2\sigma^2 \log(\sigma^{-2}T/2) - 1 - C + \varepsilon(T) \tag{5.6}$$

with $\varepsilon(T) \to 0$ as $T \to \infty$.

Remark 5.3. It is natural to try to use the processes X^{ε} instead of their modifications $\hat{X}^{\varepsilon}$ for the construction of the decisions τ_{ε}^{*}. But the proof of optimality properties of τ_{ε}^{*} based on $\hat{X}^{\varepsilon}$ is a significantly easier problem than the similar for the case of X^{ε}. Namely, directly from the definition we have the equality $\mathcal{L}(\hat{X}^{\varepsilon} | P_{\infty}^{\varepsilon}) = \mathcal{L}(\sigma W)$ which implies

$$\mathbf{E}_{\infty}^{\varepsilon} \tau_T^{*} \circ \hat{X}^{\varepsilon} = \mathbf{E}_{\infty} \tau_T^{*} = T$$

and the verification of the condition $(\tau_{\varepsilon}^{*}, \varepsilon > 0) \in \mathcal{AS}_T$ is trivial. But it is not so for a decision of the form $\tau_T^{*} \circ X^{\varepsilon}$. It is easy to obtain the weak convergence X^{ε} to σW (see §3). If τ_T^{*} as a function on $\mathbf{D}[0, \infty)$ (or on $\mathbf{D}[0, \infty)$) were bounded and continuous, then we would have

$$\lim_{\varepsilon} \mathbf{E}_{\infty}^{\varepsilon} \tau_T^{*} \circ X^{\varepsilon} = \mathbf{E}_{\infty} \tau_T^{*} = T. \tag{5.7}$$

Unfortunately, the function τ_T^{*} is neither continuous nor bounded. The problem with continuity is bypassed easily. In fact, it is not difficult to see that x is a discontinuity point for τ_T^{*} if, for some $t > 0$, $\sup_{s \leq t} \Phi_s(x) = T$, $\Phi_t(x) < T$, i.e., the path $\Phi_t(x)$, $t \geq 0$, attains the threshold T but does not cross it. But the Wiener measure of the set of all such points is zero and this is sufficient for verifying (5.7) instead of the continuity property (see (Billingsley, 1968)).
The requirement of boundness of τ_T^{*} can be replaced by the condition of uniform integrability of $\tau_T^{*} \circ X^{\varepsilon}$ as $\varepsilon \to 0$. But the verification of this condition is not easy.

Proof of Theorem 5.2 is based on the following construction.

LEMMA 5.4. *There exist* $\mathbf{F}^\varepsilon$*-adapted modifications* $\hat{X}^\varepsilon = (\hat{X}^\varepsilon_t, t \geq 0)$ *of the processes* X^ε *from (5.3) satisfying the conditions*
(i) *For each* $T > 0$ *and each* $\delta > 0$

$$\lim_\varepsilon \mathbf{P}^\varepsilon_\infty\Big(\sup_{t \leq T} |\hat{X}^\varepsilon_t - X^\varepsilon_t| > \delta\Big) = 0.$$

(ii)

$$\mathcal{L}(\hat{X}^\varepsilon | \mathbf{P}^\varepsilon_\infty) = \mathcal{L}(\sigma W), \quad \varepsilon > 0.$$

Proof. Let us fix some $\varepsilon > 0$ and let $m = [\varepsilon^{-\alpha}]$ with $\alpha = 3/2$. Define the variables $H_n, n \geq 1$, by

$$H_n = m^{-1/2} \sum_{k=1+m(n-1)}^{mn} \frac{g(Y_k)}{f(Y_k)}. \tag{5.8}$$

The r.v.'s $H_n, n \geq 1$, are i.i.d. with respect to $\mathbf{P}^\varepsilon_\infty$ with a common distribution G_ε which tends to the normal law with zero mean and covariance σ^2 as $\varepsilon \to 0$ (see §3).
At the next step we modify these r.v.'s $H_n, n \geq 1$, and define new r.v.'s $\hat{H}_n, n \geq 1$, which are closed to H_n and have the distribution $F = \mathcal{N}(0, \sigma^2)$. There are a lot of ways to do it. For example, put

$$\hat{H}_n = F^{-1} G_\varepsilon(H_n). \tag{5.9}$$

Now define $\Delta = \varepsilon^{-2} m = \varepsilon^{-2}[\varepsilon^{-3/2}]$ and

$$\hat{X}^\varepsilon_{n\Delta} = \sum_{i=0}^{n-1} \hat{H}_i.$$

It remains to cross a Wiener path over the points $\hat{X}^\varepsilon_{n\Delta}$. For this purpose the following technical result can be used.

LEMMA 5.5. *Let* ξ *be a standard normal r.v. on some probability space* $(\Omega, \mathcal{F}, \mathbf{P})$. *Then there exists a standard Wiener process* $W = (W_t)$ *on the same space* $(\Omega, \mathcal{F}, \mathbf{P})$ *such that* $W_1 = \xi$.

Proof. Let $B = (B_t, 0 \leq t \leq 1)$ be a Brownian bridge, i.e., $B_t = \tilde{W}_t - \tilde{W}_1$, $0 \leq t \leq 1$, for some Wiener process $(\tilde{W}_t)$. Put

$$W_t = t\xi + B_t\xi, \quad 0 \leq t \leq 1.$$

Direct calculation shows that $W = (W_t)$ is a Wiener process and $W_1 = \xi$.
Thus, if $t > 0$ and n is such that $0 \leq s \leq \Delta$ for $s = t - n\Delta$, then we can put

$$\hat{X}^\varepsilon_t = \hat{X}^\varepsilon_{n\Delta} + \hat{H}_n\Big(s + B^{(n)}_s\Big),$$

where $B^{(1)}, B^{(2)}, B^{(3)}, \ldots$ are independent copies of Brownian motion. Lemma 5.5 implies (ii).

To complete the proof of the lemma it remains to verify the condition (i). But the processes $\hat{X}^{\varepsilon}, X^{\varepsilon}$ coincide at the points $t = n\Delta$, and this condition readly follows from the construction.
To complete the proof of the theorem we should verify that $R(\tau_{\varepsilon}^{*}) \to R(\tau_{T}^{*})$ as $\varepsilon \to 0$. Let the number Δ be introduce in Lemma 5.4. and let $\theta = n\Delta$ for some $n \geq 1$. We show that

$$\mathbf{E}_{\theta}^{\varepsilon}\Big[\tau_{\varepsilon}^{*} - \theta | \tau_{\varepsilon}^{*} \geq \theta\Big] \longrightarrow \mathbf{E}_{\theta}\Big[\tau_{T}^{*} - \theta | \tau_{T}^{*} \geq \theta\Big] \text{ as } \varepsilon \to 0$$

uniformly in $n \geq 1$. The theorem result is easily consequence of this fact. For the process $\hat{\Phi}^{\varepsilon}$ from (5.4) the following equation holds

$$\hat{\Phi}_{t+s}^{\varepsilon} = \hat{\Phi}_{t}^{\varepsilon} Z^{\varepsilon}(t, t+s) + \int_{t}^{t+s} Z^{\varepsilon}(t+\theta, t+s)\, d\theta$$

with

$$Z^{\varepsilon}(t, t+s) = \exp\Big\{\sigma^{-1}(\hat{X}_{t+s}^{\varepsilon} - \hat{X}_{t}^{\varepsilon}) - \sigma^{-2}s/2\Big\}.$$

This means that the process $\hat{\Phi}^{\varepsilon}$ is Markovian and we get directly from the definition of the stopping times τ_{ε}^{*} that the value $\mathbf{E}_{\theta}^{\varepsilon}\Big[\tau_{\varepsilon}^{*} - \theta | \tau_{\varepsilon}^{*} \geq \theta\Big]$ depends on θ only through the conditional distribution $L_{\theta}^{\varepsilon} = \mathcal{L}\Big(\hat{\Phi}_{\theta}^{\varepsilon} | \tau_{\varepsilon}^{*} \geq \theta, \mathbf{P}_{\infty}^{\varepsilon}\Big)$. More precisely, if we put $R^{\varepsilon}(\varphi) = \mathbf{E}_{0}^{\varepsilon}\tau_{\varepsilon}^{*}$, then

$$\mathbf{E}_{\theta}^{\varepsilon}\Big[\tau_{\varepsilon}^{*} - \theta | \tau_{\varepsilon}^{*} \geq \theta\Big] = \int R^{\varepsilon}(\varphi) L_{\theta}^{\varepsilon}(d\varphi).$$

But similarly we obtain for the limit experiment

$$\mathbf{E}_{\theta}\Big[\tau_{T}^{*} - \theta | \tau_{T}^{*} \geq \theta\Big] = \int R(\varphi) L_{\theta}(d\varphi)$$

with $R(\varphi) = \mathbf{E}_{0}\tau_{T}^{*}$ and $L_{\theta} = \mathcal{L}\Big(\hat{\Phi}_{\theta} | \tau_{T}^{*} \geq \theta, \mathbf{P}_{\infty}\Big)$, and from the equality $\mathcal{L}(\hat{\Phi}^{\varepsilon} | \mathbf{P}_{\infty}^{\varepsilon}) = \mathcal{L}(\Phi | \mathbf{P}_{\infty})$ we have $L_{\theta}^{\varepsilon} = L_{\theta}$ for all $\varepsilon \geq 0$ and $\theta \geq 0$. Hence the needed result follows from uniform in $\Phi \in [0, T]$ convergence $R^{\varepsilon}(\varphi) \to R(\varphi)$ as $\varepsilon \to 0$ which is obvious. The theorem is proved.

Remark 5.6. As in §4 we can consider instead of decisions $\tau_{\varepsilon}^{*} = \tau_{T}^{*} \circ \hat{X}^{\varepsilon}$ decisions $\tilde{\tau}_{\varepsilon} = \tilde{\tau}_{T} \circ \hat{X}^{\varepsilon}$ corresponding the CUSUM method. Similarly to above one can state the following result: $(\tilde{\tau}_{\varepsilon}, \varepsilon > 0) \in \mathcal{AS}_{T}$ and

$$\lim_{\varepsilon} \sup_{\theta \in \mathbf{R}_{+}} \mathbf{E}_{\theta}^{\varepsilon}\Big[\tilde{\tau}_{\varepsilon} - \theta | \tilde{\tau}_{\varepsilon} \geq \theta\Big] = 2\sigma^{2}\log(\sigma^{-2}T/2) - 3/2 + \varepsilon(T)$$

with $\varepsilon(T) \to 0$ as $T \to \infty$.

REFERENCES

Girshich, M.A. and Rubin, H.A. (1952). A Bayes approach to a quality control model *Ann. Math. Statist.* **23**, 114–125.

Kolmogorov, A.N., Prokhorov Yu. V., and Shiryaev, A.N. (1990). Probabilistic-statistical methods of detecting spontaneously occuring effects. *Proc. Steklov Math. Inst.* 1, 1–21.

Kramkov, D.O. (1992). *On the Theory of Filtered Experiments.* Theses. Steklov Math. Inst., Moscow (in Russian).

Lai, T.L. (1975). A note on first exit times with application to sequential analysis, *Ann. Statist.* **3**, 999–1005.

Le Cam, L. (1986). *Asymptotic Method in Statistical Decision Theory.* Springer–Verlag, New York.

Lorden, G. (1970). Procedures for reacting to a change in distribution. *Ann. Math. Statist.* **42**, 1897–1908.

Nikiforov, I.V. (1983). *Sequential Selection of Changes in the Properties of Time Series.* Nauka, Moscow (in Russian).

Page, E.S. (1954). Continuous inspection schemes. *Biometrika* **41**, 100–115.

Pollak, M. and Siegmund, D. (1975). Approximations to the expected sample size of certain sequential tests. *Ann. Statist.* **3**, 1267–1282.

Pollak, M. and Siegmund, D. (1985). A diffusion process and its application to detecting a change in the drift of Brownian motion. *Biometrika* **72**, 267–280.

Shiryaev, A.N. (1961). Detection of spontaneous effects. *Soviet Math. Dokl.* **2**, 799–801.

Shiryaev, A.N. (1963a). On optimal methods in problems of quickest detection. *Theory Probab. Appl.* **8**, 26–51.

Shiryaev, A.N. (1963b). On detection of disorders in a manufactoring process. I, II. *Theory Probab. Appl.* **8**, 264–281, 431–443.

Shiryaev, A.N. (1964). Some exact formulas in the "disorder" problem. *Theory Probab. Appl.* **10**, 380–385.

Shiryaev, A.N. and Spokoiny, V.G. (1992). *Statistical Experiments and Decisions.* VINITI, Moscow (to appear).

Frontiers in Pure and Applied Probab. 1, pp. 256 – 268
H. Niemi et al. (Eds.)
1993 TVP/VSP

BRANCHING PROCESSES AND RANDOM TREES

V. A. VATUTIN
Steklov Mathematical Institute, Vavilov Street 42, 117966 Moscow, GSP-1, Russia

ABSTRACT

Let Δ be a subset of the set $\{0, 1, 2, \ldots\}$, let $\mu(t)$ be the number of particles at time t in a Bellman–Harris branching process, and let ν_Δ be the number of those particles ever existed in the process, the number of direct descendants of which belongs to Δ. Using the known theorems concerning the distribution of ν_Δ for the critical Bellman–Harris branching processes, we find the limit distribution of the degeneration moment τ of the process under the condition $\nu_\Delta = N \to \infty$. For the case $\Delta = \{0, 1, 2, \ldots\}$ we find also the distribution of the distance to the nearest ancestor of all particles existing at time t under the condition $\{\mu(t) > 0,\ \nu_\Delta = N\}$ as $t, N \to \infty$ in an appropriate way. The connection of these results with the corresponding properties of planted plane trees is discussed.

INTRODUCTION

Solving various problems of probability theory, one often encounters (in an implicit or non-implicit form) with the investigation of tree-like structures. We describe below two typical examples conforming these words.

Example 1. Let X_k, $k = 1, 2, \ldots$, be positive independent identically distributed (i.i.d.) random variables and let $S_N = \sum_{k=1}^{N} X_k$. Then the problem of investigating the limit behavior of S_N as $N \to \infty$ can be represented as the problem of finding the distribution of the height τ_N $(\equiv S_N)$ of the random "degenerate" tree with vertices numbered by $0, 1, \ldots, N$ and with edges having lengths $X_1, X_1, \ldots, X_N$:

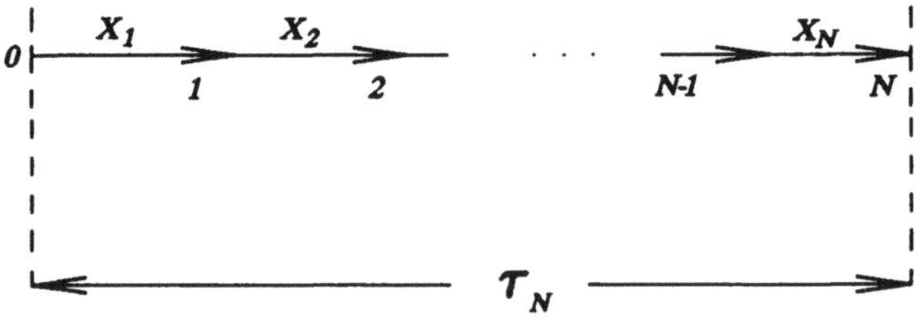

Fig. 1

Example 2. Let, as before, X_k be positive i.i.d. random variables and let $M_N = \max_{1\le k\le N} X_k$. Then the problem of investigating the limit behavior of M_N as $N \to \infty$ can be treated as the problem of finding the distribution of the height τ_N $(\equiv M_N)$ of the random bush

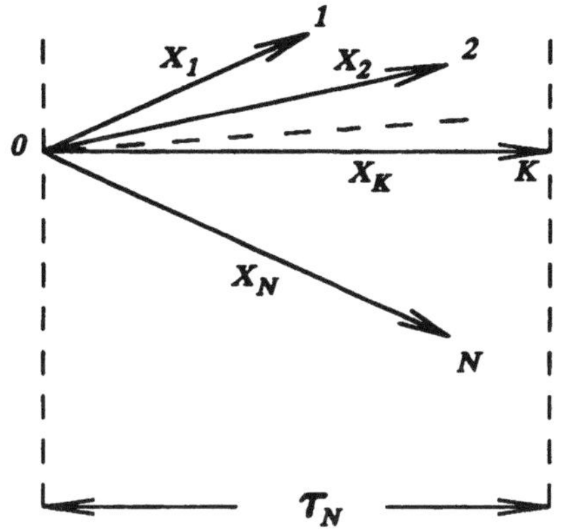

Fig. 2

as $N \to \infty$.

The examples listed above are particular cases of the following more general scheme. At first, we choose an arbitrary tree whose branches are of the unit length, then we assign positive i.i.d. random variables to each of the branches, i.e., change the lenghts of the branches, and, finally, letting the tree (or the number of its branches) grow in this or that way, we try to find the distribution of the height of the tree as the number N of its branches tends to infinity.

It is clear that to obtain a meaningful result we should pose certain restrictions on the character of trees changings. Here we use the Bellman–Harris branching processes as a model of such growth. We shall fix below not the total number of branches in trees but the total number of branches possessing these or those properties. To obtain the desired results in such situation, we use a special case of the Bellman–Harris branching processes with final types of particles which is described below.

Let Δ be a fixed subset of the set $\{0, 1, 2, \ldots\}$ and let $\chi(A)$ be the indicator of an event A. The process in question is initiated at the time $t = 0$ with a parent particle of nonfinal type which lives time l_0, $M(t) = \mathbf{P}\{l_0 \le t\}$, and then produces a random number ξ of descendants of nonfinal type and $\eta = \chi(A)$ descendants of final type. The newborn particle (if any) of final type does not change after its birth moment while the newborn particles of nonfinal type evolve independently and stochastically in the same way as the initial one and at the end of their lives produce new particles, and so on.

We introduce now some notations. Let $f(s) = \mathbf{E}s^{\xi}$,

$$F(u, s) = \mathbf{E}u^{\eta}s^{\xi} = \mathbf{E}\left[s^{\xi}; \xi \notin \Delta\right] + u\mathbf{E}\left[s^{\xi}; \xi \in \Delta\right] = F_0(s) + uF_1(s)$$

and $q = \mathbf{P}\{\xi \in \Delta\}$. Denote the number of nonfinal type particles in the process at time t by $\mu(t)$, the total number of final type particles which were born in the

process during the time of its evolution by ν_Δ. Let $\nu = \nu_\Delta$, if $\Delta = \{0, 1, 2, \ldots\}$, and let $\tau = \inf\{t > 0\colon \mu(t) = 0\}$.
The aim of this paper is to investigate the asymptotic behavior of τ under the condition $\nu_\Delta = N$ as $N \to \infty$ (§1), to find (under the same condition) the distribution of the distance to the nearest mutual ancestor of the particles living at time t (see the corresponding definition below) when t, $N \to \infty$ in an appropriate way (§2), and then to use the results obtained to find the distribution of the height of a random tree and the distribution of the distance to the root of the minimal subtree containing all vertices of a given height (§3). Here we only sketch the proofs of the main results. Their full proofs will be published elsewhere.

1. LIMIT THEOREMS FOR THE MOMENT OF DEGENERATION

We assume throughout that $f'(1) = 1$ (i.e., the process of nonfinal particles is critical),

$$f(s) = s + (1-s)^{1+\alpha} L(1-s) \equiv s + \Phi(1-s), \tag{1}$$

where $\alpha \in (0, 1]$ and the function $L(x)$ is slowly varying as $x \to +0$, and

$$\lim_{n\to\infty} \frac{n(1-M(n))}{1-f_n(0)} = 0, \tag{2}$$

where $f_n(s)$ is the n-fold iterate of $f(s)$.
Let $\{z(n),\ n = 0, 1, 2, \ldots\}$ be the embedded Galton–Watson branching process for $\mu(t)$. The process $z(n)$ is defined as the number of particles in the nth generation of $\mu(t)$.
It is known (Slack, 1968) that under condition (1) the function $a(n) = 1 - f_n(0) = \mathbf{P}\{z(n) > 0\}$ satisfies the relations

$$a(n) \sim n^{-1/\alpha} l(n), \tag{3}$$

$$a^\alpha(n) L(a(n)) \sim (n\alpha)^{-1} \tag{4}$$

as $n \to \infty$, where the function $l(n)$ varies slowly as $n \to \infty$. Let $\nu_\Delta(n+1)$ be the total number of particles in the generations $0, 1, \ldots, n$ with the number of descendants belonging to the set Δ. It was proved (Vatutin and Sagitov, 1988) that under condition (1)

$$\lim_{n\to\infty} (a(n))^{-1} \left(1 - \mathbf{E}\left[\exp\left\{-\lambda \nu_\Delta(n+1) n^{-1} a(n)\right\};\ z(n) = 0\right]\right)$$

$$= \psi(\lambda(1-q), \infty) \tag{5}$$

for any $\lambda \geq 0$, where $\psi = \psi(\lambda_1, \lambda_2)$ is the unique solution of the quasilinear partial differential equation

$$(1+\alpha)\lambda_1 \frac{\partial\psi}{\partial\lambda_1} + \lambda_2 \frac{\partial\psi}{\partial\lambda_2} = \psi - \psi^{1+\alpha} + \alpha\lambda \tag{6}$$

with the initial conditions $\psi(0,0) = 0$ and

$$\left.\frac{\partial\psi(\lambda_1,\lambda_2)}{\partial\lambda_1}\right|_{\lambda_1=\lambda_2=0} = \left.\frac{\partial\psi(\lambda_1,\lambda_2)}{\partial\lambda_2}\right|_{\lambda_1=\lambda_2=0} = 1. \tag{7}$$

The subsequent arguments use essentially Kolchin's ideas (Kolchin, 1978) and relation (5). We note also that Kolchin was the first to give a rigorious mathematical description of the correspondence between Galton–Watson branching processes and rooted labeled trees.

LEMMA 1. *If condition* (1) *holds and* $n, N \to \infty$ *in such a way that*

$$\lim N\Phi(a(n)) = x \in (0,\infty), \tag{8}$$

then

$$\lim \mathbf{E}\left[\exp\left\{i\theta\nu_{\Delta}N^{-1}\right\}|z(n) > 0\right]$$

$$= \psi\left(-\frac{i\theta(1-q)}{\alpha x},\infty\right) - \left(-\frac{i\theta(1-q)}{x}\right)^{\frac{1}{1+\alpha}} \equiv \Xi\left(\theta,\frac{x}{1-q}\right) \tag{9}$$

for any $\theta \in (-\infty,+\infty)$.

LEMMA 2. *The function* $\Xi(\theta,x)$ *is absolutely integrable with respect to* $\theta \in (-\infty,+\infty)$ *for any fixed* $x > 0$.

Owing the preceding lemma, we can calculate the density $r(y, x/(1-q))$ of the limit distribution in the right-hand side of (9) by the formula

$$r(y,x/(1-q)) = \frac{1}{2\pi}\int_{-\infty}^{+\infty} e^{-i\theta y}\Xi(\theta,x/(1-q))d\theta. \tag{10}$$

Using this representation and the usual scheme of local limit theorems proving (see, for example, (Kolchin, 1986)), one can prove the following statements.

LEMMA 3. *Let condition* (1) *be satisfied,* $0 \in \Delta$, *and let the span of the random variable* ζ *with the probability generating function* $Y(s) = F_1(s)(1-s^{-1}F_0(s))^{-1} = \mathbf{E}s^{\zeta}$ *be* h. *If* $n, N \to \infty$ *in such a way that condition* (6) *holds, then*

$$\mathbf{P}\left\{\frac{\nu_{\Delta}}{N} = y \mid z(n) > 0\right\} = \frac{h}{N}r(y,x/(1-q))(1+\mathbf{o}(1)) \tag{11}$$

uniformly with respect to integers k *within any finite interval of the form* $0 < y_0 \leq y = (kh+1)/N \leq y_1 < \infty$.

LEMMA 4. *Let condition* (1) *hold and let* h *be the span of* ζ. *Then*

$$\mathbf{P}\{\nu_{\Delta} = N\} \sim c\Phi^{-1}(1/N)\frac{1}{N}, \quad N \to \infty, \quad N-1 \equiv 0(\mathrm{mod}\, h).$$

Now we formulate two theorems concerning the distribution of the extinction moment of a critical branching process provided the total number of particles of final type ever born in the process by fixed. The first of the theorems has an auxiliary meaning and we omit its proof.
Let $\tau^* = \min\{n > 0\colon z(n) = 0\}$ and $R(x) = x^{-1}\Phi(x) = (f(1-x)-(1-x))x^{-1} = x^{\alpha}L(x)$.

THEOREM 1. *Let condition* (1) *be satisfied and let h be the span of ζ. Then*

$$\lim_{N\to\infty} \frac{\mathbf{P}\{z(x/(\alpha R(\Phi^{-1}(1/N)))) > 0;\ \nu_\Delta = N\}}{\mathbf{P}\{\nu_\Delta = N\}} = 1 - Q_q(x) \tag{12}$$

for any fixed $x > 0$ and $N = kh + 1$, $k = 0, 1, \ldots$, where

$$Q_q(x) = 1 - \left(\frac{x^{-1-1/\alpha}}{1-q}\frac{\alpha}{\Gamma(1-\alpha)}\right)^{1/(1+\alpha)} \varkappa_\alpha r\left(1, \frac{x^{-1-1/\alpha}}{1-q}\right) \tag{13}$$

and

$$\varkappa_\alpha = \pi/\left(\Gamma\left(1+\frac{1}{1+\alpha}\right)\cos\frac{\pi(1-\alpha)}{2(1+\alpha)}\right).$$

Remark 1. If $0 < B = f''(1) < \infty$, then $\alpha = 1$, $\Phi(y) \sim (B/2)y^2$ and $R(y) \sim (B/2)y$ as $y \to 0$, and $R(\Phi^{-1}(1/N)) \sim \sqrt{B/(2N)}$. It is known also (see, for instance, (Kolchin, 1978)) that in this case

$$r(1,w) = \frac{1}{\sqrt{\pi w}}\sum_{k=1}^{\infty}\left(\frac{2k^2}{w}-1\right)e^{-k^2/w}$$

and, therefore,

$$Q_q(x) = 1 - \frac{1}{x\sqrt{1-q}}\frac{\pi}{\Gamma(3/2)}\frac{1}{\sqrt{\pi}}x\sqrt{1-q}\sum_{k=1}^{\infty}\left(2k^2x^2(1-q)-1\right)e^{-k^2x^2(1-q)}$$

$$= 1 - 2\sum_{k=1}^{\infty}(2k^2x^2(1-q)-1)e^{-k^2x^2(1-q)}.$$

THEOREM 2. *If the conditions of Theorem* 1 *hold, $\rho = \mathbf{E}l_0$, and*

$$\lim_{n\to\infty}\left(\frac{n}{a(n)}\right)^2(1-M(n)) = 0, \tag{14}$$

then

$$\lim_{N\to\infty}\mathbf{P}\{\tau\alpha R(\Phi^{-1}(1/N)) \le x \mid \nu_\Delta = N\} = Q_q\left(\frac{x}{\rho}\right) \tag{15}$$

for any $x > 0$.

Proof of Theorem 2. First we would like to prove the estimates

$$-M^{*n}(t)\mathbf{P}\{z(n)>0\} \le \mathbf{P}\{\mu(t)>0, \nu_\Delta = N\}$$

$$-\mathbf{P}\{z(n)>0, \nu_\Delta = N\} \le 1 - M^{*n}(t). \tag{16}$$

With this aim in view we introduce additional notations. Let $z_t(k)$ be the number of those particles in the kth generation of the embedded process $\{z(n)\}$ which exist at time t. Put $\mu_n^+(t) = \sum_{k=n+1}^{\infty} z_t(k)$ and $\mu_n^-(t) = \sum_{k=0}^{n} z_t(k)$. It is clear that

$$\mathbf{P}\{\mu(t)>0, \nu_\Delta = N\} - \mathbf{P}\{z(n)>0, \nu_\Delta = N\}$$

$$= \mathbf{P}\{\mu(t)>0, z(n)=0, \nu_\Delta = N\} - \mathbf{P}\{\mu(t)=0, z(n)>0, \nu_\Delta = N\}. \tag{17}$$

It is easy to deduce the inequalities

$$\mathbf{P}\{\mu(t)>0, z(n)=0, \nu_\Delta = N\} \le \mathbf{P}\{\mu(t)>0, \mu_n^+(t)=0, \nu_\Delta = N\}$$

$$\le \mathbf{P}\{\mu(t)>0, \mu_n^+(t)=0\}.$$

Since $\mathbf{P}\{\mu(t)>0, \mu_n^+(t)=0\} = \mathbf{P}\{$the generation numbers of all the particles living at time t are less than $n+1$, $\mu(t)>0\} \le \mathbf{P}\{$the generation number of a fixed particle living at time t is less than $n+1$, $\mu(t)>0\} \le 1 - M^{*n}(t)$, the following inequality

$$\mathbf{P}\{\mu(t)>0, \nu_\Delta = N\} - \mathbf{P}\{z(n)>0, \nu_\Delta = N\} \le 1 - M^{*n}(t)$$

which proves the second inequality in (16) holds true. To prove the first one we use the following arguments: $\mathbf{P}\{\mu(t)=0, z(n)>0, \nu_\Delta = N\} \le \mathbf{P}\{\mu_n^-(t)=0, z(n)>0\} = \mathbf{P}\{z(n)>0$, all the particles in the nth generation of the process $\mu(t)$ have ages not exceeding $t\} \le G^{*n}(t)\mathbf{P}\{z(n)>0\}$.
Obviously,

$$\mathbf{P}\{\tau \le t \mid \nu_\Delta = N\} = 1 - \frac{\mathbf{P}\{\mu(t)>0, \nu_\Delta = N\}}{\mathbf{P}\{\nu_\Delta = N\}}. \tag{18}$$

We put $t = x/(\alpha R(\Phi^{-1}(1/N)))$ for a fixed $x>0$, choose an $\varepsilon \in (0,1)$, and set $n = [(t/\rho)(1-\varepsilon)]$. Under such a choice of t and n we have $n \sim c_1 R(\Phi^{-1}(1/N))$ or, in view of (4), the definition of R and the known properties of slowly varying functions, $a(n) \sim c_2\Phi^{-1}(1/N)$ and $\Phi(a(n)) \sim c_3\mathbf{N}^{-1}$. Using these relations and (4), we obtain once more

$$\mathbf{P}\{\nu_\Delta = N\} \sim \frac{c\Phi^{-1}(1/N)}{N} \sim \frac{c_4 a(n)}{N} \sim c_5 a(n)\Phi(a(n)) \sim c_6\frac{a^2(n)}{n}.$$

From condition (14), relation (3), and the theorem proved in (Baum and Katz, 1965) it follows that under the conditions of Theorem 2

$$1 - M^{*n}(t) = o(n^{-1}a^2(n)) = o(\mathbf{P}\{\nu_\Delta = N\})$$

for any fixed $\varepsilon > 0$. Using this relation, Theorem 1, and the second inequality in (16), we obtain

$$\liminf_{N\to\infty} \mathbf{P}\{\tau\alpha R(\Phi^{-1}(1/N)) \le x \mid \nu_\Delta = N\}$$

$$= 1 - \limsup_{N\to\infty} \frac{\mathbf{P}\{\mu(t) > 0, \nu_\Delta = N\}}{\mathbf{P}\{\nu_\Delta = N\}}$$

$$\ge 1 - \limsup_{N\to\infty} \frac{\mathbf{P}\{z\left(x(1-\varepsilon)/(\alpha\rho R(\Phi^{-1}(1/N)))\right) > 0, \nu_\Delta = N\}}{\mathbf{P}\{\nu_\Delta = N\}}$$

$$= Q_q\left(\frac{x(1-\varepsilon)}{\rho}\right). \tag{19}$$

The same arguments based upon the first inequality in (17) with $n = [t/\rho(1+\varepsilon)]$ show that

$$\limsup_{N\to\infty} \mathbf{P}\{\tau\alpha R(\Phi^{-1}(1/N)) \le x \mid \nu_\Delta = N\} = Q_q\left(\frac{x(1+\varepsilon)}{\rho}\right). \tag{20}$$

Estimates (19) and (20) prove Theorem 2, since $\varepsilon > 0$ is arbitrary and $Q_q(x)$ is continuous with respect to x.

2. THE DISTRIBUTION OF THE DISTANCE TO THE NEAREST MUTUAL ANCESTOR

In this section we consider only a single-type Galton–Watson branching process with generating function $f(s)$ governing particle production. Let $z(n)$ be the number of particles in the process under consideration at time $n = 0, 1, 2, \ldots$ and let $\lambda(n) = n - \max\{t < n\colon z(t) = 1\}$.

Definition. The random variable $\lambda(n)$ is called the distance to the nearest mutual ancestor.

To formulate the main result of this section we need some auxiliary notations. We put

$$b(\beta,\theta) = \sqrt{-2i\theta}\,\frac{\mathrm{ch}\beta\sqrt{-i\theta/2}}{\mathrm{sh}\beta\sqrt{-i\theta/2}}$$

and

$$g(\beta,y) = \frac{1}{y}r\left(1, \frac{2y}{\beta^2}\right) = \frac{\beta}{y\sqrt{2\pi y}}\sum_{k=1}^{\infty}\left(\frac{\beta^2k^2}{y} - 1\right)e^{-\frac{\beta^2k^2}{2y}}$$

for $\beta > 0$, $\theta \in (-\infty, +\infty)$, and $y > 0$.

Let us define a two-dimensional distribution function $R_\beta(z,w)$ by the relation

$$R_\beta(z,w) = \frac{1}{4\pi^2}\int^{+\infty}\int^{+\infty} e^{-i(\theta_1 z + \theta_2 w)}\varphi_\beta(\theta_1,\theta_2)\,d\theta_1 d\theta_2,$$

where

$$\varphi_\beta(\theta_1,\theta_2) = \left(\frac{\mathrm{sh}\beta\sqrt{-2i\theta_2}}{\beta\sqrt{-2i\theta_2}} - i\theta_1 \left(\frac{\mathrm{sh}\beta\sqrt{-2i\theta_2/2}}{\beta\sqrt{-2i\theta_2/2}} \right)^2 \right)^{-1} .$$

The distribution $R_\beta(z,w)$ appeared in (Kolchin, 1978) as the joint limit distribution (under the corresponding scalings and the condition $z(n) > 0$) of $z(n)$ and of the total number of particles $\nu(n)$ which were born in the time interval $[0,n]$.

THEOREM 3. *If condition* (1) *holds, the distribution of* ξ *has span* 1, *and* $n, N \to \infty$ *in such a way that*

$$n\left(\frac{B}{N}\right)^{1/2} \to \beta \in (0,\infty),$$

then

$$\lim \mathbf{P}\{\lambda(n)/n \le a \mid z(n) > 0, \nu = N\} = I_\beta(a) \equiv \frac{1}{2\pi a g(\beta,1)}$$

$$\times \int_0^\infty \int_0^1 z dR_{\beta(1-a)}(z,w) \int_0^1 g(a\beta,y)dy \int_{-\infty}^{+\infty} e^{-i\theta(1-w-y)-z(\beta(1-a)/2)b(a\beta,\theta)} d\theta$$

for any $a \in (0,1)$.

We do not prove the theorem here but apply it in the next section to obtain some results concerning the structure of random trees. To make our arguments listed below more transparent we shall describe with minor changes the well-known construction of the set of families (or genealogical trees) of a Galton–Watson branching process (Harris, 1963). We mark particles of the process in the following way. Firstly, we assign the symbol (0) to the initial particle of the process. If the particle has no offsprings, we substitute the mark (0) by (0*), and the process of marking is finished. If the initial particle has $k \ge 1$ offspring particles, then we leave the mark (0) of the particle unchanged and assign preliminary marks $(01),(02),\ldots,(0k)$ to its offspring particles in an arbitrary but fixed order. If one of these particles (say, $(0i)$) has offsprings, we leave its mark unchanged, otherwise we substitute the mark $(0i)$ by $(0i*)$. It is clear how to continue this procedure further. If the particle having preliminary mark $(0i_1\ldots i_n)$ has no offspring particles, then its final mark becomes $(0i_1\ldots i_n*)$, otherwise, its preliminary mark remains unchanged and its descendants obtain preliminary marks $(0i_1\ldots i_n1),(0i_1\ldots i_n2),\ldots$. Obviously, the following genealogical trees will be treated as distinct under such a procedure. (The daughter of my first daughter is not the same as the daughter of my second daughter.) We denote below a genealogical tree by G_N, i.e., the description of a realization of the branching process up to the moment of extinction constituted by $N+1$ particles (including the initial one) and by the set $\mathcal{G}_N = \{G_N\}$ of all possible genealogical trees having $N+1$ particles.

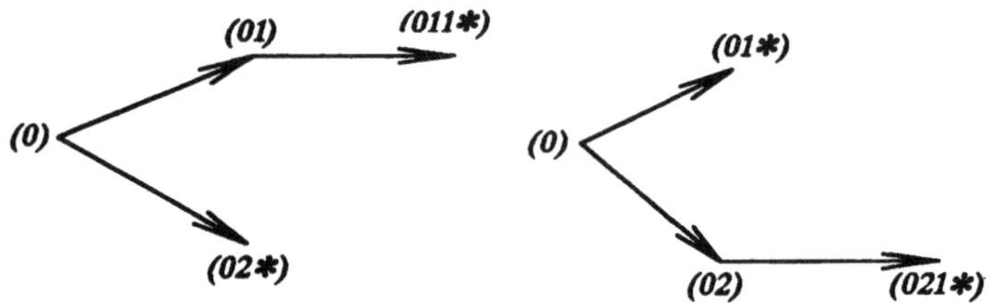

Fig. 3

Let us return now to the probability generating function $f(s) = \mathbf{E}s^{\xi} = \sum_{k=0}^{\infty} p_k s^k$ for the number of direct descendants of particles of the process under consideration and let $A = \{k\colon p_k > 0\}$. Denote the set of all admissible genealogical trees, i.e., having nonzero probability, with exactly $N+1$ particles by $\mathcal{G}_N^{(A)}$. It is clear that, if $G_N \in \mathcal{G}_N^{(A)}$ and has m_r particles producing exactly r offspring particles, $r = 0, 1, \ldots, N$, $m_0 + m_1 + \ldots + m_N = N+1$, $m_1 + 2m_2 + \ldots + Nm_N = N$, then

$$\mathbf{P}\{G = G_N | G \in \mathcal{G}_N\} = \mathbf{P}\{G = G_N | G \in \mathcal{G}_N^{(A)}\}$$

$$= \frac{\mathbf{P}\{G = G_N\}}{\mathbf{P}\{\nu = N+1\}} = \frac{p_0^{m_0} p_1^{m_1} \ldots p_N^{m_N}}{\mathbf{P}\{\nu = N+1\}}. \tag{21}$$

If, for example, $p_k = cq^k$, $k \in A$, $\sum_{k \in A} p_k = 1$, then the probability $\mathbf{P}\{G = G_N\} = c^{N+1}q^N$ does not depend on G_N and, therefore,

$$\mathbf{P}\{G = G_N \mid G \in \mathcal{G}_N^{(A)}\} = \frac{c^{N+1}q^N}{\mathbf{P}\{\nu = N+1\}} = \frac{1}{|\mathcal{G}_N^{(A)}|}. \tag{22}$$

3. LIMIT THEOREMS FOR RANDOM TREES

We recall now several notions of graph theory (Harary and Palmer, 1973).
A tree is a connected graph with no cycles. A rooted tree is a tree with one marked vertex called the root of the tree. For convenience, we assume below that edges of any rooted tree are directed from the root and call the number of edges going out of the vertex a degree of a vertex. If q is a vertex of T, then the number of edges constituting the path from the root of T to the vertex q is called a height of q and denoted by $H(q)$. The quantity $H(T) = \max_{q \in T} H(q)$ is called a height of T (we assume here that edges of T are of unit length; if it is not the case, the definition of the height of a tree needs obvious corrections).
A rooted plane tree is a rooted tree embedded into the Euclidian plane. By interchanging the position relative to the root of two different branches, different trees are obtained. A plane rooted tree is called a planted plane tree, if its root has the degree 1. We assign some marks to the vertices of a planted plane tree

using the following procedure. We assign the symbol $*$ to the root of T and the symbol (0) to the vertex adjacent to $*$. Let k be the degree of vertex (0). We assign the symbols $(01), (02), \ldots, (0k)$ to the vertices adjacent to (0) going from up to down (see Fig. 4 below).

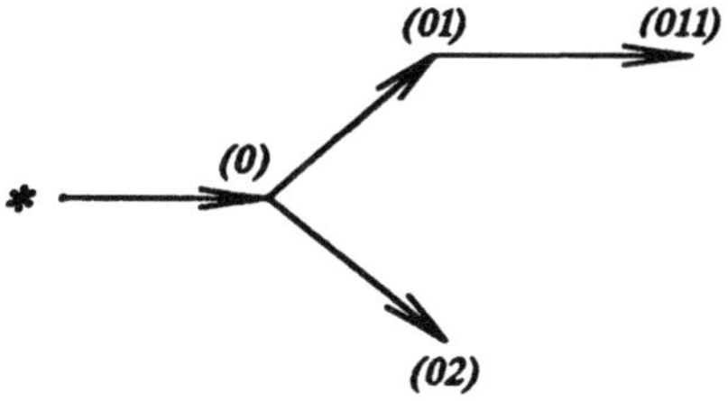

Fig. 4

We mark the yet unmarked vertices adjacent to (01) (if any) by the symbols $(011), (012), \ldots$, and so on.
Since we cannot interchange two different branches of a planted plane tree, the described procedure establishes a one-to-one correspondence between the set of numerations of such kind and the set of all planted plane trees. We denote the set of all planted plane trees having N nonrooted vertices by $\mathcal{T}_N = \{T_N\}$. From the construction given at the end of the preceding section it is clear that there exists a natural one-to-one correspondence between the elements of $\mathcal{G}_N$ and the elements of $\mathcal{T}_N$. Indeed, if we add the symbol $*$ to the marks of degree zero vertices of a tree $T_N \in \mathcal{T}_N$, $N \geq 1$, and exclude the root of the tree, we obtain the genealogical tree $G_N \in \mathcal{G}_N$, and, vice versa, if we let a fictuous parent particle marked by $*$ to be the initial particle of a genealogical tree G_N and remove the symbol $*$ from the marks of childless particles of G_N, we obtain the planted plane tree T_N. We denote the correspondence by δ.
Let $\Omega_G = \cup_{N=0}^{\infty}\{G_N \in \mathcal{G}_N\}$ and $\Omega_T = \cup_{N=0}^{\infty}\{T_N \in \mathcal{T}_N\}$ and let $\mathfrak{A}_G$ and $\mathfrak{A}_T$ be the sets of subsets of Ω_G and Ω_T, respectively. It is clear that, if $\tau^* = \tau^*(G) = k$ is the moment of degeneration of a genealogical tree G, then the height $H^* = H^*(T)$ of the tree $T = \delta(G)$ is k. Let a probability generation function $f(s)$ be given. It generates the probability measure $\mathbf{P}$ on $\mathfrak{A}_G$ which, in its turn, generates the corresponding probability measure $\mathbf{P}_\delta$ on $\mathfrak{A}_T$: $\mathbf{P}_\delta(D) = \mathbf{P}(\delta^{-1}(D))$, if $D \in \mathfrak{A}_T$. Let $\nu_\Delta = \nu_\Delta(G)$ be the total number of particles in G, the number of direct descendants of which belongs to Δ, and let $\nu_\Delta^{(\delta)} = \nu_\Delta^{(\delta)}(T) = \nu_\Delta(\delta^{-1}(T))$ be the total number of nonrooted vertices in T, the degrees of which belong to Δ. In view of the above discussion,

$$\mathbf{P}\{\nu_\Delta = N\} = \mathbf{P}_\delta\{\nu_\Delta^{(\delta)} = N\}$$

and

$$\mathbf{P}\{G \in D \mid \nu_\Delta = N\} = \mathbf{P}_\delta\{\delta(G) \in \delta(D) \mid \nu_\Delta^{(\delta)} = N\}$$

$$= \mathbf{P}_\delta\{T \in \delta(D) \mid \nu_\Delta^{(\delta)} = N\}$$

for any $D \in \Omega$.
Now Theorem 2 yields the following statement for a critical $f(s)$.

THEOREM 4. *Let the conditions of Theorem 1 be satisfied. Then*

$$\lim \mathbf{P}_\delta\{H^*\alpha R(\Phi^{-1}(1/N)) \le x \mid \nu_\Delta^{(\delta)} = N\} = Q_q(x).$$

If, in particular, $0 < f''(1) = B < \infty$, then, in view of Remark 1,

$$\lim_{N\to\infty} \mathbf{P}_\delta \left\{ H^* \sqrt{\frac{B}{2N}} \le x \mid \nu_\Delta^{(\delta)} = N \right\}$$

$$= 1 - 2\sum_{k=1}^{\infty} \left(2k^2x^2(1-q) - 1\right) e^{-k^2x^2(1-q)} = Q_q(x).$$

Example 3. Let $\Delta = \{0, 1, 2, \ldots\}$, $f(s) = (2-s)^{-1}$, i.e., $B = 2$ and $p_k = 2^{-k-1}$, $k = 0, 1, 2, \ldots$. In view of (22) with $A = \Delta$

$$\mathbf{P}\{G = G_N \mid G \in \mathcal{G}_N\} = \frac{1}{|\mathcal{G}_N|} = \frac{1}{|\mathcal{T}_N|} = \mathbf{P}_\delta\{T = T_N \mid T \in \mathcal{T}_N\}$$

and, therefore, $f(s)$ generates the uniform distribution on $\mathcal{T}_N$ given $T \in \mathcal{T}_N$ in this case (this fact was noted for the first time in (Pavlov, 1991) in a slightly different form). The following statement is an easy consequence of this fact.

COROLLARY 1. *Let a planted plane tree T be chosen randomly and equiprobably from the set $\mathcal{T}_N$. Then*

$$\lim_{N\to\infty} \mathbf{P} \left\{ \frac{H^*(T)}{\sqrt{N}} \le x \right\} = Q_1(x).$$

Example 4. Let $\Delta = \{0\}$, $f(s) = 1/2 + s^2/2$, $q = 1/2$, $A = \{0, 2\}$. Using (22), we obtain

$$\mathbf{P}\left\{G = G_{2N-1} \mid G \in \mathcal{G}_{2N-1}^{(A)}\right\} = \frac{1}{|\mathcal{G}_{2N-1}^{(A)}|}$$

$$= \mathbf{P}_\delta\left\{T = \delta(G_{2N-1}) \mid T \in \mathcal{T}_{2N-1}^{(A)}\right\} = \frac{1}{|\mathcal{T}_{2N-1}^{(A)}|},$$

where $\mathcal{T}_{2N-1}^{(A)}$ is the set of all binary rooted trees with $2N - 1$ nonrooted vertices or, what is the same, with N nonrooted vertices having the degree 0. It is clear that, if $\Delta = \{0\}$, then $\nu_\Delta = N$. Since $B = 1$ in this case, we arrive at the following statement (Gupta et al., 1990).

COROLLARY 2. *Let a binary rooted tree T be chosen randomly and equiprobably from the set $\mathcal{T}_{2N-1}^{(A)}$. Then*

$$\lim_{N\to\infty} \mathbf{P}\left\{\frac{H^*(T)}{\sqrt{N}} \leq x\right\} = Q_{1/2}\left(\frac{x}{\sqrt{2}}\right)$$

It is quite easy to understand now that, if A is a fixed subset of the set $\{0, 1, 2, \ldots\}$, $f(s)$ is chosen in such a way that $p_k = cq^k$, $k \in A$, and $p_k = 0$ otherwise, and $f'(1) = \sum_{k\in A} ckq^{k-1} = 1$, then the correspondence δ generates on $\mathcal{T}_N^{(A)}$ (the set of all planted plane trees with N nonrooted vertices, the degrees of which belong to A) the uniform distribution and, if T is a tree selected randomly from $\mathcal{T}_N^{(A)}$, then

$$\lim_{N\to\infty} \mathbf{P}\left\{H^*(T)\sqrt{\frac{B}{2N}} \leq x\right\} = Q_1(x),$$

where $B = \sum_{k\in A} ck(k-1)q^{k-2}$.

Let us consider now the following scheme. First we choose randomly and equiprobably a tree T from $\mathcal{T}_N^{(A)}$ (with edges of unit length) and then assign positive i.i.d. random variables $l_1, l_2, \ldots, l_N$ having a common distribution $M(t)$ to the edges of T. The problem is to find the distribution of the height $H(T^M)$ of the new tree T^M. Applying the arguments similar to those used above, one can deduce the following statement from Theorem 2.

COROLLARY 3. *Let $t^4(1 - M(t)) \to 0$ as $t \to \infty$. Then*

$$\lim \mathbf{P}\left\{H(T^M)\sqrt{\frac{B}{2N}} \leq x\right\} = Q_1\left(\frac{x}{\mathbf{E}l_1}\right).$$

In conclusion, we consider another problem connected with the structure of a random planted plane tree. Let $\mathcal{T}_{N,n} = \{T \in \mathcal{T}_N\colon H(T) \geq n\}$. Denote the root of such subtree of the tree T (with minimal number of vertices) which contains all the vertices of height n of T by $q(n, T)$. Let $\sigma(n, T) = n - H(q(n, T))$. It is easy to show using the correspondence δ (see §2) that

$$\mathbf{P}\{\lambda(n) \leq x \mid z(n) > 0,\ \nu = N\}$$

$$= \mathbf{P}_\delta\left\{\sigma(n, T) \leq x \mid H(T) \geq n,\ T \in \mathcal{T}_{N,n}\right\}.$$

Applying Theorem 3 with $f(s) = (2 - s)^{-1}$, we obtain

THEOREM 5. *If T is chosen randomly and equiprobably from the set $\mathcal{T}_{N,n}$ and $n, N \to \infty$ in such a way that $nN^{-1/2} \to \beta \in (0, \infty)$, then*

$$\lim \mathbf{P}\left\{n^{-1}\sigma(n, T) \leq a\sqrt{2}\right\} = I_\beta(a\sqrt{2})$$

for any $a \in (0, 1)$.

REFERENCES

Baum, L.E. and Katz, M. (1965). Convergence rates in the law of large numbers. *Trans. Amer. Math. Soc.* **120**, 108–123.

Gupta, V.K., Mesa, O., and Waymire, E. (1990). Tree-dependent extreme values: The exponential case. *J. Appl. Probab.* **27**, 124–133.

Harary, F. and Palmer, E.M. (1973). *Graphical Enumeration*. Academic Press, New York, London.

Harris, T.E. (1963). *The Theory of Branching Processes*. Springer-Verlag, New York.

Kolchin, V.F. (1978). Moment of degeneration of a branching process and height of a random tree. *Math. Notes* **24**, 954–961.

Kolchin, V.F. (1986). *Random Mappings*. Optimization Software, New York.

Pavlov, Yu.L. (1991). Planted plane trees. In: *Stochastic Processes and Their Applications*. Moskowskii Institut Elektronnogo Mashinostroeniya, Moscow, pp. 45–47 (in Russian).

Slack, R.S. (1968). A branching process with mean one and possibly infinite variance. *Z. Wahrscheinlichkeitstheor. verw. Geb.* **9**, 139–145.

Vatutin, V.A. and Sagitov, S.M. (1988). A decomposable ctitical branching process with two types of particles. *Proc. Şteklov Math. Inst.* **4**, 1–19.

Frontiers in Pure and Applied Probab. 1, pp. 269 – 286
H. Niemi et al. (Eds.)
1993 TVP/VSP

ON THE MARCINKIEWICZ WEAK LAWS OF LARGE NUMBERS IN BANACH SPACES

A. I. VOLODIN

Research Institute of Mathematics and Mechanics of Kazan University, Universitetskaya Street 17, 420008 Kazan, Russia

ABSTRACT

Some general theorems concerning the weak law of large numbers are proved. Necessary and sufficient conditions for the validity of the weak law of large numbers and the stochastic boundedness of weighted sums (under some conditions on weights are found) of independent identically distributed random elements with zero means assuming values in an arbitrary Banach space. These results are applied to certain random elements.

INTRODUCTION

The interesting problem arising in the study of the laws of large numbers is to find necessary and sufficient conditions for the validity of the weak law and the stochastic boundedness of weighted sums (under certain conditions on the weights) of independent identically distributed random elements with zero means assuming values in an Banach space. Note that in applications of the law of large numbers its weak variant and stochastic boundedness are of the utmost importance.
Let us introduce some notations and formulate the conditions ensuring the validity of the main result of this section.
Let X be a random element with zero mean assuming values in a Banach space E, $(X_k)_1^\infty$ be independent copies of X,

$$T_n = T_n(X) = \sum_{k=1}^{n} a_k(n)X_k,$$

where $\mathbf{a} = (a_k(n),\ 1 \le k \le n,\ n \in \mathbf{N})$ is a triangular array of constants which is called a weight in what follows, and let

$$S_n = S_n(X) = \sum_{k=1}^{n} X_k.$$

Let any weight $\mathbf{a}$ under consideration satisfy the following assumptions.
For all $k \in \mathbf{N}$

(i) $\lim_{n\to\infty} |a_k(n)| = 0$, and there exists $B > 0$ such that for all $m > n$: $|a_k(m)| \leq B|a_k(n)|$.

(ii) there exists $C > 1/2$ such that for all n: $|a_k(2n)| \geq C|a_k(n)|$.

Introduce the sequence $\varphi = (\varphi(n))$: $\varphi(n) = 1/\max_{1\leq k\leq n} |a_k(n)|$. Since we are interested in the law of large numbers, it is natural to use the following condition:

(iii) there exist $p < 2$ and $A > 0$ such that $\varphi(n) \geq An^{1/p}$ for all n.

Note that condition (iii) is necessary even for the random variables assuming values in $\mathbf{R}$.

Let us say a few words about condition (ii). We can assume that $C < 1$. To simplify the proofs of some technical results, we define the functions $a_k(t)$ and $\varphi(t)$, $t \geq 1$, $a_k(t) = a_k([t])$, $\varphi(t) = \varphi([t])$, where $[\cdot]$ denotes the integer part of a number. Then (ii) can be rewritten as

(ii′) $|a_k(tn)| \geq D\, t^q\, |a_k(n)|$ for all $n \in \mathbf{N}$ and $t \geq 1$, where $q = \log_2 C$ (< 0) and $D = B^2/C$.

In fact, if by $\langle t\rangle$ we denote the last integer number not smaller than t (≥ 1), then, by (i) and (ii), we have

$$|a_k(tn)| \geq B|a_k(\langle t\rangle n)| = B|a_k(2^{\log_2\langle t\rangle}n)| \geq B^2|a_k(2^{\langle\log_2\langle t\rangle\rangle}n)|$$

$$\geq B^2C^{\langle\log_2\langle t\rangle\rangle}|a_k(n)| \geq B^2C^{\log_2 t-1}|a_k(n)| \geq (B^2/C)t^q|a_k(n)|.$$

We say that a sequence of random elements (Y_n) is stochastically bounded, if $\lim_{t\to\infty}\sup_n \mathbf{P}\{\|Y_n\| > t\} = 0$.

Let us introduce the following notations:

$$\mathbf{SB_a}(E) = \{X \in L_0(E)\colon \mathbf{E}X = 0, \text{ and the sequence } (T_n(X)) \text{ is stochastically bounded}\},$$

$$\mathbf{WLLN_a}(E) = \{X \in L_0(E)\colon \mathbf{E}X = 0,\ T_n(X) \to 0 \text{ as } n \to \infty \text{ in probability}\},$$

$$\mathbf{WLLN_p}(E) = \{X \in L_0(E)\colon \mathbf{E}X = 0,\ S_n(X)/n^{1/p} \to 0 \text{ as } n \to \infty \text{ in probability}\},\quad 1 < p < 2.$$

In the sequel, we often use the following inequalities. In their formulations given below X_k are independent random elements and $t > 0$.

0.1. *Levy's inequality* (Vakhania et al., 1987, Proposition V.2.3)

If X_k are symmetric, then

$$\mathbf{P}\{\max_{1\leq k\leq n}\|X_k\| > t\} \leq 2\mathbf{P}\{\|\sum_{k=1}^{n} X_k\| > t\}.$$

0.2. *Kolmogorov's inequality* (Kwapien and Woyczynski, 1987)

If X_k have zero means, then, whenever the right-hand side below is positive,

$$\mathbf{E}\|\sum_{k=1}^{n} X_k\| \leq 3(\mathbf{E}\max_{1\leq k\leq n}\|X_k\| + 8t)/(1 - 12\mathbf{P}\{\|\sum_{k=1}^{n} X_k\| > t\}).$$

0.3. *Kwapien's inequality* (Vakhania et. al., 1987, Lemma V.4.1 (a))
If X_k are symmetric and $|\beta_k| \geq |\alpha_k|$, then

$$\mathbf{P}\{\|\sum_{k=1}^{n} \alpha_k X_k\| > t\} \leq 2\mathbf{P}\{\|\sum_{k=1}^{n} \beta_k X_k\| > t\}.$$

0.4. *Hoffmann–Jorgencen's inequality* (Hoffmann–Jorgencen, 1974)
If X_k are symmetric, then

$$\mathbf{P}\{\|\sum_{k=1}^{n} X_k\| > 3t\} \leq 4\mathbf{P}^2\{\|\sum_{k=1}^{n} X_k\| > t\} + \mathbf{P}\{\max_{1\leq k\leq n}\|X_k\| > t\}.$$

0.5. *Contraction principle* (Vakhania et al., 1987, Lemma V.4.1 (c))
If $\alpha_k \in \mathbf{R}$, then

$$\mathbf{E}\|\sum_{k=1}^{n} \alpha_k X_k\| \leq 2\max_{1\leq k\leq n}|\alpha_k|\mathbf{E}\|\sum_{k=1}^{n} X_k\|.$$

0.6. *de Acosta's inequality* (Acosta, 1981)
There exists $C > 0$ such that

$$\mathbf{E}|\,\|\sum_{k=1}^{n} X_k\|^2 - \mathbf{E}\|\sum_{k=1}^{n} X_k\|^2| \leq C\sum_{k=1}^{n}\mathbf{E}\|X_k\|^2.$$

0.7. *Inequality for means* (Araujo and Gine, 1980, Lemma III.2.7)
If $\mathbf{E}X_1 = 0$, then $\mathbf{E}\|X_2\| \leq \mathbf{E}\|X_1 + X_2\|$.
0.8. *Elementary inequality*
For all $t \leq 0$

$$\mathbf{P}\{\|\sum_{k=1}^{n} X_k\| > t\} \leq \sum_{k=1}^{n}\mathbf{P}\{\|X_k\| > t\} + \mathbf{P}\{\|\sum_{k=1}^{n} X_k I\{\|X_k\| \leq t\}\| > t\}.$$

Note that one usually regards the relation

$$\mathbf{P}\{\|\sum_{k=1}^{n} \alpha_k X_k\| > t\} \leq 2\mathbf{P}\{\max_{1\leq k\leq n}|\alpha_k|\,\|\sum_{k=1}^{n} X_k\| > t\}$$

as Kwapien's inequality.
Our inequality follows easily from this one, because

$$\mathbf{P}\{\|\sum_{k=1}^{n} \alpha_k X_k\| > t\} \leq \mathbf{P}\{\|\sum_{k=1}^{n}(\alpha_k/\beta_k)\beta_k X_k\| > t\}$$

$$\leq 2\mathbf{P}\{\max_{1\leq k\leq n}|\alpha_k/\beta_k|\,\|\sum_{k=1}^{n} \beta_k X_k\| > t\} \leq 2\mathbf{P}\{\|\sum_{k=1}^{n} \beta_k X_k\| > t\}.$$

1. CONDITIONS OF STOCHASTIC BOUNDEDNESS

In this section we state the necessary and sufficient condition for stochastic boundedness of weighed sums of i.i.d. random elements zero means. Note that in this section we do not use condition (iii).

Let us formulate and prove some technical results with the aid of which we shall prove the main statement of this section.

For $s > 0$ and $t > 0$ set $Y_{k,s}(tn) = a_k(tn)X_k I\{\|a_k(tn)X_k\| \le s\}$.

LEMMA 1.1. *Let X be a symmetric random element. Consider the conditions*

a) $\lim\limits_{s\to\infty} \sup\limits_{t>0} t\mathbf{P}\{\|X\| > s\varphi(t)\} = 0,$

b) $\lim\limits_{s\to\infty} \sup\limits_{n} \left(\mathbf{E}\|\sum\limits_{k=1}^{n} Y_{k,s}(n)\|/s\right) = 0,$

c) *there exists $u > 0$ such that*

$$\sup_{s>u}\sup_{n}\sup_{t\ge1} \frac{\mathbf{E}\|\sum\limits_{k=1}^{n} Y_{k,s}(tn)\|}{s} < \infty.$$

Conditions a) *and* b) *are sufficient, and conditions* a) *and* c) *are necessary for the inclusion $X \in \mathbf{SB}_{\mathbf{a}}(E)$.*

Proof. Sufficiency. Let a) and b) hold true. Set

$$T_n' = \sum_{k=1}^{n} a_k(n)X_k I\{\|a_k(n)X_k\| \le s\} = \sum_{k=1}^{n} Y_{k,s}(n),$$

$$T_n'' = \sum_{k=1}^{n} a_k(n)X_k I\{\|a_k(n)X_k\| > s\}.$$

In this notation, $T_n = T_n' + T_n''$, and it suffices to show that $\sup\limits_n \mathbf{P}\{\|T_n'\| > s\} \to 0$ and $\sup\limits_n \mathbf{P}\{\|T_n''\| > s\} \to 0$ as $s \to \infty$.

By Chebyshev's inequality and condition b),

$$\sup_n \mathbf{P}\{\|T_n'\| > s\} \le \sup_n \frac{\mathbf{E}\|T_n'\|}{s} = \sup_n \frac{\mathbf{E}\|\sum\limits_{k=1}^{n} Y_{k,s}(n)\|}{s} \to 0 \quad \text{as} \quad s \to \infty.$$

Furthermore, we have

$$\{\|T_n''\| > s\} \subset \{\|T_n''\| > 0\}$$
$$= \{\|\sum_{k=1}^{n} a_k(n)X_k I\{\|a_k(n)X_k\| > s\}\| > 0\} \subset \bigcup_{k=1}^{n} \{\|a_k(n)X_k\| > s\},$$

hence, by condition a),

$$\mathbf{P}\{\|T_n''\| > s\} \le \sum_{k=1}^{n} \mathbf{P}\{\|a_k(n)X_k\| > s\} \le \sum_{k=1}^{n} \mathbf{P}\{\|X_k\| > s\varphi(n)\}$$
$$= n\mathbf{P}\{\|X\| > s\varphi(n)\} \to 0 \quad \text{as} \quad s \to \infty.$$

Necessity. Now let $X \in \mathbf{SB_a}(E)$. Then $f(s) \equiv \sup_n \mathbf{P}\{\|T_n\| > s\} \to 0$ as $s \to \infty$. By Levy's inequality (0.1), for all n $f(s) \geq \mathbf{P}\{\max_{1\leq k\leq n} |a_k(n)|\,\|X_k\| > s\}/2$ or

$$1 - 2f(s) \leq \mathbf{P}\{\max_{1\leq k\leq n} \|X_k\| \leq s\varphi(n)\}$$

$$= \mathbf{P}\{\bigcap_{k=1}^{n} \{\|X_k\| \leq s\varphi(n)\}\} = \mathbf{P}^n\{\|X\| \leq s\varphi(n)\}.$$

Hence, $\mathbf{P}\{\|X\| > s\varphi(n)\} \leq 1 - (1 - f(s))^{1/n} \leq -\log(1 - 2f(s))/n$. So, $\lim_{s\to\infty} \sup_{t>0} t\mathbf{P}\{\|X\| > s\varphi(t)\} = 0$.
Furthermore, for all u, $0 < u \leq 2s$, we have

$$\mathbf{P} = \mathbf{P}\{\|\sum_{k=1}^{n} Y_{k,s}(tn)\| > u\}$$

$$= \mathbf{P}\{\|\sum_{k=1}^{n} a_k(tn)X_k - \sum_{k=1}^{n} a_k(tn)X_k I\{\|a_k(tn)X_k\| > s\}\| > u\}$$

$$\leq \mathbf{P}\{\|\sum_{k=1}^{n} a_k(tn)X_k\| > \frac{u}{2}\} + \mathbf{P}\{\|\sum_{k=1}^{n} a_k(tn)X_k I\{\|a_k(tn)X_k\| > s\}\| > \frac{u}{2}\}.$$

We estimate the first term in the right-hand side of the previous inequality by Kwapien's inequality (0.3) taking into account (i) and $t \geq 1$ and the second term by elementary inequality (0.8). As a result, we obtain

$$\mathbf{P} \leq 2\mathbf{P}\left\{\|T_n\| > \frac{u}{2B}\right\} + \sum_{k=1}^{n} \mathbf{P}\left\{\|a_k(tn)X_k I\{\|a_k(tn)X_k\| > s\}\| > \frac{u}{2}\right\}$$

$$+\mathbf{P}\left\{\|\sum_{k=1}^{n} a_k(tn)X_k I\{\|a_k(tn)X_k\| > s\} I\left\{\|a_k(tn)X_k\| \leq \frac{u}{2}\right\}\right\}.$$

By the choice of $u < 2s$, the latter event is impossible and

$$\mathbf{P} \leq 2\mathbf{P}\left\{\|T_n\| > \frac{u}{2B}\right\} + n\mathbf{P}\left\{\|X\| > \frac{u\varphi(n)}{2}\right\}.$$

Since $X \in \mathbf{SB_a}(E)$ and a) takes place, there exists $U > 0$ such that for all $s \geq U/2$

$$\sup_n \sup_{t\geq 1} \mathbf{P}\left\{\|\sum_{k=1}^{n} Y_{k,s}(tn)\| > U\right\} \leq \frac{1}{24}.$$

Then, by Kolmogorov's inequality (0.2),

$$\mathbf{E}\|\sum_{k=1}^{n} Y_{k,s}(tn)\| \le \frac{3(\mathbf{E}\max\limits_{1\le k\le n}\|Y_{k,s}(tn)\| + 8U)}{1 - 12\mathbf{P}\left\{\|\sum\limits_{k=1}^{n} Y_{k,s}(tn)\| > U\right\}} \le 6(s+8U).$$

For any $s > 0$ and any random element X define $\rho_{n,s}(X) = \sup\limits_{t\ge 0} t\mathbf{P}\{t^q\|T_n(X)\| > s\}$. Recall (see (ii)′) that $q = \log_2 C$ and C is given by (ii).

LEMMA 1.2. *There exists $H > 0$ such that $\mathbf{E}\|T_n(X)\| \le Hs(\rho_{n,s}(X) + \rho_{n,s}^{2+1/q}(X))$ for all symmetric random elements X and $s > 0$.*

Proof. For the sake of bravity, we put $\rho = \rho_{n,s}(X)$. We have

$$\mathbf{E}\|T_n(X)\| = \int_0^\infty \mathbf{P}\{\|T_n(X)\| > t\}dt$$

$$= s\int_0^\infty \mathbf{P}\{\|T_n\| > st\}dt = s\Big(\int_0^\rho \mathbf{P}\{\|T_n\| > st\}dt\Big) + \int_\rho^\infty \mathbf{P}\left\{\|\frac{T_n}{t}\| > s\right\}dt$$

$$\le s\Big(\rho + \int_\rho^\infty t^{-1/q}\mathbf{P}\left\{(t^{-1/q})^q\|T_n\| > s\right\}t^{1/q}dt\Big) \le s\Big(\rho + \rho\int_\rho^\infty t^{1/q}dt\Big).$$

Since we have $C > 1/2$ in condition (ii), it follows that $q > -1$ and the last integral converges. The straightforward computation completes the proof.
On the ground of these lemmas, we prove

THEOREM 1.1. *Let X be a symmetric random element. The inclusion $X \in \mathbf{SB_a}(E)$ holds, if and only if $\lim\limits_{s\to\infty}\sup\limits_n \rho_{n,s}(X) = 0$.*

Proof. Necessity. Let $X \in \mathbf{SB_a}(E)$, $L = 3^{1/q}$, and $0 < \varepsilon < L/2$. We shall prove that there exists an $s_0 = s_0(\varepsilon)$ such that $\rho_{n,s}(X) \le \varepsilon$ for all $s > s_0$ and $n \in \mathbf{N}$. Note that $\rho_{n,s}(X) = \sup\limits_{t\ge 0} t\mathbf{P}\{t^q\|T_n(X)\| > s\} \equiv \sup\limits_{t\ge 0} D(t) = \max(\sup\limits_{0<t\le\varepsilon} D(t), \sup\limits_{\varepsilon<t\le 1} D(t), \sup\limits_{t>1} D(t)) \le \max(\varepsilon, \sup\limits_{\varepsilon<t\le 1} D(t), \sup\limits_{t>1} D(t))$. Moreover, $\sup\limits_{\varepsilon<t\le 1} D(t) \le \mathbf{P}\{\varepsilon^q\|T_n\| > s\}$ and, since $X \in \mathbf{SB_a}(E)$, there exists $s_1 = s_1(\varepsilon)$ such that $\sup\limits_{\varepsilon<t\le 1} D(t) \le \varepsilon$ for all $s > s_1$.
Furthermore, there exists $s_2 = s_2(\varepsilon)$ such that for all $s > s_2$ and $n \in \mathbf{N}$ we have

$$\mathbf{P}\{\|T_n(X)\| > s\} < \frac{\varepsilon}{4} < \frac{L}{8} \tag{1}$$

(the last inequality follows from the choice of ε). By assertion (1) of Lemma 1.1, there exists $s_3 = s_3(\varepsilon)$ such that

$$\sup_{t>0} t\mathbf{P}\{\|X\| > s\varphi(t)\} < \frac{\varepsilon}{4} \tag{2}$$

for all $s > s_3$.

Let $s > s_0(\varepsilon) = \max(s_1, s_2, s_3)$ and $A > L$. Then, according to inequality (1),

$$I(A) \equiv \sup_{1<t<A} t\mathbf{P}\{t^q\|T_n\| > s\} = \sup_{1\le t\le L} D(t) + \sup_{L<t<A} D(t)$$

$$\le L\mathbf{P}\{\|T_n\| > s\} + \sup_{L<t<A} D(t) \le \frac{\varepsilon}{4} + \sup_{L<t<A} t\mathbf{P}\{3t^q\|T_n(X)\| > 3s\}.$$

Furthermore, by the choice of L,

$$I(A) \le \frac{\varepsilon}{4} + \sup_{1<u<A/L} \frac{u}{L}\mathbf{P}\{\|u^q T_n\| > 3s\}.$$

Applying Hoffmann–Jorgensen inequality (0.4), we obtain

$$I(A) \le \frac{\varepsilon}{4} + \frac{1}{L}\Big(4 \sup_{1<u<A/L} u\mathbf{P}^2\{\|u^q T_n\| > s\} +$$

$$\sup_{1<u<A/L} u\mathbf{P}\{\max_{1\le k\le n} \|u^q a_k(n) X_k\| > s\}\Big)$$

$$\le \frac{\varepsilon}{4} + \frac{4}{L}\Big(\mathbf{P}\{\|T_n\| > s\} \sup_{1<u<A/L} u\mathbf{P}\{\|u^q T_n\| > s\} +$$

$$\sup_{u>1} u \sum_{k=1}^{n} \mathbf{P}\{\|u^q a_k(n) X_k\| > s\}\Big).$$

Then, by condition (ii)′, in view of the inequality $L > 1$, and by virtue of (1) and (2), we have

$$I(A) \le \frac{\varepsilon}{4} + 4\mathbf{P}\{\|T_n\| > s\} D \sup_{1<u<A/L} \frac{u}{L}\mathbf{P}\{\|u^q T_n\| > s\}$$

$$+ \sup_{u>1} un\mathbf{P}\{\|X\| > s\varphi(un)\} \le \frac{\varepsilon}{4} + \frac{I(A)}{2} + \frac{\varepsilon}{4}.$$

Hence, $I(A) < \varepsilon$. Thus, $\rho_{n,s}(X) \le \varepsilon$ for $s > s_0$.

Sufficiency. By Chebyshev's inequality and Lemma 1.2,

$$\sup_n \mathbf{P}\{\|T_n\| > s\} \le \sup_n \frac{\mathbf{E}\|T_n\|}{s} \le H(\rho_{n,s}(X) + \rho_{n,s}^{2+1/q}(X)) \to 0 \quad \text{as} \quad s \to \infty.$$

Put

$$\lambda_{\mathbf{a}}(X) = \sup_n \mathbf{E}\|T_n(X)\|$$

for a random element X and introduce the space

$$\mathbf{SB}_{\mathbf{a}}^{\infty}(E) = \{X \in L_0(E)\colon\ \mathbf{E}X = 0 \ \text{and} \ \lambda_{\mathbf{a}}(X) < \infty\}.$$

This is a linear space, and $\lambda_{\mathbf{a}}(\cdot)$ is a norm. Let us prove, for example, the triangular inequality. Let $X,\ Y \in \mathbf{SB}_{\mathbf{a}}^{\infty}(E)$. If (X_k, Y_k) is a sequence of independent copies of the vector (X, Y), then

$$\lambda_{\mathbf{a}}(X+Y) = \sup_n \mathbf{E}\|T_n(X+Y)\| = \sup_n \mathbf{E}\|\sum_{k=1}^{n} a_k(n)(X_k+Y_k)\|$$

$$= \sup_n \mathbf{E}\|T_n(X)+T_n(Y)\| \le \lambda_{\mathbf{a}}(X)+\lambda_{\mathbf{a}}(Y).$$

It is easy to show that $\mathbf{SB}_{\mathbf{a}}^{\infty}(E)$ is a Banach space.

THEOREM 1.2. $\mathbf{SB}_{\mathbf{a}}(E) = \mathbf{SB}_{\mathbf{a}}^{\infty}(E)$.

Proof. If $X \in \mathbf{SB}_{\mathbf{a}}^{\infty}(E)$, then $X^s \in \mathbf{SB}_{\mathbf{a}}(E)$, where X^s is the symmetrization of X. By inequality (0.7) for means and by Lemma 1.2, we have

$$\lambda_{\mathbf{a}}(X) = \sup_n \mathbf{E}\|T_n(X)\| \le \sup_n \mathbf{E}\|T_n(X^s)\| \le sH(\rho_{n,s}(X) + \rho_{n,s}^{2+1/q}(X)) < \infty$$

for any $s > 0$ (wherein the last inequality follows from Theorem 1.1).
The reverse is evident by Chebyshev's inequality.

2. WEAK LAWS OF LARGE NUMBERS FOR THE WEIGHTED SUM

In this section we state necessary and sufficient conditions for the weak law of large numbers to hold for weighted sums of independent identically distributed random elements with zero means. We shall prove that the set of all such elements is the closure of the set of step random elements (i.e., taking on a finite number of values only) by the norm $\lambda_{\mathbf{a}}(\cdot)$ introduced at the end of the previous section. This result is an analogue of the well-known results of G. Pisier for the cases of the central limit theorem and the law of the iterated logarithm (Pisier, 1975); it is also an extension of the result of R. Norvaisa to the case of the Marcinkiewicz law of large numbers (Norvaisa, 1984).
Let us formulate and prove some technical results which enable us to obtain the main results of this section. Recall the notation

$$Y_{k,s}(tn) = a_k(tn)X_k I\|a_k(nt)X_k\| \le s\}.$$

LEMMA 2.1. *Let X be a symmetric random element. Consider the conditions*
a) $\lim_{t\to\infty} t\mathbf{P}\{\|X\| > s\varphi(t)\} = 0$,
b) $\lim_{n\to\infty} \sup_{t\ge 1} \mathbf{E}\|\sum_{k=1}^{n} Y_{k,s}(tn)\| = 0$.
In order for the inclusion $X \in \mathbf{WLLN}_{\mathbf{a}}(E)$ to hold, it is sufficient that a) *and* b) *are satisfied for at least one $s > 0$ and it is necessary that* a) *and* b) *are satisfied for all $s > 0$.*

Proof. Sufficiency. Let a) and b) be satisfied for some $s > 0$. Define

$$T_n' = \sum_{k=1}^{n} a_k(n)X_k I\{\|a_k(n)X_k\| \le s\} = \sum_{k=1}^{n} Y_{k,s}(n),$$

$$T_n'' = \sum_{k=1}^{n} a_k(n)X_k I\{\|a_k(n)X_k\| > s\}.$$

In this notation, $T_n = T_n' + T_n''$, and it is sufficient to show that $\mathbf{P}\{\|T_n'\| > \varepsilon\} \to 0$ and $\mathbf{P}\{\|T_n''\| > \varepsilon\} \to 0$ as $n \to \infty$ for all $\varepsilon > 0$.
By Chebyshev's inequality and condition b),

$$\mathbf{P}\{\|T_n'\| > \varepsilon\} \le \frac{1}{\varepsilon}\mathbf{E}\|T_n'\| = \frac{1}{\varepsilon}\mathbf{E}\|\sum_{k=1}^{n} Y_{k,s}(n)\| \to 0 \quad \text{as} \quad n \to \infty.$$

Furthermore, we have

$$\{\|T_n''\| > \varepsilon\} = \{\|T_n''\| > 0\}$$

$$= \{\|\sum_{k=1}^{n} a_k(n)X_k I\{\|a_k(n)X_k\| > s\}\| > 0\} \subset \bigcup_{k=1}^{n}\{\|a_k(n)X_k\| > s\},$$

hence, by condition a),

$$\mathbf{P}\{\|T_n''\| > \varepsilon\} \le \sum_{k=1}^{n} \mathbf{P}\{\|a_k(n)X_k\| > s\} \le \sum_{k=1}^{n} \mathbf{P}\{\|X_k\| > s\varphi(n)\}$$

$$= n\mathbf{P}\{\|X\| > s\varphi(n)\} \to 0 \quad \text{as} \quad n \to \infty.$$

Necessity. Let $X \in \mathbf{WLLN_a}(E)$. Fix some $s > 0$. Then $f(n) = \mathbf{P}\{\|T_n(X)\| > s\} \to 0$ as $n \to \infty$. By Levy's inequality (0.1), for all n we have

$$f(n) \ge \frac{1}{2}\mathbf{P}\{\max_{1\le k\le n} |a_k(n)|\, \|X_k\| > s\}$$

or

$$1 - 2f(n) \le \mathbf{P}\{\max_{1\le k\le n} \|X_k\| \le s\varphi(n)\}$$

$$= \mathbf{P}\{\bigcap_{k=1}^{n}\{\|X_k\| \le s\varphi(n)\}\} = \mathbf{P}^n\{\|X\| \le s\varphi(n)\}.$$

Hence, $\mathbf{P}\{\|X\| > s\varphi(n)\} \le 1 - (1 - f(n))^{1/n}$. So, $n\mathbf{P}\{\|X\| > s\varphi(n)\} \to 0$ as $n \to \infty$.
Furthermore, for all ε such that $0 < \varepsilon \le 2s$ and for all $t \ge 1$ we obtain

$$\mathbf{P} = \mathbf{P}\{\|\sum_{k=1}^{n} Y_{k,s}(tn)\| > \varepsilon\}$$

$$= \mathbf{P}\{\|\sum_{k=1}^{n} a_k(tn)X_k - \sum_{k=1}^{n} a_k(tn)X_k I\{\|a_k(tn)X_k\| > s\}\| > \varepsilon\}$$

$$\leq \mathbf{P}\left\{\|\sum_{k=1}^{n} a_k(tn)X_k\| > \frac{\varepsilon}{2}\right\} + \mathbf{P}\left\{\|\sum_{k=1}^{n} a_k(tn)X_kI\{\|a_k(tn)X_k\| > s\}\| > \frac{\varepsilon}{2}\right\}.$$

We estimate the first term in the right-hand side of the previous inequality by Kwapien's inequality (0.3) taking into account (i) and $t \geq 1$ and the second term — by elementary inequality (0.8). As a result, we have

$$\mathbf{P} \leq 2\mathbf{P}\left\{\|T_n\| > \frac{\varepsilon}{2B}\right\} + \sum_{k=1}^{n} \mathbf{P}\left\{\|a_k(tn)X_kI\{\|a_k(tn)X_k\| > s\}\| > \frac{\varepsilon}{2}\right\}$$

$$+\mathbf{P}\left\{\|\sum_{k=1}^{n} a_k(tn)X_kI\{\|a_k(tn)X_k\| > s\}I\left\{\|a_k(tn)X_k\| \leq \frac{\varepsilon}{2}\right\}\| > \frac{\varepsilon}{2}\right\}.$$

By the choice of $u < 2s$, the last event is impossible and

$$\mathbf{P} \leq 2\mathbf{P}\left\{\|T_n\| > \frac{\varepsilon}{2B}\right\} + n\mathbf{P}\left\{\|X\| > \frac{\varepsilon\varphi(n)}{2}\right\}.$$

Since $X \in \mathbf{WLLN_a}(E)$ and a) takes place, there exists $N = N(\varepsilon)$ such that

$$\sup_{n\geq N}\sup_{t\geq 1}\mathbf{P}\left\{\|\sum_{k=1}^{n} Y_{k,s}(tn)\| > \varepsilon\right\} \leq \frac{1}{24}.$$

Applying Hoffmann–Jorgencen's inequality (0.4) to the integral

$$I(A) = \int_0^A \mathbf{P}\left\{\|\sum_{k=1}^{n} Y_{k,s}(tn)\| > u\right\} du = \int_0^{A/3} \mathbf{P}\left\{\|\sum_{k=1}^{n} Y_{k,s}(tn)\| > 3v\right\} dv,$$

we obtain

$$I(A) \leq 3(4\int_0^{A/3} \mathbf{P}^2\left\{\|\sum_{k=1}^{n} Y_{k,s}(tn)\| > v\right\} dv$$

$$+\int_0^{A/3} \mathbf{P}\{\max_{1\leq k\leq n} |a_k(tn)|\,\|X_kI\{\|a_k(tn)X_k\| \leq s\}\| > v\}dv$$

$$\leq 12\int_0^{\varepsilon} \mathbf{P}^2\{\|\sum_{k=1}^{n} Y_{k,s}(tn)\| > v\}dv$$

$$+12\int_{\varepsilon}^{A/3} \mathbf{P}\{\|\sum_{k=1}^{n} Y_{k,s}(tn)\| > v\}\mathbf{P}\{\|\sum_{k=1}^{n} Y_{k,s}(tn)\| > v\}dv$$

$$+3\int_0^s \mathbf{P}\{\max_{1\le k\le n}|a_k(tn)|\,\|X_k\|>v\}dv.$$

By the previous reasonings, the first multiplier in the second integral is not greater than 1/24. Applying Levy's inequality (0.1) to the last integral, we have

$$I(A)\le 12\varepsilon+\frac{1}{2}\int_0^A \mathbf{P}\{\|\sum_{k=1}^n Y_{k,s}(tn)\|>v\}dv+6\int_0^s \mathbf{P}\{\|\sum_{k=1}^n a_k(tn)X_k\|>v\}dv$$

$$\le 12\varepsilon+\frac{I(A)}{2}+12B\int_0^s \mathbf{P}\{\|T_n\|>v\}dv$$

by condition (i). So, $I(A)\le 24\varepsilon+24B\int_0^s \mathbf{P}\{\|T_n\|>v\}dv$. The last integral converges to zero as $n\to\infty$ by the Lebesgue dominated convergence theorem. Hence,

$$\lim_{n\to\infty}\sup_{t\ge1}\mathbf{E}\|\sum_{k=1}^n Y_{k,s}(tn)\|=\lim_{n\to\infty}I(\infty)=0.$$

LEMMA 2.2. *Let η be a positive random variable and $b(t)$ be a positive function such that $\lim_{t\to\infty} t\mathbf{P}\{\eta>b(t)\}=0$ for any $s>0$, then $t\mathbf{E}\eta I\{\eta\le sb(t)\}/b(t)\to0$ as $t\to\infty$.*

Proof. Applying the formula of integration by parts and using the substitution $v=ub(t)$, we find

$$\frac{t}{b(t)}\mathbf{E}\eta I\{\eta\le sb(t)\}=-\frac{t}{b(t)}\int_0^{b(t)} v d\mathbf{P}\{\eta>v\}$$

$$=-t\mathbf{P}\{\eta>b(t)\}+\frac{t}{b(t)}\int_0^{b(t)}\mathbf{P}\{\eta>v\}dv$$

$$=-t\mathbf{P}\{\eta>b(t)\}+\int_0^1 t\mathbf{P}\{\eta>ub(t)\}\,du\to0$$

as $t\to\infty$ by the Lebesgue dominated convergence theorem.

LEMMA 2.3. *If $X\in\mathbf{WLLN}_{\mathbf{a}}(E)$, then for all $k\in\mathbf{N}$ and $s>0$*

$$\lim_{n\to\infty}\sup_{t\ge1} nt\mathbf{E}\|Y_{k,s}(tn)\|^2=0.$$

Proof. We have

$$d_n = \sup_{t\geq 1} nt\mathbf{E}\|Y_{k,s}(tn)\|^2 = \sup_{u\geq n} ua_k^2(u)\mathbf{E}\|X_k I\|a_k(u)X_k\| \leq s\}\|^2.$$

Let $b(u) = s^2/a_k^2(u)$ and $\eta = \|X_k\|^2$. In this notation, by Lemma 2.2, we have $d_n = \sup_{u\geq n}(u\mathbf{E}\eta I\{\eta \leq b(u)\}/b(u)) \to 0$ as $n \to \infty$, since $\mathbf{P}\{\eta I\{\eta \leq b(u)\} = u\mathbf{P}\{\|a_k(u)X_k\| > s\} \leq u\mathbf{P}\{\eta > b(u)\}$ as $n \to \infty$. With the aid of these lemmas, we prove the following theorem. Recall the notation $\rho_{n,s}(X) = \sup_{t\geq 0} t\mathbf{P}\{t^q\|T_n(X)\| > s\}$, where $q = \log_2 C$ and the constant C is taken from condition (ii).

THEOREM 2.1. *Let X be a symmetric random element. Next statements are equivalent:*
a) $X \in \mathbf{WLLN_a}(E)$,
b) $\rho_{n,s}(X) \to 0$ *as* $n \to \infty$ *for all* $s > 0$,
c) *there exists* $s > 0$ *such that* $\rho_{n,s}(X) \to 0$ *as* $n \to \infty$.

Proof. a)⇒b). Let us show that for all ε such that $0 < \varepsilon < 1$ and for all $s > 0$ there exists $N = N(\varepsilon, s)$ such that $\rho_{n,s}(X) < \varepsilon$ for all $n > N$. We have:

$$\rho_{n,s}(X) = \sup_{t>0} t\mathbf{P}\{t^q\|T_n(X)\| > s\} = \sup_{t>0} D(t)$$

$$= \max(\sup_{0<t\leq\varepsilon} D(t), \sup_{\varepsilon<t\leq 1} D(t), \sup_{t>1} D(t)) \leq \max(\varepsilon, \sup_{\varepsilon<t\leq 1} D(t), \sup_{t>1} D(t)).$$

Moreover, $\sup_{\varepsilon<t\leq 1} D(t) \leq \mathbf{P}\{\varepsilon^q\|T_n\| > s\}$ and, since $X \in \mathbf{WLLN_a}(E)$, there exists $N_1 = N_1(\varepsilon)$ such that for all $n \geq N_1$

$$\mathbf{P}\{\varepsilon^q\|T_n(X)\| > s\} < \frac{\varepsilon}{4}.$$

It remains to estimate the last supremum, that is, to prove that $A \equiv \sup_{t>1} D(t) < \varepsilon$ for sufficiently large n.
Note that, by Lemmas 2.1 and 2.3, there exists $N_2 = N_2(\varepsilon, s)$ such that for all $n \geq N_2$ and $k \leq n$

$$\sup_{u\geq n} u\mathbf{P}\{\|X\| > s\varphi(u)\} < \frac{\varepsilon}{4}. \tag{3}$$

$$\sup_{t\geq 1} \mathbf{E}\|Y_{k,s}(tn)\| \leq \frac{s}{2}, \tag{4}$$

$$\sup_{t\geq 1} t\mathbf{E}\|Y_{k,s}(tn)\|^2 \leq s^2\frac{\varepsilon}{16Cn}, \tag{5}$$

where the constant C is taken from de Acosta's inequality (0.6).
By condition (ii)′, $A \leq \sup_{t\geq 1} Dt\mathbf{P}\{\|\sum_{k=1}^n a_k(tn)X_k\| > s\}$. Now, applying elementary inequality (0.8), we get

$$A \leq 2(\sup_{t\geq 1}\sum_{k=1}^{n} t\mathbf{P}\{\|a_k(tn)X_k\| > s\}$$

$$+\sup_{t\geq 1} t\mathbf{P}\{\|\sum_{k=1}^{n} a_k(tn)X_kI\{\|a_k(tn)X_k\| \leq s\}\| > s\}).$$

Let us estimate each term separately. Using the substitution $u = tn$ and taking into account (3), we obtain the followiing estimate $I_1 \leq \sup_{u\geq n} u\mathbf{P}\{\|X\| > s\varphi(u)\} < (\varepsilon/4)$ for the first term. For the second term, we have, by (4),

$$I_2 = \sup_{t\geq 1} t\mathbf{P}\{\|\sum_{k=1}^{n} Y_{k,s}(tn)\| > s\} \leq \sup_{t\geq 1} t\mathbf{P}\{|\|\sum_{k=1}^{n} Y_{k,s}(tn)\| - \mathbf{E}\|\sum_{k=1}^{n} Y_{k,s}(tn)\|| > \frac{s}{2}\}.$$

Applying Chebyshev's inequality, de Acosta's inequality (0.6), and (5), we obtain

$$I_2 \leq 4s^{-2}\sup_{t\geq 1} t\mathbf{E}(\|\sum_{k=1}^{n} Y_{k,s}(tn)\| - \mathbf{E}\|\sum_{k=1}^{n} Y_{k,s}(tn)\|)^2$$

$$\leq 4s^{-2}C\sup_{t\geq 1}\sum_{k=1}^{n} t\mathbf{E}\|Y_{k,s}(tn)\|^2 \leq \frac{\varepsilon}{4}.$$

So, there exists $N = N(\varepsilon, s) = \max(N_1, N_2)$ such that $\rho_{n,s}(X) < \varepsilon$ for all $n > N$. Implication b)⇒c) is trivial, and c)⇒a) follows from Chebyshev's inequality and Lemma 1.2: for any $\varepsilon > 0$,

$$\mathbf{P}\{\|T_n(X)\| > \varepsilon\} \leq \frac{\mathbf{E}\|T_n(X)\|}{\varepsilon} \leq \frac{Hs}{\varepsilon}(\rho_{n,s}(X) + \rho_{n,s}^{2+1/q}(X)) \to 0 \quad \text{as} \quad n \to \infty.$$

COROLLARY. *The inclusion $X \in \mathbf{WLLN_a}(E)$ holds for a random element X with the zero mean, if and only if $\mathbf{E}\|T_n(X)\| \to 0$ as $n \to \infty$.*

Proof. If X satisfies $X \in \mathbf{WLLN_a}(E)$, then its symmetrization X^s satisfies the law of large numbers also. By inequality for means (0.7), Lemma 1.2, and condition (5) of the previous theorem, $\mathbf{E}\|T_n(X)\| \leq Hs(\rho_{n,s}(X)+\rho_{n,s}^{2+1/q}(X)) \to 0$ as $n \mapsto \infty$. The convergence follows easily from Chebyshev's inequality.
From the Corollary and Theorem 1.2 we conclude that

$$\mathbf{WLLN_a}(E) \subset \mathbf{SB_a}(E) = \mathbf{SB_a}^{\infty}(E).$$

THEOREM 2.2. *The linear set $\mathbf{WLLN_a}(E)$ is closed in the Banach space $(\mathbf{SB_a}(E), \lambda_{\mathbf{a}}(\cdot))$ or, equivalently, $X \in \mathbf{WLLN_a}(E)$, if and only if for any $\varepsilon > 0$ there exists $Y \in \mathbf{WLLN_a}(E)$ such that $\lambda_{\mathbf{a}}(X - Y) < \varepsilon$.*

Proof. Take any $X \bar{\in} \overline{\mathbf{WLLN_a}(E)}^{\lambda_{\mathbf{a}}(E)}$. Then for any $\varepsilon > 0$ and $\delta > 0$ there exists $Y \in \mathbf{WLLN_a}(E)$ such that $\lambda_{\mathbf{a}}(X - Y) \leq \varepsilon\delta/2$. By the corollary to Theorem 2.1, there exists $N = N(\varepsilon, \delta)$ such that $\mathbf{E}\|T_n(Y)\| \leq \varepsilon\delta/2$. By Chebyshev's inequality,

$$\mathbf{P}\{\|T_n(X)\| > \varepsilon\} \le \frac{1}{\varepsilon}\mathbf{E}\|T_n(X)\| \le \frac{1}{\varepsilon}(\mathbf{E}\|T_n(X-Y)\| + \mathbf{E}\|T_n(Y)\|)$$
$$\le \frac{1}{\varepsilon}(\lambda_{\mathbf{a}}(X-Y) + \mathbf{E}\|T_n(Y)\|) \le \delta$$

and, since ε and δ are arbitrary, $X \in \mathbf{WLLN_a}(E)$.
In the next section we shall need the following result which is originally due to R. Norvaisa (Norvaisa, 1984, Lemma 3.6).
We use the notation

$$\mathbf{WLLN_p}(E) = \{X \in L_0(E)\colon \mathbf{E}X = 0 \text{ and } S_n(X)/n^{1/p} \to 0$$
$$\text{as } n \to \infty \text{ in probability}\}, \quad 1 < p < 2.$$

COROLLARY. *The space $\mathbf{WLLN_p}(E)$ is closed in the sense of the norm $\lambda_p(X) = \sup_n \mathbf{E}\|S_n(X)\|$ or, equivalently, $X \in \mathbf{WLLN_p}(E)$, if and only if for any $\varepsilon > 0$ there exists $Y \in \mathbf{WLLN_p}(E)$ such that $\lambda_{\mathbf{p}}(X - Y) < \varepsilon$.*

Let $\mathbf{E}(E)$ be the set of all step random elements (i.e., those assuming only a finite number of values) with zero means. It is easy to note that condition (iii) implies the inclusion $\mathbf{E}(E) \subset \mathbf{WLLN_a}(E)$. In fact, an element from $\mathbf{E}(E)$ is supported on a finite-dimensional and bounded space, hence, it satisfies the law of large numbers (Marcus and Woyczynski, 1987, Theorem 3.1).

THEOREM 2.3. *The space $\mathbf{WLLN_a}(E)$ is the closure of the set $\mathbf{E}(E)$ in $(\mathbf{SB_a}(E), \lambda_{\mathbf{a}}(\cdot))$.*

Proof. Since the space $\mathbf{WLLN_a}(E)$ is closed, it is sufficient to find for any $X \in \mathbf{WLLN_a}(E)$ a sequence from $\mathbf{E}(E)$ which converges to X in the sense of the norm $\lambda_{\mathbf{a}}(\cdot)$.
Let $\mathcal{F}_1 \subseteq \mathcal{F}_2 \subseteq \ldots$ be finite σ-algebras such that the σ-algebra generated by the random element X is contained in the σ-algebra generated by $\bigcup_{k=1}^{n} \mathcal{F}_k$. Consider the martingale $M_k = \mathbf{E}(X|\mathcal{F}_k) \in \mathbf{E}(E)$ and show that $\lambda_{\mathbf{a}}(X - M_k) \to 0$ as $k \to \infty$.
We have

$$T_n(X) - T_n(M_k) = (T_n(X) - T_n(M_1)) - (T_n(M_k) - T_n(M_1))$$
$$= (T_n(X) - T_n(M_1)) - \mathbf{E}(T_n(X) - T_n(M_1)|\mathcal{F}_k).$$

Hence, $\mathbf{E}\|T_n(X - M_k)\| = \mathbf{E}\|T_n(X) - T_n(M_k)\| \le 2\mathbf{E}\|T_n(X) - T_n(M_1)\|$.
By the corollary to Theorem 2.1, for any $\varepsilon > 0$ there exists $N = N(\varepsilon)$ such that $\mathbf{E}\|T_n(X) - T_n(M_1)\| < \varepsilon/2$ for all $n \ge N$, i.e., for all $k \in \mathbf{N}$ and $n \ge N$

$$\mathbf{E}\|T_n(X - M_k)\| \le \varepsilon. \tag{6}$$

Moreover, since $M_k \to X$ as $k \to \infty$ in $L_1(E)$, (Vakhania et al., 1987, Theorem II.4.1) there exists $K = K(\varepsilon)$ such that for all n: $1 \le n \le N$ and $k \ge K$: $\mathbf{E}\|T_n(X - M_k)\| \le \varepsilon$.
Combining this inequality and (6), we obtain

$$\lambda_{\mathbf{a}}(X - M_k) = \max(\sup_{1\le k\le n} \mathbf{E}\|T_n(X - M_k)\|, \sup_{n\ge N} \mathbf{E}\|T_n(X - M_k)\|) \le \varepsilon$$

for $k \ge K$.
Note that the case of the classical law of large numbers ($a_k(n) = 1/n$ for all $k \in \mathbf{N}$) is not covered by our results. This is caused by the fact that $C > 1/2$ in condition (ii) while in the classical case $C = 1/2$. It is evident that the existence of $\mathbf{E}X$ is not necessary for the proof of the classical weak law of large numbers (even the strong law takes place).

3. AN APPLICATION

Using Pisier's inequality (see below) and the results of the previous section, we examine the weak law of large numbers for certain random elements with values in an arbitrary Banach space.
Let $1 < s < \infty$ and let $(b_k)_{k\ge 1}$ be a sequence of real numbers. Set $\|(b_k)_{k\le n}\|_{s,\infty} = \max_{k\le n} k^{1/s} b_k^*$, where $(b_k^*)_{k\le n}$ is the nonincreasing permutation of $(|b_k|)_{k\le n}$. Define $\Lambda_s(\xi) = (\sup_{t>0} t^s \mathbf{P}\{|\xi| > t\})^{1/s}$ for a random variable ξ.
To facilitate the formulation of the theorem, we introduce the random element $X = \sum_{i=1}^{\infty} \eta_i x_i$ (with the series converging almost surely), where (x_i) is a nonrandom sequence of elements in E and (η_i) is a sequence of independent random variables with zero means. Note that we do not require the sequence (η_i) to be identically distributed. It should be mentioned that I. K. Matsak (Matsak, 1988) proved the Central Limit Theorem for such random elements.
By β_q denote the symmetric random variable whose absolute value has the Weibull distribution with a parameter $q > 0$, i.e., $\mathbf{P}\{|\beta_q| > t\} = \exp\{-t^q\}$, and let $(\beta_{q,i})_{i\ge 1}$ be independent copies of β_q.

THEOREM 3.1. *Let $X = \sum_{i=1}^{\infty} \eta_i x_i$ and let the weight satisfy condition (iii) only. If there exists s: $2 > s > p$ such that*
a) $W = \sum_{i=1}^{\infty} \beta_{q,i} x_i$ converges a.s. for $\frac{1}{p} + \frac{1}{s} = 1$,
b) $\Lambda_s(\sup_{i\ge 1} |\eta_i|) < \infty$,
then $X \in \mathbf{WLLN}_{\mathbf{a}}(E)$.

In order to prove this theorem, we need some lemmas. In the sequel, we shall use the same notation as in the formulation of the theorem.

LEMMA 3.1. *If $W = \sum_{i=1}^{\infty} \beta_{q,i} x_i$ converges a.s., then it converges in $\mathbf{L}_p(E)$.*

Proof. It is sufficient to show that $W \in \mathbf{L}_p(E)$ (Vakhania et al., 1987, corollary to Theorem V.3.2). It suffices to prove the following inequality for all $t > 0$ (Vakhania et al., 1987, Theorem V.5.1):

$$\int_t^{\infty} \mathbf{P}\{|\beta_q| > u\} u^{p-1} du \le C t^p \mathbf{P}\{|\beta_q| > t\},$$

where the constant C does not depend on t. In the particular case β_q of a Weibull random variable this inequality can be rewritten as

$$\int_t^\infty \exp\{-u^q\}u^{p-1}du \le Ct^p \exp\{-t^q\}.$$

In fact, even a more precise estimation holds:

$$\int_t^\infty \exp\{-u^q\}u^{p-1}du \le Ct^{p-q} \exp\{-t^q\}, \tag{7}$$

since

$$t^{p-1}\exp\{-t^q\} \le t^{p-1}\exp\{-t^q\}\left(q + \frac{q-p}{p}t^{-q}\right) \tag{8}$$

and $p < 2 < q$. The terms of this inequality are the derivatives of the corresponding terms of (7) taken with the opposite sign, hence, (7) follows from (8) after integration of (8) on the half-line (t, ∞).

Pisier's inequality (Pisier, 1986, Lemma 2.7). Suppose $1 < s < 2$, q is the conjugate of s: $(1/q) + (1/s) = 1$, $(b_k)_{k\le n} \subset \mathbf{R}$, and let $(\varepsilon_k)_{k\le n}$ be independent Bernoulli random variables (i.e., $\mathbf{P}\{\varepsilon_1 = \pm 1\} = 1/2$). Then, for $k_s = q(q-2)/2$

$$\mathbf{P}\{|\sum_{k=1}^n b_k\varepsilon_k| > t\} \le 2\exp\left\{-\frac{t^q}{(k_s\|(b_k)_{k\le n}\|_{s,\infty})^q}\right\}.$$

The following lemma is just a reformulation of Pisier's inequality.

LEMMA 3.2. *Let* $\tau = \sum_{k=1}^n b_k\varepsilon_k$ *and* $\nu = \|(b_k)_{k\le n}\|_{s,\infty}\beta_q$, *where* $(b_k)_{k\le n} \subset \mathbf{R}$, *and let* (ε_k) *be a sequence of independent Bernoulli random variables. Then, for any* $t > 0$

$$\mathbf{P}\{|\tau| > t\} \le 2\mathbf{P}\{k_s|\nu| > t\}.$$

Let $(\eta_{k,i})_{k\ge 1}$ be independent copies of the sequence (η_i). We introduce the random variables $\mu(i,s,n) = \|(\eta_{k,i})_{k\le n}\|_{s,\infty}$.

LEMMA 3.3. *Let* η_i *be a sequence of symmetric random variables,* $X_k = \sum_{i=1}^\infty \eta_{k,i}x_i$, *and* $Y = \sum_{i=1}^\infty \beta_{q,i}\mu(i,s,n)x_i$. *Then the inequality* $\mathbf{E}\|\sum_{k=1}^n X_k\|^p \le C\mathbf{E}\|Y\|^p$ *holds, where* $C = 2^{1+p}k_s$.

Proof. First consider the case of $\eta_{k,i} = \varepsilon_{k,i}b_{k,i}$, where $b_{k,i} \in \mathbf{R}$ and $\varepsilon_{k,i}$ are independent Bernoulli random variables. Define

$$\tau_i = \sum_{k=1}^n b_{k,i}\varepsilon_{k,i} \quad \text{and} \quad \nu_i = \beta_{q,i}\|(b_{k,i})_{k\le n}\|_{s,\infty}.$$

It follows from Lemma 3.2 that $\mathbf{P}\{|\tau_i| > t\} \le 2\mathbf{P}\{k_s|\nu_i| > t\}$ for all $i \in \mathbf{N}$. Utilizing Theorem V.4.5 (Vakhania et al., 1987), we conclude that

$$\mathbf{E}\|\sum_{k=1}^{n} X_k\|^p = \mathbf{E}\|\sum_{i=1}^{\infty} \tau_i x_i\|^p \le 2^{1+p}k_s^p\mathbf{E}\|\sum_{i=1}^{\infty} \tau_i x_i\|^p = C\mathbf{E}\|Y\|^p.$$

In the general case, the desired result follows from this relation by a standard procedure. Replace the symmetric random variables η_i by the random variables $\eta_i\varepsilon_i$ which have the same distributions. Here ε_i are independent Bernoulli random variables which are independent of η_i. The left-hand side of the inequality from the statement of Lemma 3.3 is averaged over ε_i for a fixed η_i, and, finally, Fubini's theorem is applied (see, for example, the proof of Lemma V.2.1 (Vakhania et al., 1987)).

Proof of Theorem 3.1. First consider the case of a symmetric η_i. Let $X_k = \sum_{i=1}^{\infty} \eta_{k,i}x_i$ be independent copies of X. Applying Kwapien's inequality (0.3), we see that

$$\mathbf{P}\{\|T_n(X)\| > \varepsilon\} \le 2\mathbf{P}\left\{\frac{\|S_n(X)\|}{\max\limits_{k\le n}|a_k(n)|} > \varepsilon\right\} \le 2\mathbf{P}\left\{\frac{\|S_n(X)\|}{n^{1/p}} > A\varepsilon\right\}$$

by condition (iii). Hence, it is sufficient to prove that $S_n(X)/n^{1/p} \to 0$ in probability as $n \to \infty$. Let $Z_m = \sum_{i=1}^{m} \eta_i x_i$ and let $Z_{k,m} = \sum_{i=1}^{m} \eta_{k,i}x_i$ be independent copies of Z_m. By Lemma 3.3 and inequality from (Vakhania et al., 1987, V.4, exercise 1(a)):

$$\begin{aligned}\mathbf{E}\left\|\frac{S_n(X-Z_m)}{n^{1/p}}\right\|^p &\le C\mathbf{E}\left\|\frac{\sum_{i>m}\beta_{q,i}\mu(i,s,n)}{n}\right\|^p \\ &\le \frac{C2^p}{n}\mathbf{E}(\sup_{i\ge1}\mu(i,s,n))^p\mathbf{E}\|\sum_{i>m}\beta_{q,i}x_i\|^p.\end{aligned}$$

Note that, by Lemma 3.1, $\mathbf{E}\|\sum_{i>m}\beta_{q,i}x_i\|^p \to 0$ as $m \to \infty$. Furthermore,

$$\begin{aligned}\mathbf{E}(\sup_{i\ge1}\mu(i,s,n))^p &\le (1 + \int_1^{\infty} \mathbf{P}\{\sup_{i\ge1}\|(\eta_{k,i})_{k\le n}\|_{s,\infty} > t\}dt^p) \\ &\le 1 + 2en\int_1^{\infty}\frac{\Lambda_s(\sup_{i\ge1}|\eta_i|)}{t^s}dt^p\end{aligned}$$

by Marcus–Pisier's result (Pisier, 1986, Lemma 4.11). Since $s > p$, we have

$$\frac{1}{n}\mathbf{E}(\sup_{i\ge1}\mu(i,s,n))^p \le C\Lambda_s(\sup_{i\ge1}|\eta_i|) + 1,$$

whence $\sup_n \mathbf{E}\|S_n(X-Z_m)/n^{1/p}\|^p \to 0$ as $m \to \infty$.

Note that the random vector Z_m takes on its values in the finite-dimensional space Span $(x_1, \dots, x_m)$ and $\Lambda_p(\|Z_m\|) < \infty$, since $\Lambda_s(\sup_{i\geq 1} |\eta_i|) < \infty$ and $s > p$. Then $S_n(Z_m)/n^{1/p} \to 0$ in probability as $n \to \infty$ (Marcus and Woyczynski, 1979, Theorem 3.1). By the corollary to Theorem 2.2, we conclude that $S_n(X)/n^{1/p} \to 0$ in probability as $n \to \infty$.

If X is an arbitrary not necessarily symmetric centered random element satisfying the hypotheses of Theorem 1, then its symmetrization X^s also satisfies the same hypotheses and inequality for means (0.7):

$$\sup_n \mathbf{E} \left\| \frac{T_n(X - Z_m)}{n^{1/p}} \right\| \leq \sup_n \mathbf{E} \left\| \frac{T_n(X^s - Z_m)}{n^{1/p}} \right\| \to 0$$

as $m \to \infty$. It remains to apply the corollary to Theorem 2.2.

Mention should be made of an important particular case, namely, the case of random variables (η_i) which are nondegenerate, i.e., $\inf_i \mathbf{E}|\eta_i| > 0$. Then condition a) of Theorem 1 is not necessary, since the convergence of the series $\sum_{i=1}^{\infty} \eta_i x_i$ follows from the convergence of the series $\sum_{i=1}^{\infty} \beta_{q,i} x_i$ (Vakhania et al., Lemma V.5.1 and Proposition V.5.5).

REFERENCES

Acosta, A. de (1981). Inequalities for B-valued random vectors with applications to the strong law of large numbers. *Ann. Probab.* **9**, 157–161.

Araujo, A. and Gine, E. (1980). *The Central Limit Theorem for Real and Banach Space Valued Random Variables.* John Wiley, New York.

Hoffmann-Jorgensen, J. (1974). Sums of independent Banach space valued random variables. *Stud. Cerc. Math.* **52**, 159–186.

Kwapien, S. and Woyczynski, W. A. (1987). Double stochastic integrals, random quadratic forms, and random series in Orlicz spaces. *Ann. Probab.* **15**, 1072–1096.

Marcus, M.B. and Woyczynski, W. A. (1979). Stable measures and central limit theorem in spaces of stable type. *Trans. Amer. Math. Soc.* **251**, 71–102.

Matsak, I.K. (1988). A remark on the central limit theorem in Banach spaces. *Ukrainian Math. J.* **40**, 649–652 (in Russian).

Norvaisa, R. (1984). The law of large numbers for identically distributed random variables taking values in Banach spaces. *Lithuanian Math. J.* **24**, 133–150.

Pisier, G. (1975). Le theoreme limite centrale et la loi du logarithme iteree dans les espaces de Banach. In: *Seminaire Maurey–Schwartz* 1975–1976, exp III.

Pisier, G. (1986). Probabilistic methods in the geometry of Banach spaces, probability, and analysis. *Lect. Notes Math.* **1206**, 167–241.

Vakhania, N.N., Tarieladze, V.I., and Chobanyan, S.A. (1987). *Probability Distributions on Banach Spaces.* Reidel, Dordrecht.

Frontiers in Pure and Applied Probab. 1, pp. 287 - 295
H. Niemi et al. (Eds.)
1993 TVP/VSP

ESTIMATING FUNCTIONS AND EFFICIENCY IN A FILTERED MODEL

W. WEFELMEYER
Department of Mathematics, University of Cologne, Weyertal 86, 5000 Cologne 41, Germany

ABSTRACT

Suppose we observe a stochastic process which is a special semimartingale with respect to the filtration generated by the process. Assume that its compensator and the predictable quadratic variation of the corresponding martingale are known up to a (one-dimensional) parameter. This model was introduced by Thavaneswaran and Thompson (Thavaneswaran and Thompson, 1986). They consider estimators based on estimating functions which are stochastic integrals with respect to the underlying martingale; they judge these estimating functions by the criterion of Godambe (Godambe, 1985). Two questions arise: Can one relate Godambe's criterion to properties of the estimator? If so, are there better estimators than the one based on the best estimating function? We discuss conditions under which estimating functions lead to asymptotically normal estimators with variance determined by the criterion. Such estimators can be judged, at least, asymptotically by the criterion. Then we sketch conditions under which the model is locally asymptotically normal and we determine an asymptotic bound for the concentration of regular estimators around the true parameter. It turns out that the best estimating function does not lead to an efficient estimator. This fact is demonstrated on a simple example.

1. INTRODUCTION

Suppose we observe a (special) semimartingale on some time interval. Three different types of models have been suggested: Specify a parametric form for (1) the compensator of the process, (2) the compensator of the process and the predictable quadratic variation of the associated martingale, (3) the predictable characteristics of the process. We shall mainly be concerned with model 2 but occasionally it will be useful to compare it with models 1 and 3. The discussion will be heuristic throughout. We shall refer to papers of Greenwood and Wefelmeyer for regularity conditions (Greenwood and Wefelmeyer, 1989, 1990, 1991, 1992).

Specifically, let X be a cadlag process on $[0, \infty)$ with $X_0 = 0$ for simplicity. We observe X at state t on the time interval $[0, t]$. Later we shall consider the asymptotic behavior as t tends to infinity. Let $(\mathcal{F}_t)_{t \geq 0}$ be the filtration generated by X. Let F be nondecreasing and predictable. To exclude a statistically trivial situation, the quadratic characteristic of X is assumed known, say $C = c \bullet F$.

Let Θ denote the parameter space. For simplicity, let Θ be one-dimensional. In *model* 1, the process X is a special semimartingale with a compensator

$$A_\vartheta = a_\vartheta \bullet F, \quad \vartheta \in \Theta. \tag{1}$$

In *model* 2 we assume, in addition, a parametric form for the predictable quadratic variation of $M_\vartheta = X - A_\vartheta$:

$$\langle M_\vartheta \rangle = m_\vartheta \bullet F, \quad \vartheta \in \Theta. \tag{2}$$

This model was introduced in (Thavaneswaran and Thompson, 1986).

Model 3 is, perhaps, the most common one: We specify the predictable characteristics $(A_\vartheta, C, \nu_\vartheta)$, where ν_ϑ is the compensator of the random jump measure μ associated with X.
Model 1 contains model 2. Model 2 contains model 3, if $\langle M_\vartheta \rangle = C + x^2 * \nu_\vartheta$.
In general, all the three models are *partially specified*: If $\mathcal{P}_\vartheta^i$ denotes the family of all distributions compatible with the specifications of model i, then the restrictions $\mathcal{P}_{\vartheta t}^i$ to $\mathcal{F}_t$ do not reduce to a single probability measure. The underlying model is the union of the families $\mathcal{P}_\vartheta^i$, $\vartheta \in \Theta$. The parameter ϑ can be interpreted as a *functional* assigning the value ϑ to each $P \in \mathcal{P}_\vartheta^i$.
However, model 3 often differs from models 1 and 2 in this respect: Under appropriate conditions, the predictable characteristics determine the underlying distribution locally uniquely. In other words, model 3 is *parametrized* by ϑ: The family $\mathcal{P}_{\vartheta t}^3$ reduces to a single distribution $P_{\vartheta t}^3$ for each $t \geq 0$. Moreover, the likelihood process admits a representation in terms of the characteristics. The best estimating function is given by the logarithmic derivative of the likelihood. The corresponding estimator, the maximum likelihood estimator, is known to be asymptotically efficient under certain technical conditions.
Additional structural assumptions (not depending on the parameter) are occasionally made in all the three models. For example, the process may be assumed to have continuous paths or to be a pure step process. More specifically, we may assume that the process is a diffusion process or a counting process. In these two cases, the characteristics are determined by the compensator of the process, and models 1–3 coincide and are fully specified. (Recall that C is fixed.) But model 2 is larger than model 3, and model 1 is larger than model 2 when the process has different jump sizes or both jump and continuous parts.
For model 1, Thavaneswaran and Thompson (Thavaneswaran and Thompson, 1986) suggest estimating functions of the form

$$Q_\vartheta = q_\vartheta \bullet M_\vartheta. \tag{3}$$

Godambe ((Godambe, 1985, 1987) for discrete time) and Thavaneswaran and Thompson ((Thavaneswaran and Thompson, 1986) for continuous time) suggest the better judging an estimating function Q_ϑ the smaller the *criterion*

$$E_P Q_\vartheta^2 / (E_P Q_\vartheta')^2, \quad P \in \mathcal{P}_\vartheta^1, \tag{4}$$

with prime denoting derivative with respect to ϑ. We have

$$\langle Q_\vartheta \rangle_P = (q_\vartheta^2 m_P) \bullet F. \tag{5}$$

The compensator, say $\overline{Q}_\vartheta$, of Q'_ϑ under $P \in \mathcal{P}'_\vartheta$ depends on P through ϑ only,

$$\overline{Q}_\vartheta = -(q_\vartheta a'_\vartheta) \bullet F. \tag{6}$$

Heyde (Heyde, 1987) suggests a *second*, random *criterion*,

$$\langle Q_\vartheta \rangle_P / \overline{Q}_\vartheta^2, \quad P \in \mathcal{P}_\vartheta^1. \tag{7}$$

For $P \in \mathcal{P}_\vartheta^2$ this criterion depends on P through ϑ only.
In general, the smaller the model the larger the class of sensible estimating functions and the better the optimal estimating function. But even though model 2 is smaller than model 1, no larger class of estimating functions suggests itself.
As noted by Sørensen (Sørensen, 1990), we have a larger class of estimating functions for model 3 than for model 1 or 2,

$$Q_\vartheta = b_\vartheta \bullet X_\vartheta^c + y_\vartheta * (\mu - \nu_\vartheta), \tag{8}$$

where X_ϑ^c is the continuous martingale part of X for $P \in \mathcal{P}_\vartheta^3$. (Here we have tacitly assumed that the distributions in $\mathcal{P}_\vartheta^3$ are locally mutually absolutely continuous.) We obtain estimating function (3) with $q_\vartheta = b_\vartheta$ for $y_\vartheta(t,x) = b_{\vartheta t} x$.
For model 2, Thavaneswaran and Thompson (Thavaneswaran and Thompson, 1986) show that the optimal estimating function $Q_\vartheta^* = q_\vartheta^* \bullet M_\vartheta$ for the first criterion is given by

$$q_\vartheta^* = a'_\vartheta / m_\vartheta. \tag{9}$$

A generalization to finite-dimensional parameters is given in (Hutton and Nelson, 1986). As shown by Godambe and Heyde (Godambe and Heyde, 1987), the estimating function Q_ϑ^* is optimal for the second criterion, too. See also (Heyde, 1988, 1989). The proofs are simple enough to reproduce them here. By (6),

$$\langle Q_\vartheta, Q_\vartheta^* \rangle_\vartheta = -\overline{Q}_\vartheta. \tag{10}$$

The Schwarz inequality shows now that Q_ϑ^* is optimal with respect to the second criterion. The optimal value of the second criterion is

$$\langle Q_\vartheta^* \rangle_\vartheta / \overline{Q}_\vartheta^{*2} = 1/(a_\vartheta'^2 / m_\vartheta) \bullet F. \tag{11}$$

For the first criterion, take expectations in (6) and apply the Schwarz inequality again. Analogous results hold for estimating functions of form (8); see (Sørensen, 1990; Jacod, 1990).
Consider now an estimator $\hat{\vartheta}_t$ which solves an estimating equation of the form

$$q_\vartheta \bullet M_{\vartheta t} = 0.$$

The *first question* is: Under which conditions are the criteria related to asymptotic properties of such an estimator? This problem is solved in (Greenwood and Wefelmeyer, 1991) for counting processes; the arguments generalize immediately.

Greenwood and Wefelmeyer give conditions under which $\delta_t^{-1}(\hat{\vartheta}_t - \vartheta)$ is asymptotically normal for $P \in \mathcal{P}_\vartheta^1$. They assume, in particular, a deterministic limit for $\varepsilon_t^2(q_\vartheta^2 m_P) \bullet F_t$ and $\varepsilon_t \delta_t (q_\vartheta a'_\vartheta) \bullet F_t$. The asymptotic variance of the estimator is the limit of the expression

$$\delta_t^{-2}(q_\vartheta^2 m_P) \bullet F_t / ((q_\vartheta a'_\vartheta) \bullet F_t)^2. \tag{12}$$

Relations (5) and (6) show that this expression equals the second criterion (7) up to the norming constant δ_t^{-2}. In particular, both criteria normed appropriately converge to the asymptotic variance of the estimator. This implies that small values of the criteria correspond to good estimators, at least, asymptotically.
The usual proof of this result starts with the expansion

$$Q_{\hat{\vartheta}_t} = Q_\vartheta + (\hat{\vartheta}_t - \vartheta) Q'_\vartheta + \dots,$$

leading to criteria of form (4) and (7). See (Hutton and Nelson, 1986; Sørensen, 1989; Barndorff-Nielsen and Sørensen, 1989). The proof differs here as it does not require *differentiability* of the estimating function. It suggests writing the criterion as (12). Then optimality of Q_ϑ^* follows immediately.
In general, of course, it is impossible to describe the quality of an estimator by a real number like the first criterion or even by the second criterion even for a fixed ϑ and $P \in \mathcal{P}_\vartheta^1$ and even asymptotically. Different loss functions will favour different estimators.
One motive for restricting attention in models 1 and 2 to estimating functions is that it seems difficult to get asymptotic variance bounds for *arbitrary* (regular) estimators. We shall see in §2 that this requires, in fact, some effort. Specifically, our *second question* is: Does the optimal estimating function in model 2 lead to an asymptotically efficient estimator? We have reasons to doubt it. A rather vague one is that model 2 is infinite-dimensional. Efficient estimators for functionals in such models are usually not obtained from a simple estimating function. A more specific reason is that the estimating functions do not use the second specification of model 2 on the predictable quadratic variation of the martingale associated with the process. In other words: They are estimating functions for the larger model 1. Hence the corresponding estimators cannot be better than the best estimator for model 1.
We show that the estimator based on optimal estimating function (9) is asymptotically as good as we can hope for: It achieves within model 2 the asymptotic variance bound for estimators for model 1. We argue that the asymptotic variance bound for model 2 is smaller. This implies that the optimal estimating function gives an inefficient estimator. A simple example, with independend identically distributed (i.i.d.) observations, illustrates the results.

2. ASYMPTOTIC VARIANCE BOUNDS

The results mentioned so far refer to estimating functions. They do not require looking at the underlying model. However, an asymptotic variance bound *depends* on the underlying model. This is why the present section is more involved. The bound for model i is obtained in two steps. Firstly, we construct a local model with *linear* local parameter space around the true distribution and obtain an infinite-dimensional version of *local asymptotic normality*. This induces an inner product, the *acuity*, on the local parameter space. Secondly, we show that the functional mapping $P \in \mathcal{P}^i_\vartheta$ into ϑ is differentiable. Then the squared length of the gradient of the functional with respect to the acuity is an asymptotic variance bound for regular estimators of the functional. For i.i.d. observations, this approach is due to (Levit, 1974, 1975; Koshevnik and Levit, 1976); see the monographs (Pfanzagl and Wefelmeyer, 1982; Bickel et al., 1992). The approach is generalized to counting processes by Dzhaparidze (Dzhaparidze, 1985) and to general semimartingales by (Greenwood and Wefelmeyer, 1989, 1992).

Firstly we determine an asymptotic variance bound for regular estimators of ϑ in model 1. Without loss of generality, write $\nu_P(dt, dx) = K_{Pt}(dx)\, dF_t$.

Fix $\vartheta \in \Theta$ and $P \in \mathcal{P}^1_\vartheta$. Choose $\tau \in \Theta$ and $Q \in \mathcal{P}^1_\tau$. Write (A_τ, C, ν_Q) for the characteristics of X under Q. According to the Girsanov theorem (Jacod and Shiryaev, 1987, p. 159, Theorem 3.24), we can write

$$\nu_Q(dt, dx) = Y_Q(t, x) K_{Pt}(dx)\, dF_t,$$

$$A_\tau - A_\vartheta = \left(\beta_Q c + \int (Y_Q(\cdot, x) - 1) x K_P(dx)\right) \bullet F.$$

Note that $\beta_P = 0$ and $Y_P = 1$. Model specification (1) entails the side condition

$$\beta_Q c + \int (Y_Q(\cdot, x) - 1) x K_P(dx) = a_\tau - a_\vartheta. \tag{13}$$

on β_Q and Y_Q. Introduce a *local model* around P as follows. Consider all predictable processes b and $y = y(t, x)$ such that there exists a one-dimensional submodel of model 1 for each t, say $P^{tuby} \in \mathcal{P}^1_{\vartheta+\delta_t u}$ with (local) parameter $u \in \mathbf{R}$, which passes through P and is *asymptotically differentiable* at $u = 0$ with the derivative (b, y). More specifically, the corresponding predictable processes β^{tub} and Y^{tuy} are asymptotically differentiable at $u = 0$ with derivatives b and y, respectively, as $t \to \infty$:

$$\beta^{tub} - \delta_t u b \to 0, \quad Y^{tuy} - 1 - \delta_t u y \to 0.$$

(See (Greenwood and Wefelmeyer, 1989) for the appropriate differentiability concept.) Restrict attention to b and y with asymptotically *deterministic* Hellinger limits: For certain real numbers $\|b\|_c$ and $\|y\|_j$,

$$\delta_t^2 (b^2 c) \bullet F_t \to \|b\|_c^2, \tag{14}$$

$$\delta_t^2 \int y(t, x)^2 K_{Pt}(dx)\, dF_t \to \|y\|_j^2. \tag{15}$$

The subscripts c and j refer to the continuous and the jump parts, respectively. Assume that the model specification is smoothly parametrized. More specifically, let a_τ be asymptotically as $t \to \infty$ differentiable at $\tau = \vartheta$ with derivative a' in the appropriate sense: $\delta_t^{-1}(a_{\vartheta+\delta_t u} - a_\vartheta) \to u a'$.

Then side condition (13) on β_Q and Y_Q translates into the side condition

$$bc + \int y(\cdot, x) x K_P(dx) \doteq a' \tag{16}$$

on the local parameters b and y. Write H_1^1 for all such (b, y). Assume that the set H_1 of all $u(b, y)$ with $u \in \mathbf{R}$ is a linear space. We call it a *local parameter space.* Under an appropriate Lindeberg condition, the likelihood process is *locally asymptotically normal* (Greenwood and Wefelmeyer, 1989):

$$\log dP_t^{tuby}/dP_t = u\delta_t^{-1}(b \bullet X_{Pt}^c + y * (\mu - \nu_P)_t)$$
$$-\frac{1}{2}u^2(\|b\|_c^2 + \|y\|_j^2) + o_P(1). \tag{17}$$

Here we have assumed, for simplicity, that the fixed jump probabilities do not depend on the underlying distribution. To represent the likelihood process in terms of the characteristics (Jacod and Shiryaev, 1987, p. 186, Theorem 5.32), we also are to assume that the characteristics determine the underlying distribution locally uniquely.

Now we can calculate the asymptotic variance bound for estimators of ϑ in model 1. We apply the version of Hájek's convolution theorem given in (Greenwood and Wefelmeyer, 1990). Consider the parameter ϑ as a *functional* $f(P)$ mapping $P \in \mathcal{P}_\vartheta^1$ into ϑ. We determine its *gradient* with respect to the inner product induced by local asymptotic normality (17) on the local parameter space H_1. It is convenient to write the gradient as $I^{-1}(\bar{b}, \bar{y})$, $(\bar{b}, \bar{y}) \in H_1^1$ being the *score function.* The gradient is characterized by

$$1 = \delta_t^{-1}(\vartheta + \delta_t 1 - \vartheta) = \delta_t^{-1}(f(P^{t1by}) - f(P)) = I^{-1}((\bar{b}, b)_c + (\bar{y}, y)_j) \tag{18}$$

for all $(b, y) \in H_1^1$. For $(b, y) = (\bar{b}, \bar{y})$ we obtain $I = \|\bar{b}\|_c^2 + \|\bar{y}\|_j^2$.
Hence we can rewrite (18) as

$$\|\bar{b}\|_c^2 + \|\bar{y}\|_j^2 = (\bar{b}, b)_c + (\bar{y}, y)_j. \tag{19}$$

According to definitions (14) and (15) of the norms, this means that

$$\bar{b}^2 c + \int \bar{y}(\cdot, x)^2 K_P(dx) = \bar{b} bc + \int \bar{y}(\cdot, x) y(\cdot, x) K_P(dx). \tag{20}$$

The solution $(\bar{b}, \bar{y})$ is easy to determine. Write $\langle M_\vartheta \rangle_P = m_P \bullet F$. We have

$$\langle M_\vartheta \rangle_P = C + x^2 * \nu_P = (c + \int x^2 K_P(dx)) \bullet F.$$

Hence,

$$m_P = c + \int x^2 K_P(dx).$$

Define $\bar{b}_t = a'_{\vartheta t}/m_{Pt}$, $\bar{y}(t, x) = a'_{\vartheta t} x / m_{Pt}$.

Then $(\bar{b}, \bar{y})$ satisfies (16) and, consequently, is in H_1^1. It solves (20), since, on the one hand,

$$\bar{b}^2 c + \int \bar{y}(\cdot\,, x)^2 K_P(dx) = a_\vartheta'^2 / m_P \tag{21}$$

and, on the other hand,

$$\bar{b} b c + \int \bar{y}(\cdot\,, x)\, y(\cdot\,, x) K_P(dx) = a_\vartheta'^2 / m_P$$

by side condition (16) on b and y. The *convolution theorem* can now be stated as follows. Let $\hat{\vartheta}_t$ be *regular* at ϑ with the *limit* L: For all $(b, y) \in H_1^1$, $\delta_t^{-1}(\hat{\vartheta}_t - \vartheta - \delta_t u) \Rightarrow L$ under P^{tuby}. Then $L = N + M$ with M independent on N and N normal with the mean 0 and the variance equal to the squared length of the gradient I^{-1}.

We see that I can be interpreted as (quasi-) *Fisher information*. By definition and (21), it is the limit of

$$\delta_t^2 (\bar{b}^2 c + \int \bar{y}(x, \cdot)^2 K_P(dx)) \bullet F_t = \delta_t^2 (a_\vartheta'^2 / m_P) \bullet F_t. \tag{22}$$

Assume now that model 2 is true. Let $P \in \mathcal{P}_\vartheta^2$. By model specification (2), we have $m_P = m_\vartheta$. We have seen in §1 that the optimal estimating function $(a'_\vartheta / m_\vartheta) \bullet M_\vartheta$ for model 2 gives an estimator with the asymptotic variance equal to the variance bound I^{-1} for model 1. Compare (11) and (22). The estimator is regular; see (Greenwood and Wefelmeyer, 1991).

Of course, this does not mean that the estimator is efficient for model 1. We have not specified a parametric form for m_P in model 1. The expression a'_ϑ / m_P depends on $P \in \mathcal{P}_\vartheta^1$ not only through ϑ and cannot be used as an estimating function for ϑ. It remains to show that the asymptotic variance bound for model 2 is smaller than the bound for model 1. However, the calculation of the gradient is less straightforward than for model 1, and we restrict ourselves to indicating how to modify the above approach. We can write

$$\langle M_\tau \rangle_Q - \langle M_\vartheta \rangle_P = \int (Y_Q(\cdot\,, x) - 1) x^2 K_P(dx) \bullet F.$$

Second model specification (2) entails the second side condition on Y_Q:

$$\int (Y_Q(\cdot\,, x) - 1) x^2 K_P(dx) = m_\tau - m_\vartheta.$$

Assume that the specification is smoothly parametrized: $\delta_t^{-1}(m_{\vartheta + \delta_t u} - m_\vartheta) \to u m'$. Then the side condition translates into the (second) side condition on the local parameter y:

$$\int y(\cdot\,, x) x^2 K_P(dx) = m'. \tag{23}$$

Write H_2^1 for all (b, y) satisfying both side conditions (16) and (23). Assume that the set H_2 of all $u(b, y)$ with $(b, y) \in H_2^1$ and $u \in \mathbf{R}$ is a linear space. Now proceed as for model 1 with H_1 replaced by H_2. The problem is to find an explicit solution of (20) with $(\bar{b}, \bar{y})$ not in H_1 but in H_2. This gives the gradient for model 2. Its squared length is the asymptotic variance bound for model 2. We omit the calculation. Instead, we illustrate by the simple example that the bound for model 2 is, in general, smaller than the bound for model 1.

Example. Let x_i be i.i.d. with variance equal to expectation. Fix a distribution with (variance and) expectation $\vartheta > 0$. Then

$$X_t = \sum_{i \le t} x_i = \sum_{i \le t} (x_i - \vartheta) + \vartheta[t] = M_{\vartheta t} + A_{\vartheta t}$$

is a special semimartingale satisfying specifications (1) and (2) of model 2:

$$A_{\vartheta t} = \langle M_\vartheta \rangle_t = \vartheta[t].$$

According to (9), the optimal estimating function is

$$Q^*_{\vartheta t} = \frac{1}{\vartheta} \sum_{i \le t} (x_i - \vartheta).$$

The corresponding estimator is the sample mean. It is not efficient, since it ignores information about ϑ in the sample variance. In this example, however, the estimator happens to be efficient for model 1. Note that specification (1) here is not a restriction at all.

REFERENCES

Barndorff-Nielsen, O. and Sørensen, M. (1989). Asymptotic likelihood theory for stochastic processes. A review. *Int. Statist. Rev.* (to appear).

Bickel, P.J., Klaassen, C.A., Ritov, Y., and Wellner, J.A. (1992). *Efficient and Adaptive Estimation in Semiparametric Models*. Johns Hopkins Univ. Press, Baltimore (to appear).

Dzhaparidze, K. (1985). On asymptotic inference about intensity parameters of a counting process. *Bull. Intern. Statist. Inst.* **51**, 23.2–1 — 23.2–15.

Godambe, V.P. (1985). The foundations of finite sample estimation in stochastic processes. *Biometrika* **72**, 419–428.

Godambe, V.P. (1987). The foundations of finite sample estimation in stochastic processes — II. In: *Proceedings of the First World Congress of the Bernoulli Society,* Ed. by Yu.V. Prohorov, V.V. Sazonov. Vol. 2, Mathematical Statistics. Theory and Applications, VNU Sci. Press, Utrecht, pp. 49–54.

Godambe, V.P. and Heyde, C.C. (1987). Quasi-likelihood and optimal estimation. *Intern. Statist. Rev.* **55**, 231–244.

Greenwood, P.E. and Wefelmeyer, W. (1989). *Partially and Fully Specified Semimartingale Models and Efficiency*. Preprints in Statistics 124, Department of Mathematics, University of Cologne, Cologne.

Greenwood, P.E. and Wefelmeyer, W. (1990). Efficiency of estimators for partially specified filtered models. *Stoch. Proc. Appl.* **36**, 353–370.

Greenwood, P.E. and Wefelmeyer, W. (1991). On optimal estimating functions for partially specified counting process models. In: *Estimating Functions*, Ed. by V.P. Godambe. Oxford Univ. Press, Oxford, pp. 147–160.

Greenwood, P.E. and Wefelmeyer, W. (1992). *Efficient Estimation for Stochastic Processes*. Johns Hopkins Univ. Press, Baltimore (to appear).

Heyde, C.C. (1987). On combining quasi-likelihood estimating functions. *Stoch. Proc. Appl.* **25**, 281–287.

Heyde, C.C. (1988). Fixed sample and asymptotic optimality for classes of estimating functions. *Contemporary Math.* **80**, 241–247.

Heyde, C.C. (1989). On efficiency for quasi-likelihood and composite quasi-likelihood methods. In: *Statistical Data Analysis and Inference*, Ed. by Y. Dodge, North-Holland/Elsevier, Amsterdam, pp. 209–213.

Hutton, J.E. and Nelson, P.I. (1986). Quasi-likelihood estimation for semimartingales. *Stoch. Proc. Appl.* **22**, 245–257.

Jacod, J. (1990). Regularity, partial regularity, partial information process, for a filtered statistical model. *Probab. Theory Rel. Fields* **86**, 305–335.

Jacod, J. and Shiryaev, A.N. (1987). *Limit Theorems for Stochastic Processes*. Springer-Verlag, Berlin.

Koshevnik, Yu.A. and Levit, B.Ya. (1976). On a nonparametric analogue of the information matrix. *Theory Probab. Appl.* **21**, 738–753.

Levit, B.Ya. (1974). On optimality of some statistical estimates. In: *Proceedings of the Prague Symposium on Asymptotic Statistics*, Ed. by J. Hájek. Vol. 2, Charles Univ. Press., Prague, pp. 215–238.

Levit, B.Ya. (1975). On the efficiency of a class of nonparametric estimates. *Theory Probab. Appl.* **20**, 723–740.

Pfanzagl, J. and Wefelmeyer, W. (1982). *Contributions to a General Asymptotic Statistical Theory*. Springer-Verlag, New York.

Sørensen, M. (1989). Some asymptotic properties of quasi-likelihood estimators for semimartingales. In: *Proceedings of the Fourth Prague Symposium on Asymptotic Statistics*, Ed. by P. Mandl, M. Hušková. Charles Univ. Press, Prague, pp. 469–479.

Sørensen, M. (1990). On quasi-likelihood for semimartingales. *Stoch. Proc. Appl.* **35**, 331–346.

Thavaneswaran, A. and Thompson, M.E. (1986). Optimal estimation for semimartingales. *J. Appl. Probab.* **23**, 409–417.

LIST OF CONTRIBUTORS

Asmussen, S.	1 - 6
Borodin, A. N.	7 - 21
Borovkov, K. A.	22 - 28
Foss, S. G.	1 - 6
Gjessing, H.	29 - 67
Höglund, R.	68 - 85
Holden, H.	29 - 67
Jacobsen, M.	86 - 94
Kaj, I.	95 - 102
Kolchin, V. F.	103 - 111
Konakov, V. D.	112 - 134
Koski, T.	135 - 144
Kramkov, D. O.	145 - 164
Lindstrøm, T.	29 - 67
Melnikov, A. V.	165 - 182 183 - 196
Øksendal, B.	29 - 67
Östermark, R.	68 - 85
Penttiten, A.	197 - 205
Prokhorov, Yu. V.	206 - 208
Rodkina, A. E.	165 - 182 183 - 196
Salminen, P.	7 - 21 95 - 102
Sazonov, V. V.	209 - 222
Shiryaev, A. N.	223 - 227
Skorokhod, A. V.	228 - 237
Spokoiny, V. G.	223 - 227 238 - 255
Ubøe, J.	29 - 67
Valkeila, E.	183 - 196
Vatutin, V. A.	256 - 268
Volodin, A. I.	269 - 286
Wefelmeyer, W.	287 - 295
Zhang, T.-S.	29 - 67

ORGANIZING COMMITTEE

G. Högnas (Åbo Academy, Turku)
H. Niemi (Helsinki University, Helsinki)
A.V. Melnikov (Steklov Mathematical Institute, Moscow)
A.N. Shiryaev (Steklov Mathematical Institute, Moscow)
E. Valkeila (Helsinki University, Helsinki)

Bei Fragen zur Produktsicherheit wenden Sie sich bitte an:
If you have any questions regarding product safety,
please contact:

Walter de Gruyter GmbH
Genthiner Straße 13
10785 Berlin
productsafety@degruyterbrill.com